Crop Yield and Management

Crop Yield and Management

Edited by Corey Aiken

SYRAWOOD
PUBLISHING HOUSE

New York

Published by Syrawood Publishing House,
750 Third Avenue, 9th Floor,
New York, NY 10017, USA
www.syrawoodpublishinghouse.com

Crop Yield and Management
Edited by Corey Aiken

© 2017 Syrawood Publishing House

International Standard Book Number: 978-1-68286-379-4 (Hardback)

Cataloging-in-publication Data

Crop yield and management / edited by Corey Aiken.
    p. cm.
Includes bibliographical references and index.
ISBN 978-1-68286-379-4
1. Crop yields. 2. Agronomy. 3. Crop science. 4. Agricultural pests--Integrated control. I. Aiken, Corey.
SB91 .C76 2017
633--dc23

Printed in the United States of America.

TABLE OF CONTENTS

PREFACE

Crop yield and management deals with the study of the diverse methods involved in crop production and crop management. Increasing interest and access to various types of ecosystems have taken research forward and has ensured that the topics of genetic modification, pest management and plant breeding are discussed in detail. This book is a complete source of knowledge on the present status of this important field. It traces the progress of this field and highlights some of its key concepts and applications. This book presents researches and studies performed by experts across the globe. While understanding the long term perspectives of the topics, the book makes an effort in highlighting their impact as a modern tool for the growth of the discipline. This book with its detailed analysis and data will prove immensely beneficial to professionals and students involved in this area at various levels.

This book has been an outcome of determined endeavour from a group of educationists in the field. The primary objective was to involve a broad spectrum of professionals from diverse cultural background involved in the field for developing new researches. The book not only targets students but also scholars pursuing higher research for further enhancement of the theoretical and practical applications of the subject.

It was an honour to edit such a profound book and also a challenging task to compile and examine all the relevant data for accuracy and originality. I wish to acknowledge the efforts of the contributors for submitting such brilliant and diverse chapters in the field and for endlessly working for the completion of the book. Last, but not the least; I thank my family for being a constant source of support in all my research endeavours.

                                                                                             Editor

Establishing a Regional Nitrogen Management Approach to Mitigate Greenhouse Gas Emission Intensity from Intensive Smallholder Maize Production

Liang Wu, Xinping Chen, Zhenling Cui*, Weifeng Zhang, Fusuo Zhang

Center for Resources, Environment and Food Security, China Agricultural University, Beijing, People's Republic of China

Abstract

The overuse of Nitrogen (N) fertilizers on smallholder farms in rapidly developing countries has increased greenhouse gas (GHG) emissions and accelerated global N consumption over the past 20 years. In this study, a regional N management approach was developed based on the cost of the agricultural response to N application rates from 1,726 on-farm experiments to optimize N management across 12 agroecological subregions in the intensive Chinese smallholder maize belt. The grain yield and GHG emission intensity of this regional N management approach was investigated and compared to field-specific N management and farmers' practices. The regional N rate ranged from 150 to 219 kg N ha^{-1} for the 12 agroecological subregions. Grain yields and GHG emission intensities were consistent with this regional N management approach compared to field-specific N management, which indicated that this regional N rate was close to the economically optimal N application. This regional N management approach, if widely adopted in China, could reduce N fertilizer use by more than 1.4 MT per year, increase maize production by 31.9 MT annually, and reduce annual GHG emissions by 18.6 MT. This regional N management approach can minimize net N losses and reduce GHG emission intensity from over- and underapplications, and therefore can also be used as a reference point for regional agricultural extension employees where soil and/or plant N monitoring is lacking.

Editor: Shuijin Hu, North Carolina State University, United States of America

Funding: The work has been funded by the National Basic Research Program of China (973, Program: 2009CB118606) (website: http://www.973.gov.cn/AreaAppl. aspx). National Maize Production System in China (CARS-02-24)(website: http://119.253.58.231/). Special Fund for Agro-scientific Research in the Public Interest (201103003)(website: http://www.hymof.net.cn/webapp/login.asp). The funders had no role in study design, data collection and analysis, decision to publish, or preparation of the manuscript.

Competing Interests: The authors have declared that no competing interests exist.

* E-mail: cuizl@cau.edu.cn

Introduction

The need to increase global food production while also increasing nitrogen (N) use efficiency and limiting environmental costs [e.g., greenhouse gas (GHG) emissions] have received increasing public and scientific attention [1–6]. Coordinated global efforts are particularly critical when dealing with N-related GHG emissions because such emissions and their impacts recognize no borders. The most rapidly developing countries, such as China and India, are becoming central to the issue, not only because these countries consume the most chemical N fertilizer [7,8], but they have also become dominating forces in the production of new N fertilizers in recent decades [7,8]. From 2001 to 2010, global N fertilizer consumption increased from 83 to 105 MT, with 83% of this global increase originating from five rapidly developing countries, specifically China (9.9 MT), India (5.2 MT), Pakistan (0.8 MT), Indonesia (1.1 MT), and Brazil (1.1 MT). In comparison, chemical N fertilizer consumption decreased by 6.5% (0.7 MT) in Western Europe and Central Europe, and increased by only 7.1% (0.8 MT) in the United States over this period [8]. Optimizing N management in these rapidly developing countries clearly has important implications worldwide.

In the past 30 years, the N application rate in many developed economies has been optimized based on recommended systems, and have included soil nitrate (NO_3) and plant testing [9,10], and more recently, remote sensing [11]. However, in rapidly developing countries, small-scale farming with high variability between fields and poor infrastructure in the extension service makes the use of many advanced N management technologies difficult. Fox example, the average area per farm in China is only 0.6 ha, and individually managed fields are generally 0.1–0.3 ha [12]. Therefore, the challenge is to develop agronomically effective and environmentally friendly practices that are applicable to hundreds of millions of smallholder farmers, while producing high yields and reducing N losses.

Decisions regarding the optimal N fertilizer application rate require knowledge of existing soil N supplies, crop N uptake, and the expected crop yield in response to N application [13]. Optimal N rates often vary depending on soil-specific criteria and/or crop management variables such as soil productivity, producer management level, and geographic location [14]. However, the optimal N rate will become more uniform under geographically similar soil and climatic conditions, and when the main factors causing the variation in optimal N rates are either addressed or removed [14].

Our hypothesis is that a regional N management approach could be adopted to accommodate hundreds of millions of small farmers and reduce variation among farms, increase crop yield, and lower the GHG emission intensity of maize production. In China, maize (*Zea mays* L.) is the largest food crop produced, accounting for 37% of Chinese cereal production and 22% of the global maize output in 2011 [15]. Chinese maize production results in some of the most intensive N applications globally, and the resulting enrichment of N in soil, water, and air has created serious environmental problems.

In the present study, we developed a regional N management approach across major maize agroecological regions in China. We also compared grain yield and GHG emissions between the regional N management approach and site-specific N management, and evaluated the potential for increasing grain yields and mitigating GHG emission intensity using this regional N management approach when compared to farmers' practices across each region.

Materials and Methods

Description of China's agroecological maize regions

In China, maize is grown primarily in 4 main agroecological regions and 12 agroecological subregions, including Northeast China (NE1, NE2, NE3, NE4), North China Plain (NCP1, NCP2), Northwest China (NW1, NW2, NW3), and Southwest China (SW1, SW2, SW3) (Fig. 1) [16]. These agroecological subregions were divided based on climatic conditions, terrains, agricultural management practices (e.g., irrigation), and soil types. Detailed information on each of these subregions is provided in Table S1 and Text S1.

Farmers' survey

A multistage sampling technique was used to select representative farmers for a face-to-face, questionnaire-based household survey conducted once a year between 2007 and 2009 [17]. In this study, 5,406 farmers from 66 counties in 22 provinces were surveyed (Table 1). In each province, three counties were randomly selected, three townships were randomly selected in each county, two to five villages were randomly selected in each township, and 20 farmers from the villages were randomly surveyed to collect information on N fertilizer use and grain yield in each farmer's household. This study was approved by a research ethics review committee at the College of Resources and Environmental Science (CRES), China Agricultural University, Beijing, China. Data was collected through an in-house survey, which was conducted by research staff at the College of Resources and Environmental Science. Before beginning the survey, an informed consent information sheet was given to the farmer to read (or in some cases was read to the farmer), and verbal informed consent was requested. Because this study was considered anonymous and each participating household could not be identified directly or indirectly, the research ethics review committee of CRES waived the need for written informed consent from the participants.

On-farm field experiments

In total, 1,726 on-farm maize N fertilizer experiments in 181 counties of 22 provinces were conducted from 2005 to 2010 in the NE ($n = 397$) and NW ($n = 416$) spring maize areas, and in the NCP ($n = 407$) and SW ($n = 506$) summer maize areas. All 66 counties where farm surveys were conducted were included in these 181 counties.

All experimental fields received four treatments without replication: without N fertilizer (N0), medium N rate (MN), 50% and 150% of MN. The amount of N fertilizer for the MN treatment was recommended by local agricultural extension

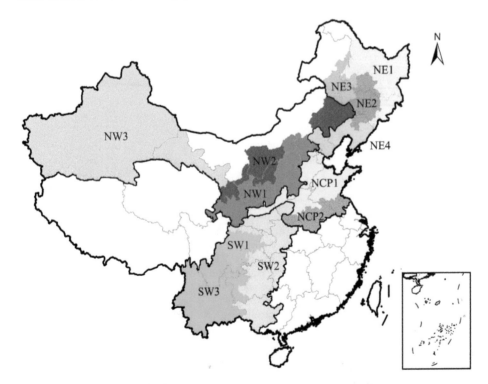

Figure 1. Map showing the four major maize-planting agroecological regions (thick lines, NE, NCP, NW, SW) and their subregions in China (different colors). Northeast China (NE1, NE2, NE3, NE4), North China Plain (NCP1, NCP2), Northwest China (NW1, NW2, NW3), and Southwest China (SW1, SW2, SW3). Here, we show the distribution of maize production in China; the total maize sowing area in the 12 subregions is approximately 32 million hectares, which represents 96% of the total maize production in China.

Table 1. N fertilizer application rate, maize grain yield, N balance, and GHG emission intensity of N fertilizer use, N fertilizer production and other sources in different agro-ecological subregions.

Region & Subregion	n [a]	N rate (kg ha⁻¹)	Grain yield (Mg ha⁻¹)	N balance (kg N ha⁻¹)	GHG emission intensity (kg CO_2 eq Mg⁻¹ grain)			
					N fertilizer use	N fertilizer production	Other sources	Total
NE	1263	195±61	8.91±1.19	32±30	115±38	182±60	50±17	347±131
NE1	361	156±43	8.59±0.79	3±13	96±27	151±40	51±13	298±85
NE2	411	201±59	9.03±1.16	40±28	116±34	183±54	49±15	348±124
NE3	311	226±76	8.80±1.52	68±41	137±46	213±71	51±17	402±164
NE4	180	205±77	8.68±1.50	49±39	124±51	196±74	52±21	373±183
NCP	1983	208±72	7.42±1.24	61±43	148±46	233±76	55±19	436±178
NCP1	1460	206±71	7.68±1.27	54±38	141±49	223±77	54±19	418±180
NCP2	523	217±66	7.14±1.15	76±46	161±45	252±75	57±17	471±174
NW	882	238±107	7.58±1.91	95±68	170±77	261±119	56±27	487±240
NW1	394	234±103	6.93±1.70	98±73	182±80	280±123	58±26	520±282
NW2	289	246±128	8.22±2.50	91±64	163±86	248±129	54±29	466±171
NW3	199	234±83	7.15±1.48	106±47	176±62	272±96	60±20	508±257
SW	1278	250±91	5.45±1.17	144±86	251±93	381±140	78±30	710±319
SW1	427	257±83	5.41±1.13	151±71	263±85	394±127	75±24	732±225
SW2	447	232±90	5.36±1.08	129±93	232±90	358±134	84±34	675±359
SW3	404	272±101	5.59±1.32	160±103	274±100	403±152	75±28	752±375

[a] n: number of observations.

employees based on experience and target yield (1.1 times the average yield of the past 5 years). The median N application rates for the 1,726 sites are shown in Table 2. Approximately one-third of the granular urea was applied by broadcasting at sowing, while the remainder was applied as a side-dressing at the six-leaf stage. All experimental fields received 30–150 kg P_2O_5 (P) ha^{-1} as triple superphosphate and 30–135 kg K_2O (K) ha^{-1} as potassium chloride, based on experience and target yield. All P and K fertilizers were applied by broadcasting before sowing. No manure was used, which is common for maize production in China. Detailed information regarding the N application rate and selected soil chemical properties before maize planting at 1,726 on-farm experimental sites is provided in Table S2.

Individual plots were approximately 40 m^2 (5 m wide and 8 m long). All experiments were managed (including maize variety, density, planting, harvesting, herbicide and insecticide for pests, diseases, and weeds) by local farmers based on a field manual provided by local agricultural extension employees, whereas for the treatments, local agricultural extension employees conducted fertilizer applications. The time of planting and harvest were determined by farmers and differed among sites. Generally, in NE and NW, maize was planted in early May and harvested in late September. Maize was planted from June to October in NCP and from April to August in SW. Plant densities were 50,000–65,000 plants ha^{-1} in NE, 70,000–75,000 plants ha^{-1} in NCP, 65,000–75,000 plants ha^{-1} in NW, and 45,000–50,000 plants ha^{-1} in SW. The locations of the 1,726 experiments were not privately-owned or protected in any way. No specific permits were required for the field studies. The farming operations employed during the experiment were similar to the operations routinely employed on rural farms and did not involve endangered or protected species. All operations were approved by the CRES, China Agricultural University.

Sampling and laboratory procedures

Prior to the experiments, five chemical soil properties were examined. Values were determined based on soil samples from a combined soil sample of the 10–20 cores from depths of 0–20 cm. Soil samples collected before planting were air-dried and sieved through a 0.2-mm mesh. Soil samples were used to measure organic matter content (OM) [18], alkaline hydrolyzable N (AN) [19], Olsen-P [20], NH_4OAc-K [21], and pH [22]. Upon harvest, approximately 2.5×8-m^2 sections of each plot were assessed, and ears were harvested from all plants by hand. The grain yield was adjusted to a moisture content of 15.5%.

A regional N management approach

A guideline for regional N rate was calculated for each subregion through several steps. First, yield data were collected from a large number of N response trials ($n = 1,726$). Grain yield responses to N fertilizer curves were fit using a quadratic model with PROC NLIN (SAS Institute Inc., Cary, NC, USA) to generate yield function equations (the yield significantly ($P<0.05$) responded to N) [23,24]. Next, from the response curve equation at each experimental site, the yield increase (above the yield in the N0 treatment), gross Chinese yuan return at that yield increase (maize grain price times yield), N fertilizer cost (N fertilizer price times N fertilizer rate), and net return to N ratio (gross yuan return minus N fertilizer cost) were calculated for each 1 kg N fertilizer rate increment from 0 to 270 kg N ha^{-1}. Finally, for each incremental N rate, the net return was averaged across all trials in the subregional data set to generate an estimated ratio of the maximum return to N rate, and the corresponding yield across all trials at an N fertilizer:maize grain price ratio [14,25]. In recent

years, the fertilizer:maize grain price ratio has remained relatively stable, and a value of 2.05 was used in this study.

Field-specific N management

In total, grain yield responses to N fertilizer curves were fit for 1,726 on-farm sites, using a quadratic model with PROC NLIN (SAS Institute Inc.) to generate yield function equations (the yield significantly ($P<0.05$) responded to N) [23,24]. The minimum N rate for the maximum net return was calculated from the selected model based on an N:maize price ratio of 2.05.

Nitrogen use efficiency and N balance

Nitrogen use efficiency for each treatment using the partial factor productivity (PFP_N) indices.

$$PFP_N = \frac{Y_N}{F_N} \qquad (1)$$

Where Y_N = Crop yield with N applied;
F_N = Amount of N applied.
Soil surface N balance was calculated as described in the Organization for Economic Co-operation and Development (OECD) [26].

$$N\ balance = N\ input - N\ uptake \qquad (2)$$

where N input is N applied as chemical fertilizer, and N uptake is N in the harvested yield.

$$N\ uptake = Aboveground\ N\ uptake \times Yield \qquad (3)$$

The maize aboveground N uptake requirement per million grams (Mg) grain yield in China was determined previously; spring maize grain yield was <7.5 Mg ha^{-1}, 7.5–9.0 Mg ha^{-1}, 9.0–10.5 Mg ha^{-1}, and 10.5–12.0 Mg ha^{-1}, and N uptake requirements per Mg grain yield were 19.8, 18.1, 17.4 and 17.1 kg, respectively [27]. Summer maize N uptake requirements per Mg grain yield were 20 kg [28].

Estimation of GHG emissions and emission intensity

Total GHG emissions during the entire life cycle of maize production, including CO_2, CH_4, and N_2O, consisted of three components: (1) emissions during N fertilizer application, production and transportation, (2) emissions during P and K fertilizer production and transportation, and (3) emissions from pesticide and herbicide production (delivered to the gate) and diesel fuel consumption during sowing, harvesting, and tillaging operations [29].

$$GHG = (GHGm + GHGt) \times N\ rate + total\ N_2O \times 44/28 \times 298 + GHGothers \qquad (4)$$

where GHG (kg CO_2 eq ha^{-1}) is the total GHG emission, and GHGm is the GHG emission originating from fossil fuel mining as the industry's energy source to N product manufacturing, and was 8.21 kg CO_2 eq kg N^{-1} (Table S3) [30]. GHGt is the N fertilizer transportation emission factor, and was 0.09 kg CO_2 eq kg N^{-1} (Table S3) [30]. N rate is the N fertilizer application rate (kg N ha^{-1}). GHG$_{others}$ represents GHG emission of P and K fertilizer

Table 2. The number of on-farm experiments, maize yield without N, medium N rate, grain yield at the medium N rate and N rate, grain yield, GHG emission intensity of N fertilizer use, N fertilizer production and other sources for regional N management approach and field-specific N management.

Subregion	n[a]	Yield without N (Mg ha^{-1})	Medium N rate (kg ha^{-1})	Yield for medium N rate (Mg ha^{-1})	Regional N management approach							Field-specific N management					
					N rate (kg ha^{-1})	Grain yield (Mg ha^{-1})	GHG emission intensity (kg CO$_2$ eq Mg^{-1} grain)				N rate (kg ha^{-1})	Grain yield (Mg ha^{-1})	GHG emission intensity (kg CO$_2$ eq Mg^{-1} grain)			
							N fertilizer use	N fertilizer production	Other sources	Total			N fertilizer use	N fertilizer production	Other sources	Total
NE1	132	6.40±1.01[b]	153±6	8.98±1.09	150	8.85	91	141	49	280	158±26	8.87±1.11	91±19	149±32	49±7	289±55
NE2	62	6.82±1.23	147±21	9.05±1.51	150	9.18	87	136	49	272	155±25	9.13±1.48	87±14	143±23	51±8	281±42
NE3	126	6.50±1.26	162±15	9.48±1.38	164	9.01	96	151	50	298	165±20	9.10±1.25	92±18	153±29	50±8	295±53
NE4	77	6.92±1.66	204±28	8.93±1.59	188	8.76	113	178	55	346	191±49	8.84±1.48	117±37	183±53	57±9	356±94
NCP1	348	6.58±1.13	194±22	8.23±2.13	178	8.13	115	182	58	355	179±27	8.14±1.19	113±23	185±36	59±9	357±64
NCP2	59	6.91±1.13	213±30	8.67±0.95	177	8.37	111	176	55	342	185±33	7.59±1.12	129±30	208±45	62±9	399±80
NW1	100	6.30±1.06	190±34	8.35±1.03	181	8.13	117	185	59	360	180±44	8.13±1.12	115±29	184±45	59±9	357±77
NW2	309	8.12±1.83	190±20	10.53±1.71	176	10.38	89	141	47	277	182±34	10.48±1.82	91±25	148±37	48±9	288±68
NW3	7	7.23±1.74	221±9	10.33±1.59	219	9.83	118	185	46	349	215±16	9.85±1.56	116±20	185±30	46±7	347±57
SW1	78	5.70±1.30	217±22	7.63±1.20	174	7.46	123	194	63	379	191±39	7.56±1.19	134±34	214±53	63±11	412±93
SW2	368	5.59±1.13	195±22	7.72±1.18	183	7.71	125	197	66	387	184±38	7.77±1.26	125±33	202±52	67±12	394±92
SW3	60	6.00±1.09	207±25	8.29±1.33	186	8.10	121	191	65	376	191±37	8.38±1.33	120±28	192±44	65±11	376±78
National[c]	-	6.60	187	8.69	174	8.56	108	171	56	334	178	8.55	109	185	57	343

[a]n: number of observations.
[b]Mean ± SD.
[c]National values are computed from the regional values weighted by area. The regional weights are as follows:
NE1, 4.5%; NE2, 14.9%; NE3, 4.7%; NE4, 6.4%; NCP1, 25.6%; NCP2, 6.0%; NW1, 10.4%; NW2, 7.3%; NW3, 2.6%; SW1, 3.5%; SW2, 7.9%; SW3, 6.2%.

production and transportation, pesticide and herbicide production and transportation, and diesel fuel consumption (Table S3).

Total N_2O emission included direct and indirect N_2O emissions. Indirect N_2O emissions were estimated with a method used by the International Panel on Climate Change [31], where 1% and 0.75% of ammonia (NH_3) volatilization and nitrate (NO_3^-) leaching, respectively, is lost as N_2O. N_2O emission is calculated based on empirical models. Based on previous reports, the final data set consisted of 10 (30 observations) and 22 (117 observations) studies on direct N_2O emissions for spring maize and summer maize, respectively. Detailed information is provided in Table S4 and Figure S1.

$$\text{Direct } N_2O \text{ emission for spring maize} = 0.576\exp(0.0049 \times N \text{ rate}) \tag{5}$$

$$\text{Direct } N_2O \text{ emission for summer maize} = 0.593\exp(0.0045 \times N \text{ rate}) \tag{6}$$

NH_3 volatilization and N leaching employs the following equation (Cui *et al* 2013, Global Change Biology, main text, Fig. 2) [6].

$$NH_3 \text{ volatilization} = 0.24 \times N \text{ rate} + 1.30 \tag{7}$$

$$N \text{ leaching} = 4.46\exp(0.0094 \times N \text{ rate}) \tag{8}$$

The system boundaries were set as the periods of the life cycle from the production inputs (such as fertilizers, pesticides, and herbicides), delivery of the inputs to the farm gates, and farming operations. We calculated total GHG emissions expressed as kg CO_2 eq ha^{-1} and the GHG emission intensity expressed as kg CO_2 eq Mg^{-1} grain. The change in soil organic carbon content was also not included in our analysis, because it was difficult to detect the small magnitude of the changes that occurred over a short time [32]. The soil CO_2 flux as a contributor to global warming potential (GWP) was also not included in this study because the net flux was estimated to contribute less than 1% to the GWP of agriculture on a global scale [33].

To calculate total GHG emissions and emission intensity, the N rate and corresponding yield of each farm were used for farmers' N practices. The regional N rate and corresponding yield of each subregion were used for the regional N management approach, and the optimal N rate and corresponding yield of each field were used for field-specific N management.

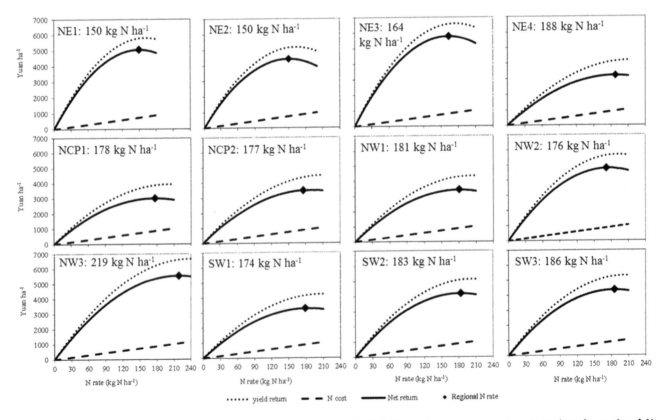

Figure 2. Maize grain yield and fertilizer economic components of calculated net return across N rates using the regional N management approach indicated at the 2.05 price ratio (N price 4.87 yuan kg^{-1} and maize price 2.37 yuan ha^{-1}) in the 12 agroecological subregions. In total, 1,726 N responses trials were used to estimate the regional N rate. The net return is the increase in yield times the grain price at a particular N rate, minus the cost of that amount of N fertilizer. The maximum return is the N rate at which the net return is greatest.

Results

Farmers' Practice

Across all 5,406 farms, maize grain yield averaged 7.56 Mg ha^{-1}, the corresponding N application rate averaged 220 kg ha^{-1}, and the N balance averaged was 69 kg N ha^{-1} (Table S5). Calculated GHG emission intensity averaged 482 kg CO_2 eq Mg^{-1} grain (Table S5), including the contributions of 155, 242, and 85 kg CO_2 eq Mg^{-1} grain from N fertilizer use, N fertilizer production, and other sources, respectively (data not shown).

Large variations were observed in grain yield and N fertilizer application rates across the four main agroecological regions. The N application rates followed the order SW (250 kg N ha^{-1}) ≈ NW (238 kg N ha^{-1}) > NCP (208 kg N ha^{-1}) ≈ NE (195 kg N ha^{-1}). In contrast, the maize grain yields were highest in NE (8.91 Mg ha^{-1}) followed by NW (7.58 Mg ha^{-1}), NCP (7.42 Mg ha^{-1}) and SW (5.45 Mg ha^{-1}). The GHG emission intensity averaged 347, 436, 487, and 710 kg CO_2 eq Mg^{-1} grain for NE, NCP, NW, and SW, respectively (Table 1).

Regional N management approach

Across all 1,726 on-farm experiments, the average grain yield under the N0 treatment, weighted by maize area in each subregion, was 6.60 Mg ha^{-1} and ranged from 5.59 Mg ha^{-1} (SW2) to 8.12 Mg ha^{-1} (NW2) (Table 2). The average medium N rate (MN) recommended by local extension employees, weighted by maize area in each subregion, was 187 kg N ha^{-1} and ranged from 147 kg N ha^{-1} (NE2) to 221 kg N ha^{-1} (NW3). The corresponding grain yield under MN treatment averaged 8.69 Mg ha^{-1} and ranged from 7.63 Mg ha^{-1} (SW1) to 10.53 Mg ha^{-1} (NW2) (Table 2).

Considering all on-farm experiments, the calculated regional N rate based on the cost response to N application rate for the subregions, weighted by maize area in each subregion, averaged 174 kg N ha^{-1} and ranged from 150 kg N ha^{-1} (NE1 & NE2) to 219 kg N ha^{-1} (NW3) (Table 2, Fig 2). The corresponding grain yield averaged 8.56 Mg ha^{-1} and ranged from 7.46 Mg ha^{-1} (SW1) to 10.38 Mg ha^{-1} (NW2) (Table 2). Calculated GHG emission intensity, weighted by maize area in each subregion, averaged 334 kg CO_2 eq Mg^{-1} grain and ranged from 272 kg CO_2 eq Mg^{-1} grain (NE2) to 387 kg CO_2 eq Mg^{-1} grain (SW2).

Based on the maize grain yield response to N application rates in all 1,726 on-farm experiments, the calculated field-specific N rate, weighted by maize area in each subregion, averaged 178 kg N ha^{-1} (Table 2) and ranged from 53 kg N ha^{-1} to 271 kg N ha^{-1} (Table S2), with a coefficient of variation (CV) of 18% (data not shown). The corresponding grain yield averaged 8.63 Mg ha^{-1} (Table 2) and ranged from 4.29 Mg ha^{-1} to 14.91 Mg ha^{-1} (Table S2), with a CV of 19% (data not shown). The calculated GHG emission intensity averaged 343 kg CO_2 eq Mg^{-1} grain (Table 2). The similar N rate, grain yield and GHG emission intensity between the regional N management approach and field-specific N management supported the notion that the regional N rate was close to an economic and environmentally optimal N application (Table 2).

Opportunities to reduce the GHG emission intensity

Compared to farmer's practices, the regional N management approach proposed reducing N fertilizer by 20.9% (220 vs. 174 kg N ha^{-1}). The grain yield would increase by 13.2% (7.56 vs. 8.56 Mg ha^{-1}). The GHG emission intensity would decrease by 30.7%, from 482 to 334 kg CO_2 eq Mg^{-1} grain. The overuse and high variability of N use by farmers has resulted in a high variability in GHG emission intensity, ranging from 364 to

1,399 kg CO_2 eq Mg^{-1} grain (Table S5) with a CV of 43% (data not shown).

Of the 12 agroecological subregions, NE2, NE3, NW1, NW2, SW1, SW2, and SW3 showed the highest potential for N-reduction (>20%), ranging from 21.0% to 31.5% and accounting for 55% of the total maize-sown area. Reduced N rates in other subregions ranged from 3.8% to 18.4% and accounted for 45% of the total maize-sown area. The subregions with a high yield increase potential (>15%; Fig. 3) were NCP2, NW1, NW2, NW3, SW1, SW2, and SW3, with increases ranging from 17.2% to 44.9% and accounting for 44% of the total maize-sown area. Grain yield in other regions ranged from 0.5% to 5.9%, accounting for 56% of the total maize-sown area. Subregions with a high potential to decrease GHG emission intensity (>20%) included NE2, NE3, NCP2, NW1, NW2, NW3, SW1, SW2, and SW3, ranging from 21.8% to 50.0% and accounting for 64% of the total maize-sown area. Reduced GHG emission intensity in other regions ranged from 6.0% to 15.1%, accounting for 36% of the total maize-sown area.

This regional N management approach, if widely adopted in China, regional N fertilizer consumption would be reduced by 1.4 MT (−20.3%), and 91% of this reduction would occur in the NE2, NE3, NCP1, NW1, NW2, SW1, SW2, and SW3 subregions (Table 3). At the same time, Chinese maize production could be increased by 31.9 MT (13.1%), from 244.1 MT to 276.0 MT, when undertaking this regional N management approach (Table 3). Total GHG emissions would be reduced by 18.6 MT eq CO_2 year^{-1} (−16.9%) (from 110.2 to 91.5 MT eq CO_2 year^{-1}) (Table 3), with 91% of this reduction occurring in the NE2, NE3, NCP1, NW1, NW2, SW1, SW2, and SW3 subregions.

Discussion

The current intensive maize system used in farmers' practices in China results in a median yield, high N application, and GHG emission intensity of 7.56 Mg ha^{-1}, 220 kg N ha^{-1}, and 482 kg CO_2 eq Mg^{-1} grain, respectively. These yields and N application rates are higher than the reported global averages (4.81 Mg ha^{-1} and 104.9 kg N ha^{-1}, the N rate calculated based on maize N fertilizer consumption and maize area harvested) for these crops in 2006 [8,15,34] and are similar to the previously reported Chinese averages for maize [35,36]. In comparison, grain yield in central Nebraska, USA, averaged 13.2 Mg ha^{-1} with only 183 kg N ha^{-1}. GHG emission intensity in this region was only 231 kg CO_2 eq Mg^{-1} grain, which was 48% lower than the average for China [37] and 109% lower than the 482 kg CO_2 eq Mg^{-1} grain for individual farmer's practices in China. The median yield and large GHG emission intensity for Chinese maize systems were attributable to the large variation in N application rates among fields. Considering 5,406 farms, N application rates ranged from 46 (only 56% of crop N uptake) to 615 kg N ha^{-1} (414% of crop N uptake). Similar results were reported by Wang et al (2007), showing that one-third of farmers apply too little N, while another one-third of farmers apply too much (n = 10,000) [38].

In small-scale farming, a lack of basic knowledge and information on crop responses to N fertilizer often results in the over- and underapplication of N fertilizer [39,40]. We developed and assessed regional N management approach using large pools of response trial data that have been grouped according to criteria that indicate differing N responses for regions with similar management, climates, and soil. Our guide provides a N application rate that can be used to reduce the potential for N-deficiency or N-surplus, lowers the likelihood of reduced yields and profits, and lessens GHG emissions intensity (particularly N_2O

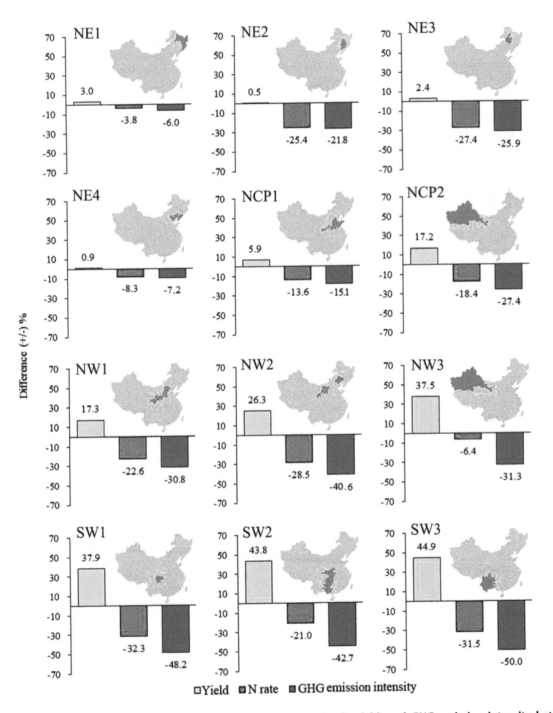

Figure 3. Regional differences (±%) in N application rates, grain yield, and GHG emission intensity between the regional N management approach and farmers' practice in the 12 agroecological subregions. Regional difference (±%) = (regional approach minus farmers' practice)/farmers' practice ×100.

emissions associated with N fertilization). Using a regional N management approach, potential for crop productivity increases and the mitigation of GHG emission intensity are likely to be achieved through a combination of increased N application in regions with a low N input and improved PFP_N in regions where N fertilizer application is already high. Meanwhile, crop N uptake and N use efficiency can improve the ratio split application, with one-third for base dressing and two-thirds for top dressing [36]. Currently, typical farmers' practices apply 50% of the total N fertilizer before planting or at the early growth stage [36,41]. Some

recent practices have indicated that the amount of basal application should be added to the ratio of the top dressing to improve N use efficiency and increase grain yield [36].

The gains in yield and reduced GHG emissions achieved using regional N management approach are significant. Moreover, we believe these benefits can be further improved by applying other best-management strategies to fertilizer (e.g., slow-release N fertilizer, N transformation inhibitors, and fertigation) [42] and related practices that enhance the crop recovery of applied N (e.g., rotation with N fixing crops, precision agriculture management

Table 3. Maize production, N fertilizer consumption and total GHG emission between the regional N rate and farmers' practice in 12 agro-ecological subregions.

Subregion	Area (million ha)	N fertilizer consumption (MT)			Maize production (MT)			Total GHG emission (MT eq CO_2 yr^{-1})		
		Farmers' practice	Regional N rate	Difference [a]	Farmers' practice	Regional N rate	Difference [a]	Farmers' practice	Regional N rate	Difference [a]
NE1	1.45	0.23	0.22	−0.01	12.5	12.8	0.4	3.7	3.6	−0.1
NE2	4.80	0.96	0.72	−0.24	43.8	44.1	0.2	15.2	12.0	−3.3
NE3	1.50	0.34	0.25	−0.09	13.2	13.5	0.3	5.3	4.0	−1.3
NE4	2.06	0.42	0.39	−0.04	17.9	18.0	0.2	6.7	6.2	−0.4
NCP1	8.24	1.70	1.47	−0.23	63.3	67.0	3.7	26.4	23.8	−2.7
NCP2	1.94	0.42	0.34	−0.08	13.9	16.2	2.4	6.5	5.6	−1.0
NW1	3.35	0.78	0.61	−0.18	23.2	27.2	4.0	12.1	9.8	−2.3
NW2	2.36	0.58	0.42	−0.17	19.4	24.5	5.1	9.0	6.8	−2.2
NW3	0.85	0.20	0.19	−0.01	6.1	8.4	2.3	3.1	2.9	−0.2
SW1	1.14	0.29	0.20	−0.09	6.2	8.5	2.3	4.5	3.2	−1.3
SW2	2.54	0.59	0.46	−0.12	13.6	19.6	6.0	9.2	7.6	−1.6
SW3	1.99	0.54	0.37	−0.17	11.1	16.1	5.0	8.4	6.1	−2.3
National [b]	32.23	7.06	5.62	−1.43	244.1	276.0	31.9	110.2	91.5	−18.6

[a]Different mean the different of maize production, N fertilizer consumption, and total GHG emission between regional N rate and farmer's practice.
[b]National values are computed from the regional values weighted by area. The regional weights are as follows:
NE1, 4.5%; NE2, 14.9%; NE3, 4.7%; NE4, 6.4%; NCP1, 25.6%; NCP2, 6.0%; NW1, 10.4%; NW2, 7.3%; NW3, 2.6%; SW1, 3.5%; SW2, 7.9%; SW3, 6.2.

techniques) [42]. While this approach for N fertilizer management should be extended to farmers throughout the entire Chinese cereal production area, it is also relevant to other high-yield cropping systems outside of China. The economic approach to N rate recommendations based on multiple N rate trials has been applied for two to three decades in the U.S. Midwest, and has been more recently "formalized" with the Iowa State MRTN approach for seven Midwestern states [43].

This regional N management approach, if widely adopted in China, could reduce fertilizer N consumption by 20.3%, increase Chinese maize production by 13.1%, and reduce total GHG emissions by 16.9%. Moreover, the recommendations provide reasonable N rates and high net return, and can be easily adopted in rural areas of China where no available soil and/or plant N monitoring facilities exist [44]. The regional N rate can also be used as a reference point for agricultural extension employees without any soil and/or plant N monitoring. In practice, some factors also affect these suggested regional N rates, such as timing of crop rotation, tillage system, and soil productivity [14]. For example, the recommended N rate for soybean following maize rotations is lower than maize following maize rotations [14]. No-till management can delay or reduce residue breakdown, or mineralization, thereby reducing the N supplied from crop residue [14]. Soils where productivity is limited frequently require higher rates of fertilizer N to reach optimum yield. Conversely, lower rates of fertilizer N may be needed to reach optimum yield on highly productive soils [14].

Although this regional N management approach can easily be adopted in rural areas, delivering this technology to millions of farmers is challenging due to the lack of effective advisory systems and knowledgeable farmers. For example, educated young male farmers tend to leave the farming sector for more profitable jobs, leaving farmwork to the older and less educated individuals, especially in low income or remote areas [45]. In addition, adding more N fertilizer based on the regional N rate is difficult for farmers with low incomes or in remote areas. The Chinese central government has been aware of this problem and has attempted to provide agricultural technologies to these areas. For example, China has launched national programs for soil testing and fertilizer recommendations since 2005. In 2009, 2,500 counties in China were involved in the programs, receiving a total of 1.5 billion yuan from the Chinese central government [40].

Although the on-farm trials were conducted by local farmers in the same counties as the farmers' surveys (including experimental counties), the management and environment is not always the same for on-farm trials and farmers' surveys. While gains in grain yield and GHG were achieved by farmers using the trials, we believe that the majority of these gains can be realized in practice in many counties if improved agronomic and N management techniques are adopted. The management and environment differed among four maize regions; thus, N losses may also differ. For example, the annual direct N_2O emission accounted for 0.92% of the applied N with an uncertainty of 29%. The highest N_2O fluxes occurred in East China as compared with the lowest

fluxes in West China [46]. In this study, we use the different exponential relationships of the N application rate and N_2O fluxes for spring maize and summer maize, respectively. However, developing N loss models at the regional or subregional scale is difficult due to insufficient field measurement data in China. Long-term field observations covering all subregions are required to accurately assess farming potential and mitigate GHG emissions.

Supporting Information

Figure S1 Relationships between the N application rate and direct N_2O emissions for spring maize (A) and summer maize (B) production in China based on a meta-analysis. The direct N_2O emission data was taken from Table S4.

Table S1 The criteria and values for the sub-regional divisions.

Table S2 The site, year, soil type, irrigation, crop rotations, soil organic matter (SOM) content, alkaline hydrolyzable N (AN), Olsen-P (AP), NH_4OAc-K (AK), pH, medium N rate (MN), recommended P_2O_5 rate (RP), recommended K_2O rate (RK), grain yield without N fertilizer, yield at 50% MN, yield at 100% MN, yield at 150% MN, economic optimal N rate (EONR), yield at EONR, and GHG emissions intensity at EONR for all 1,726 on-farm experiments.

Table S3 GHG emission factors of agricultural inputs.

Table S4 The site, year, annual mean precipitation, temperature, soil organic matter (SOM), total N content, pH, N rate, grain yield, and direct N_2O emissions at different experimental sites.

Table S5 Descriptive statistics of the surveyed farms N fertilizer application rate, maize grain yield, PFP_N, N balance and GHG emission intensity for 5,406 farmed fields between 2007 and 2009 in China.

Text S1 Detailed information for each of these regions.

Author Contributions

Conceived and designed the experiments: FsZ XpC. Performed the experiments: FsZ XpC. Analyzed the data: LW ZlC WfZ. Contributed reagents/materials/analysis tools: LW ZlC. Wrote the paper: LW. Designed the NH3 volatilization and N leaching models used in analysis: ZlC.

References

1. Tilman D, Fargione J, Wolff B, D'Antonio C, Dobson A, et al. (2001) Forecasting agriculturally driven global environmental change. Science. 292: 281–284.

2. Tilman D, Cassman KG, Matson PA, Naylor R, Polasky S (2002) Agricultural sustainability and intensive production practices. Nature. 418: 671–677.

3. Conley D J, Paerl H W, Howarth R W, Boesch D F, Seitzinger S P, et al. (2009) Controlling eutrophication: nitrogen and phosphorus. Science. 323: 1014–1015.

4. Tilman D, Balzer C, Hill J, Befort BL (2011) Global food demand and the sustainable intensification of agriculture. Proc. Natl. Acad. Sci. USA.108: 20260–20264.

5. Zhang F, Cui Z, Fan M, Zhang W, Chen X, Jiang R (2011) Integrated soil-crop system management: reducing environmental risk while increasing crop productivity and improving nutrient use efficiency in China J Environ. Qual. 40: 1051–1057.

6. Cui Z, Yue S, Wang G, Meng Q, Wu L, Yang Z, et al. (2013) Closing the yield gap could reduce projected greenhouse gas emissions: a case study of maize production in China. Global Change Biol. 19: 2467–2477.

7. Zhang F, Cui Z, Chen Z, Ju X, Shen J, et al. (2012) Chapter one-Integrated nutrient management for food security and environmental quality in China. In Adv. Agron. ed Donald L S (Academic Press) 1–40 p.

8. IFA IFA Statistics (Paris: International Fertilizer Industry Association). Available at: www.fertilizer.org/ifa/HomePage/STATISTICS. Accessed 2013 Sept 6.

9. Soper R, Huang P (1963) The effect of nitrate nitrogen in the soil profile on the response of barley to fertilizer nitrogen Can. J. Soil Sci. 43: 350–358.

10. Wehrmann J, Scharpf HC, Kuhlmann H (1988) The Nmin method – an aid to improve nitrogen efficiency in plant production, In Nitrogen Efficiency in Agricultural Soils ed Jenkinson D S, Smith K A (Netherlands: Elsevier Applied Science) 38–45 p.

11. Gebbers R, Adamchuk VI (2010) Precision agriculture and food security. Science. 327: 828–831.

12. Chen X P, Cu Z L, Vitousek P M, Cassman K G, Matson P A, et al. (2011) Integrated soil-crop system management for food security. Proc. Natl. Acad. Sci. USA 108 6399–6404.

13. Dobermann A, Witt C, Abdulrachman S, Gines H, Nagarajan R, et al. (2003) Estimating indigenous nutrient supplies for site-specific nutrient management in irrigated rice. Agron. J. 95: 924–35

14. Sawyer J, Nafziger E, Randall G, Bundy L, Rehm G, Joern B (2006) Concepts and rationale for regional nitrogen rate guidelines for corn. Iowa: Iowa State University, University Extension. 15–24 p.

15. FAO FAOSTAT–Agriculture Database. Available: http://faostat.fao.org/site/339/default.aspx. Accessed 2013 Sept 6.

16. National Bureau of Statistics of China. China Statistical Yearbook. Available: http://www.stats.gov.cn/tjsj/ndsj/. Accessed 2013 Sept 6.

17. Etimi N A, Solomon VA (2010) Determinants of rural poverty among broiler farmers in Uyo, Nigeria: implications for rural household food security. J. Agric. Soc. Sci. 6: 24–28.

18. Walkley A (1947) A critical examination of a rapid method for determining organic carbon in soils-effect of variations in digestion conditions and of inorganic soil constituents. Soil Sci. 63: 251–264.

19. Khan S, Mulvaney R, Hoeft R (2001) A simple soil test for detecting sites that are nonresponsive to nitrogen fertilization. Soil Sci. Soc. Am. J. 65: 1751–1760.

20. Olsen SR (1954) Estimation of available phosphorus in soils by extraction with sodium bicarbonate (Washington, DC: US Department of Agriculture)

21. van Reeuwijk LP (1993) Procedures for soil analysis (International Soil Reference and Information Centre).

22. Richards LA (ed) (1954) Diagnosis and improvement of saline and alkali soils (Washington, DC: US USDA. U.S. Gov. Print. Office).

23. Wallach D, Loisel P (1949) Effect of parameter estimation on fertilizer optimization Appl. Stat. 641–651.

24. Magee L (1990) R^2 measures based on Wald and likelihood ratio joint significance tests. American Statistician. 44: 250–253.

25. Hoben J, Gehl R, Millar N, Grace P, Robertson G (2011) Nonlinear nitrous oxide (N_2O) response to nitrogen fertilizer in on-farm corn crops of the US Midwest. Global Change Biol. 17: 1140–1152.

26. OECE. Environmental indicators for agriculture: Methods and results (Paris: Organisation for Economic Co-operation and Development). Available: www.oecd.org/greengrowth/sustainable-agriculture/1916629.pdf. Accessed 2013 Sept 6.

27. Hou P, Gao Q, Xie R, Li S, Meng Q, et al. (2012) Grain yields in relation to N requirement: Optimizing nitrogen management for spring maize grown in China. Field Crop Res. 129: 1–6.

28. Meng Q F (2012) Strategies for achieving high yield and high nutrient use efficiency simultaneously for maize (Zea mays L.) and wheat (Triticum aestivum L.), Ph.D. Diss. China Agriculture University.

29. Forster P, Ramaswamy V, Artaxo P, Berntsen T, Betts R, et al. (2007) Changes in atmospheric constituents and in radiative forcing In Climate Change. The Physical Science Basis Contribution of Working Group I to the Fourth Assessment Report of the Intergovernmental Panel on Climate Change. ed Solomon S, Qin D, Manning M, Chen Z, Marquis M, Averyt K B, Tignor M, Miller H L 2007(Cambridge: Cambridge University Press).

30. Zhang WF, Dou ZX, He P, Ju XT, Powlson D, et al. (2013) New technologies reduce greenhouse gas emissions from nitrogenous fertilizer in China Proc. Natl. Acad. Sci. USDA 110: 8375–8380

31. Klein CD, et al. (2006) IPCC Guidelines for National Greenhouse Gas Inventories Chapter 11: N_2O emissions from managed soils, and CO_2 emissions from lime and urea application avaluable at: www.ipcc-nggip.iges.or.jp/public/2006gl/pdf/4_Volume4/V4_11_Ch11_N2O&CO2.pdf

32. Conant RT, Ogle SM, Paul EA, Paustian K (2010) Measuring and monitoring soil organic carbon stocks in agricultural lands for climate mitigation Front. Ecol. Environ 9: 169–173.

33. IPCC 2007Climate Change 2007: Mitigation. Contribution of Working Group III to the Fourth Assessment Report of the Intergovernmental Panel on Climate Change. ed Smith P, et al(Cambridge: Cambridge University Press).

34. Heffer P (2009) Assessment of fertilizer use by crop at the global level 2006/07–2007/08. International Fertilizer Industry Association. (Paris, France).

35. Cui Z (2005) Optimization of the nitrogen fertilizer management for a winter wheat-summer maize rotation system in the North China Plain - from field to regional scale Ph.D. Diss. China Agriculture University. (Chinese with English abstract).

36. Cui Z, Chen X, Miao Y, Zhang F, Sun Q, et al. (2008) On-farm evaluation of the improved soil N-based nitrogen management for summer maize in North China. Plain Agron. J. 100: 517–525.

37. Grassini P, Cassman KG (2012) High-yield maize with large net energy yield and small global warming intensity Proc. Natl. Acad. Sci. USA 109: 1074–1079.

38. Wang JQ (2007) Analysis and evaluation of yield increase of fertilization and nutrient utilization efficiency for major cereal crops in China Ph.D. Diss. China Agriculture University.

39. Huang J, Hu R, Cao J, Rozelle S (2008) Training programs and in-the-field guidance to reduce China's overuse of fertilizer without hurting profitability J. Soil Water Conserv. 63: 165A–167A.

40. Cui Z, Chen X, Zhang F (2010) Current nitrogen management status and measures to improve the intensive wheat–maize system in China AMBIO. 39: 376–384.

41. Chen XP (2003) Optimization of the N fertilizer management of a winter wheat/summer maize rotation system in the Northern China Plain Ph.D. diss. Univ.of Hohenheim.

42. Good AG, Beatty PH (2011) Fertilizing nature: a tragedy of excess in the commons. Plos Biol. 9: e1001124.

43. Iowa State University – Agronomy Extension. Corn Nitrogen Rate Calculator. Available: extension.agron.iastate.edu/soilfertility/nrate.aspx. Accessed 2013 Sept 6.

44. Zhu Z, Chen D (2002) Nitrogen fertilizer use in China – Contributions to food production, impacts on the environment and best management strategies Nutr. Cycl. Agroecos. 63: 117–127.

45. Barning R (2008) Economic evaluation of nitrogen application in the North China Plain Ph.D. diss. Univ. of Hohenheim.

46. Lu Y, Huang Y, Zou J, Zheng X (2006) An inventory of N_2O emissions from agriculture in China using precipitation-rectified emission factor and background emission. Chemosphere. 65: 1915–1924.

2

Impacts of Organic and Conventional Crop Management on Diversity and Activity of Free-Living Nitrogen Fixing Bacteria and Total Bacteria Are Subsidiary to Temporal Effects

Caroline H. Orr[1][¤], Carlo Leifert[2], Stephen P. Cummings[1]*, Julia M. Cooper[2]

1 Faculty of Health and Life Sciences, Northumbria University, Newcastle-Upon-Tyne, United Kingdom, **2** Nafferton Ecological Farming Group, Newcastle University, Nafferton Farm, Stocksfield, Northumberland, United Kingdom

Abstract

A three year field study (2007–2009) of the diversity and numbers of the total and metabolically active free-living diazotophic bacteria and total bacterial communities in organic and conventionally managed agricultural soil was conducted using the Nafferton Factorial Systems Comparison (NFSC) study, in northeast England. Fertility management appeared to have little impact on both diazotrophic and total bacterial communities. However, copy numbers of the *nifH* gene did appear to be negatively impacted by conventional crop protection measures across all years suggesting diazotrophs may be particularly sensitive to pesticides. Impacts of crop management were greatly overshadowed by the influence of temporal effects with diazotrophic communities changing on a year by year basis and from season to season. Quantitative analyses using qPCR of each community indicated that metabolically active diazotrophs were highest in year 1 but the population significantly declined in year 2 before recovering somewhat in the final year. The total bacterial population in contrast increased significantly each year. It appeared that the dominant drivers of qualitative and quantitative changes in both communities were annual and seasonal effects. Moreover, regression analyses showed activity of both communities was significantly affected by soil temperature and climatic conditions.

Editor: A. Mark Ibekwe, U. S. Salinity Lab, United States of America

Funding: This work was funded by the following: European Community financial participation under the Seventh Framework Programme for Research, Technological Development and Demonstration Activities, for the Integrated Project NUE CROPS EU-FP7 222–645. The Yorkshire Agricultural Society. www.yas.co.uk. Grant NPD/JMD/08/72. Nafferton Ecological Farming Group. http://research.ncl.ac.uk/nefg. Institutional funding was provided by Northumbria University Research Development Fund. http://www.northumbria.ac.uk. The funders had no role in study design, data collection and analysis, decision to publish, or preparation of the manuscript.

Competing Interests: The authors have declared that no competing interests exist.

* E-mail: stephen.cummings@unn.ac.uk

¤ Current address: School of Science and Engineering, Teesside University, Middlesbrough, United Kingdom

Introduction

Yields of arable crops depend on sufficient reservoirs of plant available nitrogen in agricultural soils. However, as conventional fertility management using inorganic nitrogen fertiliser is becoming increasingly expensive and is recognised as having significant detrimental effects on the environment [1], there is growing interest in more sustainable systems that can exploit biologically fixed nitrogen or use inorganic nitrogen as efficiently as possible.

One microbiological approach to improve sustainability is to enhance the activity of the nitrogen fixing bacteria in soil [2] and to optimise the effects of land use [3], crop management [4,5], N management [4], and seasonal variations [6] on N fixation processes, especially by free-living diazotrophs.

Fertility management, crop protection and crop rotation have all been shown to exert significant effects on the soil microbial communities present in organically and conventionally managed agricultural soils. Previous studies that looked at the impact of farm management on the function and diversity of these communities report the most significant factor affecting soil

microbial communities is the fertility management regime [7,8]. However, results are equivocal and studies have mostly focussed on comparing farming systems over a single season. Here we extend these studies by exploring the effects of different organic and conventional farm management practices on the total bacterial and free-living N fixing community using a factorial design that allows us to investigate the individual effects of crop protection practices and fertility management over several seasons. In general, organic fertility management systems, that include the application of farmyard manure, and the use of diverse crop rotations have been shown to have a positive effect on microbial biomass, diversity and activity [9,10,11,12], when compared with conventional systems. These differences are mainly attributed to; the increased organic C added as manure; lower background levels of readily-available nitrogen, and pH values that are, on average, closer to neutral in organically managed soils [13,14]. As nitrogen fixation is energy-expensive, it is reliant on carbon sources that are more abundant and are retained longer in organically managed compared to conventional soils [15]. Therefore, organic soils are more likely to offer optimal conditions for nitrogen fixation and it is perhaps unsurprising that increases in soil organic carbon have been shown to stimulate nitrogen fixation, although results have been inconsistent [16,17,18,2]. Additional drivers of the activity of the diazotrophic community have been identified as the soil microbial biomass and total nitrogen [3] both of which are typically higher in organic systems.

Other secondary effects of fertility management could also be significant, in particular, changes in soil pH, which is considered a predictor of soil microbial community composition [19]. Hallin et al, [20] found that pH affected total bacterial community composition among soils treated with different fertilizers. Phosphorus can also stimulate nitrogen fixation as it is required for microbial energy production. Reed et al, [21] observed doubling of nitrogen fixation in response to the addition of phosphorus. Most studies discussing free-living N fixing bacteria have described results from a single season (e.g. [22,23]. Since free-living N fixers are known to be very sensitive to environmental conditions, it is important to establish whether crop management effects are consistent across dates within a given year, and over several growing seasons.

Crop protection measures could also potentially affect the soil microbial community. Conventional farmers can use a complex mixture of pesticides to protect their crops [24]; whereas, organic farmers rely on the diversification of crops in the field (intercropping) and over time (crop rotation), use of resistant varieties, optimal timing of tillage for weed control, and use of a limited range of organically approved pesticides [25]. While conventional crop protection measures include the use of chemicals that are toxic to specific organisms, the majority of studies into the effects of chemical pesticides on the soil microbial community have found that they do not significantly affect microbial populations when used at the correct dose [26,27,28]. Nitrogen fixing bacteria are thought to be especially sensitive to pesticides [29]. However, most of the work looking at the effects of pesticides has been carried out on symbiotic diazotrophs. For example, it was shown both *in vitro* and *in vivo* that around 30 different pesticides have a negative effect on the relationship between *S. meliloti* and alfalfa probably due to a disruption in the chemical signalling between the bacteria and its host [30,31].

In this study we utilise an existing factorial field trial that enables comparisons between elements of organic and conventional systems, including fertility management, crop protection and crop rotation, to be analysed over several years. Previously, we have used this trial to demonstrate how organic and conventional crop rotations affect the bulk soil microbial community with emphasis on free-living diazotrophs [23] within a single growing season. In this paper the effects of fertility management and crop protection as well as sample date and sample year, on the general total bacterial and diazotrophic communities over three years are reported.

Materials and Methods

Soil Sampling

The soil (sandy loam; Stagnogley) used in this study was taken from the Nafferton Factorial Systems Comparison (NFSC) study, a field trial based at Nafferton Farm in the Tyne Valley, northeast England. The NFSC was established in 2001 and consists of a series of four field experiments established within four replicate blocks. Each experiment is a split split-plot design with three factors. The main factor is crop rotation: an eight year, conventional cereal intensive rotation is compared to an eight year, diverse legume intensive organic crop rotation. Each crop rotation main plot is split to compare two levels of crop protection:

organic (ORG CP; according to Soil Association organic farming standards [32]) and conventional (CON CP; following British Farm Assured practice). Each crop protection subplot is further split into two fertility management sub-subplots: organic (ORG FM) and conventional (CON FM). Each of the four field experiments has the same design, but was begun in a different year to allow a diversity of crops to be grown in the trial in any given year. In this study soils were taken from potato plots (6 x 24 m in size) grown in 2007, 2008 and 2009 on three dates in each year (March, July, September). Soil samples in March were taken from bare soil which had been mouldboard ploughed the previous autumn, prior to the application of any fertility or crop protection treatments. Before samples were taken in June, the soil had undergone secondary tillage and potato planting, as well as frequent ridging for weed control. Pesticides had been applied to CON CP treatments and mineral fertilisers to CON FM treatments. Compost was applied to ORG FM treatments in April prior to potato planting. Final samples were taken post harvest. Prior to potato harvest CON CP treatments were treated with a chemical defoliant, while ORG CP plots were mechanically defoliated. All potato crops followed a winter cereal the previous year; however, in 2007, the potatoes were in a conventional crop rotation following a crop of winter barley that followed two previous years of winter wheat following a grass/clover ley. In contrast, both the 2008 and 2009 potato crops were grown in an organic crop rotation following a preceding crop of winter wheat that followed a grass/clover ley.

Full details of the organic and conventional fertility management and crop protection practices used in the potato crop and for the preceding year are shown in Table 1.

Soils were sampled and a standard set of soil properties (pH, soil organic C, soil total N, NO_3-N, NH_4-N, soil basal respiration (SBR) and Mehlich 3-extractable P, K, and Fe) were analysed as described in Orr *et al*, [23]. Weather conditions at the experimental site were monitored on an hourly basis using a Delta-T GP1 datalogger with sensors. Mean results for soil temperature and rainfall in the 14 days prior to each soil sampling occasion are shown in Table S1.

Nucleic Acid Extraction and PCR

RNA was extracted from 0.25 g of soil with the UltraClean microbial RNA isolation kit (MoBio) and reverse transcribed with the Superscript II reverse transcriptase kit (Invitrogen). DNA was extracted from 0.25 g of soil using the UltraClean Soil DNA extraction kit (MoBio). The *nifH* gene was amplified using an adapted method first described by Wartiainen et al [33]. Initially a 360 bp fragment is amplified using using PolF and PolR primers [34] followed by nesting with PolFI and AQER-GC30 primers [33]. To amplify the total bacterial community, the V3 variable region of 16S rRNA was amplified using V3FC and V3R primers [35]. Full PCR conditions and primer sequences are described in Orr et al, [23].

DGGE

DGGE was carried out using the D-Code system (Bio-Rad Laboratories) as described by Baxter & Cummings [35]. PCR products were electrophoresed through gels containing 35–55% denaturing gradient at 200 V for either 6 (*nifH*) or 4.5 (16S rRNA) hours. Bands were identified and relative intensities were calculated with Quantity One software (Bio-Rad). Shannon's diversity index (H') was calculated by the formula $H' = -\Sigma pi \ln(pi)$, where pi is the ratio of intensity of band i compared with the total intensity of the lane.

Table 1. Crop protection protocols and fertility management used in the NFSC experiments for 2006, 2007, 2008 and 2009 under organic crop protection (ORG CP) or conventional crop protection (CON CP) and organic fertility management (ORG FM) or conventional fertility management (CON FM).

Current crop	
Potatoes (2007–9)	
ORG CP	mechanical weeding (ridging); copper-oxychloride[b] (23 kg/ha)
CON CP	aldicarb[d] (33.5 kg/ha); linuron[a] (3.5 L/ha); fluazinam[c] (1.5 L/ha); mancozeb and metalaxyl-M[c] (4.7 kg/ha); oiquat[e] (2 L/ha)
ORG FM	composted cattle manure (equivalent to 180 kg total N/ha with 2–9% of total N in plant available forms; 2–17 kg total P$_2$O$_5$/ha; 5–149 kg total K$_2$O/ha)
CON FM	0:20:30 (134 kg P$_2$O$_5$/ha; 200 kg K$_2$O/ha); Nitram (180 kg N/ha)
Previous crop	
Winter barley (2006)	
ORG CP	mechanical weeding (finger weeder)
CON CP	Pendimethalin[a] (2.5 L/ha); isoproturon[a] (1.5 L/ha); Duplosan[a] (1 L/ha); Acanto[b] (0.4 L/ha); Proline[b] (0.4 L/ha); Corbel[b] (0.5 L/ha); Fluroxypyr[b] (0.75L/ha); Amistar[b] (0.25 L/ha); Bravo 500[b] (0.5 L/ha); Cleancrop EPX[b] (0.4 L/ha)
ORG FM	no amendment
CON FM	0:20:30 (64 kg P2O5/ha; 96 kg K2O/ha); Nitram (170 kg N/ha)
Winter Wheat (2007–2008)	
ORG CP	mechanical weeding (finger weeder)
CON CP	isoproturon[a] (6 L/ha); Optica[a] (1 L/ha); Pendimethalin[a] (1.5 L/ha); Corbel[b] (0.2 L/ha); Cleancrop EPX[b] (1.25 L/ha); Bravo 500[b] (1.75 L/ha); chlormequat[c] (2.3 L/ha); Tern[b] (0.15 L/ha); Twist[b] (0.25 L/ha)
ORG FM	no amendment
CON FM	0:20:30 (64 kg P2O5/ha; 96 kg K2O/ha); Nitram (210 kg N/ha)

[a]herbicide;
[b]fungicide;
[c]growth regulator;
[d]nematicide;
[e]desiccant.

qPCR

Reactions were set up using SYBR green (Thermo Fisher Scientific) according to Orr et al, [23] with the Rotor-Gene RG 3000 (Corbett Research). PolF and PolR primers were used for *nifH* qPCR, and Eub338 and V3R were used for total bacteria qPCR. A standard curve was set up using 10-fold dilutions of pGEM-T Easy vector plasmid DNA containing either the *nifH* gene of *Rhizobium* sp. strain IRBG74 bacterium [36] or the 16S rRNA gene of *Pseudomonas aeruginosa* NCTC10662. Each soil extraction, no-template control, and standard curve dilution was replicated three times. Average copy number was converted into copies of the gene per g of oven dry soil.

Standard deviation was determined (by the Rotor-Gene 6 software [Corbett Research]) on the replicate threshold cycle (*CT*) scores. qPCR was repeated if the deviation was above 0.4. Samples were considered to be below reasonable limits of detection if the *CT* score was above 30 [37]. In the system used in this study, this would equate to results below 1.0×10^4 copies per g of soil being rejected. All no-template control results fell below this threshold (35.4±2.8). The standard curve produced was linear ($r^2 = 0.98$), and the PCR efficiency was 0.9.

Sequencing

All sequencing was carried out on a 3130 genetic analyzer (Applied Biosystems). DGGE bands of interest were excised from the gel. DNA was eluted into 10 µl of sterile water and used as the template in the *nifH* PCR. The process was repeated until the band of interest was isolated. The PCR product was then cleaned up using ExoSAP-IT. PCR products were then purified using ethanol precipitation. Sequence data was analyzed using the NCBI BLAST tool.

Statistical Analysis

In all tests, significant effects/interactions were those with a *P* value of 0.05. Treatment effects were analyzed by analysis of variance of a linear mixed effects model, using the lme function in R version 2.6.1 [38] with the maximum likelihood method and the random error term (block/year/date/crop protection) specified to reflect the nested structure of the design [39]. The combined data for all years were analyzed first, and where interaction terms were significant, further analyses were conducted at each level of the interacting factor. Where analysis at a given level of a factor was carried out, that factor was removed from the random error term. The normality of the residuals of all models was tested with QQ plots, and data were log-transformed when necessary to meet the criteria of normal data distribution. Differences between main effects were tested by analysis of variance (ANOVA). Differences between years or sample dates (both across years or within a year) were tested with Tukey contrasts in the general linear hypothesis testing (glht) function of the multcomp package in R. A linear mixed effects model was used for the Tukey contrasts, containing a year or sample date main effect with the random error term specified as described above.

Step-wise regression was carried out in Minitab [40] using the results over the three years and three sample dates for *q*PCR and Shannon's diversity index data as response variables and the measured soil parameters listed above (pH, NO_3^-, NH_4^+, soil basal respiration, total N and organic C) as well as environmental variables (average soil temperature and total rainfall in the 14 days prior to sampling) as explanatory variables.

DGGE data were analyzed by detrended correspondence analysis (DCA) on relative intensities followed by direct ordination with Monte Carlo permutation testing. Direct ordination was either by canonical correspondence analysis (CCA) or redundancy discriminate analysis (RDA), depending on the length of the DCA axis (where an axis of $>3.5 =$ CCA and an axis of $<3.5 =$ RDA) using CANOCO 4.5 and CANODRAW for Windows [41].

Results

Diversity and Expression of *nifH* and 16S mRNA Transcripts and Genes

A single band of 360 bp, corresponding to the expected *nifH* mRNA transcript, was successfully amplified from RNA extracted from all 2007 and 2009 plots. However, the *nifH* mRNA transcript could not be detected in certain plots in September 2008. When using the qPCR approach the CT score for the *nifH* mRNA transcript was below the reasonable limits of detection for all plots at all sample dates in 2008. In contrast, acceptable copy numbers of the 16S mRNA transcript were successfully amplified from all 2008 samples suggesting that the *nifH* gene was not being expressed at certain dates in 2008 rather than a problem with the extraction and amplification protocol.

When *nifH* RNA diversity indices (H') from 2007 and 2009 were analyzed (Table 2), the year, sample date and sample date × year interaction terms were all significant, while crop protection and fertility management factors did not contribute to a significant proportion of the variation in results. H' was significantly higher overall in 2007 and generally, the June sample date had the lowest diversity across the three years. However, when separate analyses were conducted for each year, sample date was highly significant for both 2007 and 2009 (P = 0.002 in both years). In both years June samples had the lowest *nifH* mRNA transcript diversity, although H' values for June 2009 were extremely low compared with June 2007 (Fig. 1). In addition, September 2007 *nifH* mRNA transcript diversity was significantly higher than the other two sample dates in that year, whereas in 2009, there was no difference in *nifH* mRNA transcript diversity between March and September sample dates.

In contrast to the RNA results, analysis of the *nifH* DNA-DGGE diversity results showed that year was not a significant factor but there was a significant interaction between sample date and year (Table 2); again, crop protection and fertility management were not significant factors in the model. Since the year × sample date term was significant, a separate analysis was conducted for each year for both RNA and DNA extractions (Fig. 1). This showed that the ranking of dates for DNA-DGGE diversity was not the same in each year. In 2007 and 2009 H' was highest for the March sampling date, while in 2008 it was highest in June.

The diversity of the active bacterial community was also analyzed (Table 2). ANOVA indicated that overall diversity was highest in 2009, and within a given year was greatest in March; however, there were significant year by date interactions. These are illustrated in Fig. 1 which shows that sample date had no effect on bacterial community diversity in 2009, while on the other two dates there were some differences among sample dates.

In contrast to *nifH* community diversity, the overall bacterial community diversity was affected by crop management practices. There was also a significant interaction between sample date, fertility management and crop protection. When only the March

samples were analyzed across all three years, there was a significant crop protection by fertility management interaction (P = 0.007), although neither factor had a significant main effect. Fig. 2 shows that in March of every year, highest bacterial diversity was measured in the fully conventionally managed plots. However, on the other two sampling dates, crop management had no effect on bacterial community diversity and year was the only significant factor in the model. For all sample dates, highest bacterial community diversity was measured in 2009.

The diazotrophic and total bacterial community composition were further analysed using constrained ordination, for each sample date, to determine how soil biochemical properties measured on the sample dates and environmental conditions in the two weeks prior to sampling may be driving community composition. The importance of the fertility management and crop protection treatments as drivers of community composition were also investigated in the constrained ordinations. For *nifH*, although fertility management and crop protection did not represent a significant portion of the variance on any of the sample dates, factors that were significantly affected by fertility management on all sample dates did contribute significantly to the variation in *nifH* community structure. Specifically, soil basal respiration (greater under organic FM, P<0.001) and nitrate and ammonium (both greater under conventional FM, P<0.001 and = 0.003 respectively) were correlated with changes in *nifH* diversity at certain dates. Factors associated with fertility management also contributed to a much greater proportion of the variance when analysed as separate factors rather than grouped as one management factor (Table 3). The constrained ordination did, however show that crop management significantly affected total bacterial community composition in June 2007 and June 2009 and that fertility management significantly affected total bacterial community composition in September 2008. The significant effect of fertility management in 2008 coincides with pH also significantly affecting total bacterial community composition (Table 3).

For the *nifH* community attempts were made to sequence all bands on the gels. This resulted in 22 bands being sequenced and identified from the DGGE gels. The sequences were around 200 bp in length and enabled gross taxonomic resolution but were too short for higher level phylogenetic affiliation to be determined. Sequence data is available at the GenBank database under accession numbers JQ618105–JQ618126. Table S2 shows the closest match from the NCBI database. Of the 22 sequences, 17 were from uncultured taxa; 10 sequences belonged to *Alpha-Proteobacteria*, 9 belonged to *Beta-Proteobacteria*, 2 belonged to *Gamma-Proteobacteria* and 1 belonged to the order Clostridia. The remaining 3 bands were identified as belonging to *Rhizobium huautlense*. By analyzing the relative intensities of the sequenced bands using ANOVA (data not shown) it was found that management type did not significantly affect the presence or intensity of any of the sequenced bands.

Quantification of the *nifH* and 16S mRNA Transcripts and Genes with qPCR

The predominant factors affecting *nifH* mRNA transcripts and DNA copy numbers and the 16S ribosomal mRNA transcript copy numbers were Year and Sample date, although in some cases crop management practices also affected these parameters (Table 4). On average there were significantly more copies of the *nifH* mRNA transcript detected in 2007 compared with 2009. September samples also had more copies of *nifH* mRNA; however, there were strong interactions between Year and Sample date, and Year and Fertility Management. For this reason a separate year by year analysis was conducted. In both years Sample date was the

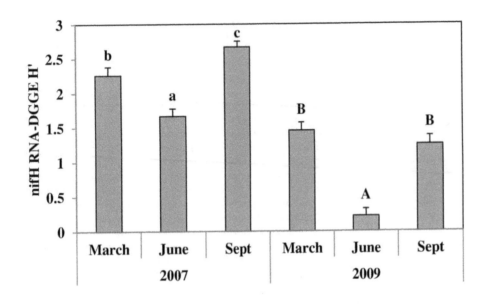

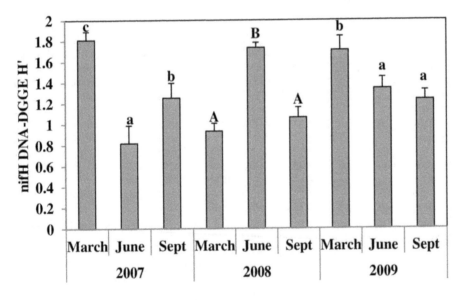

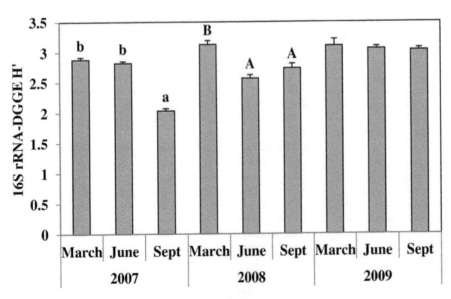

Figure 1. Showing Shannon's diversity indices of the metabolically active and total diazotrophic bacteria and the total bacterial communities derived from the DGGE analyses. The *nifH* mRNA transcripts are represented by the top, the *nifH* gene by the middle and the 16S mRNA transcript by the bottom panels respectively. Bars labelled with the same letter in the same year are not different ($P = 0.05$). Bars are standard errors (n = 16).

predominant factor affecting *nifH* mRNA transcript copy numbers. Highest numbers were detected in September, although in 2009 significantly lower numbers were detected in June while in 2007 March and June results did not differ (Fig. 3). In both years the use of organic fertility management always resulted in higher levels of *nifH* gene expression than conventional fertilisation.

Quantities of the *nifH* gene were also strongly affected by Year with highest copy numbers observed in 2008, but in contrast to the *nifH* mRNA transcript, crop protection practices were also significant. The use of organic crop protection practices resulted in significantly higher quantities of the *nifH* gene compared with conventional crop protection (Table 4). Year by year analysis of the *nifH* gene shows that sample date is only a significant factor in 2007 with *nifH* gene copy number increasing throughout the year (Table 5). Year by year analysis also shows that in 2007 organic fertility management results in increased *nifH* gene copy number (Table 5).

Although the ANOVA results indicated that sample date had a significant effect on copies of the 16S mRNA transcript (Table 4) there were no significant differences among the months identified using Tukey's HSD test. Year was not a significant factor affecting numbers of 16S mRNA transcript but a significant Year by Sample date interaction was observed. When each year was analysed individually (Table 5) it was found that Sample date was a significant factor in 2008, where highest numbers of 16S mRNA

transcript were observed in September, and in 2009, where highest numbers were observed in March. Stepwise regression was used to determine how soil biochemical properties may be driving *nifH* and 16S RNA transcript and gene activity (qPCR) and diversity (DGGE H') (Table S3). This analysis indicated that soil temperature had a slightly negative effect on *nifH* diversity (for both transcript and gene) and a positive effect on *nifH* transcript activity. Rainfall was negatively correlated with *nifH* transcript diversity (RNA) and positively related to *nifH* gene diversity. In addition, the diversity of the *nifH* mRNA transcript was negatively related to soil C and soil basal respiration. Whereas for activity of the *nifH* gene measured using DNA extracts, pH had a positive effect while soil basal respiration was negatively correlated with expression (Table S3). In general there was a positive correlation between *nifH* RNA diversity and copy number and likewise a positive correlation between 16S rRNA diversity and copy number. For the 16S rRNA gene, copy numbers were also negatively correlated with rainfall. Negative correlations were observed between the DGGE H' data set and average soil temperature with average rainfall positively correlated with both *nifH* and 16S DNA DGGE H'. 16S rRNA DGGE H' was also affected by available nitrate; total carbon and available ammonium. These correlations to environmental parameters are distinct from those of 16S expression, suggesting *nifH* expression did not

Table 2. Summary of Shannon diversity analysis of all DGGE results from all sample years and nucleic acid types.

	H' for *nifH* DGGE (RNA) band data (mean+SE)	H' for nifH DGGE band data (DNA) (mean+SE)	H' for 16S rRNA DGGE band data (mean+SE)
Year (Y)			
2007	2.20±0.08 a	1.29±0.10 a	2.58±0.06 a
2008		1.24±0.07 a	2.81±0.05 a
2009	0.98±0.10 b	1.43±0.07 a	3.06±0.04 b
Sample Date (SD)			
March	1.86±0.11 b	1.48±0.08[1] a	3.04±0.05 b
June	0.95±0.15 a	1.30±0.09 a	2.82±0.04 a
September	1.97±0.15 b	1.18±0.07 a	2.60±0.07 a
Crop protection (CP)			
ORG	1.62±0.13 a	1.37±0.06 a	2.82±0.05 a
CON	1.56±0.13 a	1.27±0.07 a	2.82±0.05 a
Fertility management (FM)			
ORG	1.57±0.13 a	1.28±0.06 a	2.79±0.05 a
CON	1.61±0.13 a	1.37±0.07 a	2.85±0.05 a
ANOVA P-values			
Y	0.001		<0.001
SD	<0.001	0.012	<0.001
Y*SD	0.040	<0.001	<0.001
CP*FM			0.006

[1]Although date was a significant factor in the ANOVA, means comparison tests did not indicate any significant differences among dates.
P-values are only shown for terms with P<0.05. Means followed by the same letter for a given factor are not significantly different (P<0.05; Tukey's HSD test where there are more than two treatment levels).

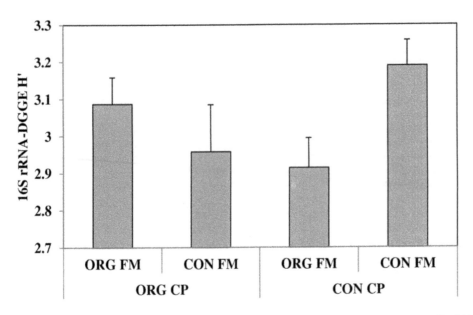

Figure 2. The interaction between organic and conventional crop protection (ORG CP, CON CP) and organic and conventional fertility management (ORG FM, CON FM) on March sample dates only for three years (2007, 2008 and 2009) for Shannon's diversity index of the 16S mRNA transcript. Bars are standard errors (n = 12).

simply mirror the response of the broader bacterial community (Table S3).

Discussion

The NFSC trial enables studies to be conducted with spatial and temporal replication of each system of interest allowing for robust analyses of the impact of management and environmental effects on the microbial communities [42]. Previously we have shown that soils in a conventional crop rotation had a significantly greater diversity and number of free-living diazotrophic bacteria than those within an organic rotation [23]. Here we compared the effect of organic versus conventional fertility management and crop protection activities on the total and free-living N fixing bacterial communities in three different years, on three sampling dates in each year. We hypothesised that the predominant factor affecting diazotrophic and total bacterial diversity, biomass, activity and community structure would be enhanced under organic fertility management, as a result of increased levels of organic carbon, phosphorus and higher soil pH, as has been previously observed [10,43–47].

However, although overall activity of soil organisms was enhanced under ORG FM (e.g. higher soil basal respiration 1.14 mg CO_2 kg^{-1} h^{-1} for CON FM versus 1.38 mg CO_2 kg^{-1} h^{-1} for ORG FM), and pH was significantly reduced in conventional fertility management (6.35 for CON FM versus 6.58 for ORG FM on average) while the availability of P, nitrate and ammonium was increased; (Table S4) our data demonstrated that the most significant explanatory variables for quantitative changes in both the diversity and numbers of free-living diazotrophic and total bacterial populations in agricultural soil in a multiple year study were temporal and seasonal effects. These observations contrast with previous work, where fertility source (farmyard manure versus mineral or no fertilizer) was the dominant factor driving bacterial community structure [8,11], indicating that an increase in organic carbon, associated with organic fertility management activities, had a positive correlation with bacterial soil diversity [48,49]. However, other studies that have more

resonance with our data, indicate that changes to bacterial structure and diversity due to management practices are often subtle [50] and seasonal and plant growth effects often have a greater influence than those due to management processes [51]. One explanation for our findings may be that, although the NFSC trial has been ongoing since 2001, there were no significant differences in soil organic C or total N between the soil management treatments in any of the study years.

Although overall fertility management had no effect on the diversity of the diazotrophs, the factors soil basal respiration and available nitrate were associated with changes in *nifH* diversity and activity (Table 3). There are very few studies on the impact of organic farming on the free-living diazotrophic communities in agricultural soil. DeLuca et al, [22] compared the use of cattle manure and urea fertilizers and found that both fertilizer types inhibited nitrogen fixation (measured by acetylene reduction assay) and that pH was correlated with nitrogen fixation ability. Previous studies, looking at the effect of individual attributes of farm management on the rhizospheric nitrogen fixing community, suggested that the application of increased amounts of nitrogen fertilizer (normally associated with conventional fertility management) would result in decreased diazotrophic diversity and activity [4,52]. Rather our data suggests that many different factors affect the nitrogen fixing community (Table 3 and S3). Although our results were not as conclusive as previous studies, organic fertility management was observed to correlate with increased *nifH* mRNA transcripts in 2007 and 2009, and increased *nifH* gene copy number in 2007 (Table 5). Soil nitrate levels were also negatively correlated with *nifH* qPCR data (Table S3), which corresponds to other studies which have reported inhibition of nitrogenase activity in free-living N_2 fixing bacteria [53–56]. The interacting effects of nitrogen level, carbon availability and crop protection practices, make it difficult to recommend one suite of management practices that can be expected to enhance N fixation by diazotrophs in agricultural soils.

It was hypothesised that conventional crop protection would have a negative effect on *nifH* diversity, and expression, as studies into the environmental impacts of pesticides have shown that they

Table 3. Summary of CCA and RDA analysis of RNA DGGE results showing significant variables.

Gene of interest	Sample date	Variables tested	Significant variables selected by forward selection			Variance of DGGE data explained by the model (%)		
			2007	2008	2009	2007	2008	2009
nifH	March	FM				8.0		4.8
		CP				5.8		5.1
		Associated variables[1]				12.7		13.2
		Associated variables[1], FM, CP				14.4		15.6
	June	FM				6.1		8.0
		CP				6.4		10.1
		Associated variables[1]			NH_4^+	11.1		23.6
		Associated variables[1], FM, CP	SBR		NH_4^+	23.4		36
	September	FM				7.2		9.1
		CP				5.2		5.5
		Associated variables[1]			NO_3^-, NH_4^+	14.3		20.6
		Associated variables[1], FM, CP			NO_3^-, NH_4^+	15.2		22.6
16S rRNA	March	FM				3.1	5.8	4.7
		CP				6.2	6.1	5.9
		Associated variables[1]				9.9	4.5	4.2
		Associated variables[1], FM, CP				10	8.8	9.1
	June	FM				6.2	4.5	4.4
		CP	CP		CP	14.1	8.0	11.7
		Associated variables[1]				9.1	9.2	6.8
		Associated variables[1], FM, CP	CP		CP	17.8	10.5	13.5
	September	FM	FM			11.3	9.3	5.8
		CP				4.7	3.9	5.3
		Associated variables[1]		pH	NO_3^-	11.2	12.9	11.2
		Associated variables[1], FM, CP		FM, pH	NO_3^-	22.1	18.5	12.9

FM = fertility management, CP = crop protection.

[1]Associated variables measured at the time of sampling pH, soil basal respiration (SBR), ammonium (NH_4^+) and nitrate (NO_3^-). Soil basal respiration was measured in June samples only and pH was measured in September samples only.

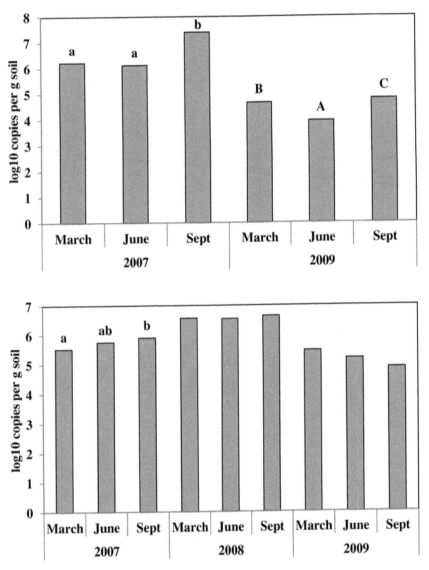

Figure 3. Showing copy numbers of the *nifH* mRNA transcript and the *nifH* gene. *nifH* mRNA transcripts shown in the top and *nifH* gene in the bottom panels respectively. Bars labelled with the same letter in the same year are not different ($P = 0.05$). Unlabelled bars in the same year are not significantly different.

can significantly affect the bacterial community as a whole and that diazotrophs could be particularly affected [57,58,51]. Our results suggest that conventional crop protection did in fact exert a negative effect on the diazotrophic activity when *nifH* copy numbers derived from the DNA data set were considered (Table 4) but appeared not to impact on the diversity or structure of the community. The DNA results suggest that the size of the *nifH* population in plots under conventional crop protection has been significantly reduced due to the long-term application of pesticides. That levels of *nifH* expression (RNA-qPCR results) did not mirror the DNA-qPCR results suggests that activity of the diazotrophic community is not limited by its size, but rather by other controlling factors. A range of pesticides are applied to the potato crops in the NFSC experiment (Table 1) some of which have been shown to have some inhibitory effect on diazotrophs at high concentrations [57,59,60]. Many previous studies looking at the effect of pesticides on the diazotrophic community have focussed on nitrogen fixers which are symbiotic with legumes. *Bradyrhizobium japonicum*, for example, has been found to be particularly

susceptible to the effects of glyphosphate due to the sensitivity of its phosphate synthase enzyme [61,62]. Other studies have found that herbicides will affect nitrogenase activity, nodule formation, nodule biomass and leghaemoglobin concentrations [61–63]. However, it is unclear whether this is due to direct changes in the rhizobia, indirect physiological changes in the plant, or both and does not explain why we see significant changes in the free-living nitrogen fixing community [64,61].

In contrast crop protection strategy had no significant effect on the activity of the total bacterial population, although, it was a significant driver of community structure in June of both 2007 and 2009. To our knowledge this is the first study which fully investigates the effects of crop protection protocols in the field on the activity of both diazotrophic and total bacterial communities.

The temporal effects observed on both diversity and number of the diazotrophic, and total bacterial communities, were primarily affected by the recent environmental conditions. On most occasions, rainfall and soil temperature were significant factors affecting activity and diversity according to stepwise regression

Table 4. Summary of qPCR analysis across all years and sample dates for all genes and nucleic acid types.

	qPCR average copy numbers for *nifH* RNA data set (mean ± SE)	qPCR average copy numbers for *nifH* DNA data set (mean ± SE)	qPCR average copy numbers for 16S rRNA gene data set (mean ± SE)
year (Y)			
2007	$9.29 \times 10^6 \pm 2.85 \times 10^6$b	$5.69 \times 10^5 \pm 6.81 \times 10^4$b	$8.05 \times 10^7 \pm 1.17 \times 10^7$a
2008		$3.89 \times 10^6 \pm 2.62 \times 10^5$c	$5.26 \times 10^7 \pm 2.30 \times 10^7$a
2009	$3.92 \times 10^4 \pm 5.37 \times 10^3$a	$1.77 \times 10^5 \pm 6.34 \times 10^4$a	$2.96 \times 10^8 \pm 1.29 \times 10^8$a
sample date (SD)			
March	$8.46 \times 10^5 \pm 4.99 \times 10^5$b	$1.45 \times 10^6 \pm 2.65 \times 10^5$a	$2.99 \times 10^8 \pm 1.29 \times 10^8$
June	$6.62 \times 10^5 \pm 2.33 \times 10^5$b	$1.42 \times 10^6 \pm 2.51 \times 10^5$a	$4.31 \times 10^7 \pm 7.29 \times 10^6$a
September	$1.25 \times 10^7 \pm 4.16 \times 10^6$a	$1.76 \times 10^6 \pm 3.45 \times 10^5$a	$8.66 \times 10^7 \pm 2.41 \times 10^7$a
Crop protection (CP)			
org	$6.25 \times 10^6 \pm 2.72 \times 10^5$a	$1.74 \times 10^6 \pm 2.44 \times 10^5$a	$1.06 \times 10^7 \pm 5.20 \times 10^7$a
con	$3.07 \times 10^6 \pm 1.18 \times 10^6$a	$1.35 \times 10^6 \pm 2.27 \times 10^5$b	$1.81 \times 10^8 \pm 8.13 \times 10^7$a
fertility management			
org	$4.97 \times 10^6 \pm 2.59 \times 10^6$a	$1.64 \times 10^6 \pm 2.31 \times 10^5$a	$1.23 \times 10^8 \pm 3.92 \times 10^7$a
con	$4.35 \times 10^6 \pm 1.53 \times 10^6$a	$1.45 \times 10^6 \pm 2.42 \times 10^5$	$1.63 \times 10^8 \pm 7.99 \times 10^7$a
ANOVA P values			
Y	<0.001	<0.001	
SD	<0.001		0.032
CP		0.013	
Y*SD	0.005		<0.001
Y*FM	0.001	0.032	
CP*FM	0.037		
Y*SD*FM		0.022	

P-values are only shown for terms with P<0.05; all data were log-transformed before analysis. Means followed by the same letter for a given factor are not significantly different (P<0.05; Tukey's HSD test where there are more than two treatment levels).

(Table S3), although the effects were not always positive. Diversity tended to be higher in March for both *nifH* (mRNA transcripts and genes) and 16S mRNA transcripts. Optimum temperature for growth and activity of diazotrophs is between 10 and 25°C (similar to the temperature range in the field between June and September) [65,66]. Activities of the general bacterial community were largely unaffected by sample date, suggesting that this community included species with a wide range of optimal temperatures that were able to adapt to the environmental conditions throughout the growing season. As expected, diversity and copy number of the 16S rRNA gene were always higher than diversity and copy number of the *nifH* gene. Ratios of the *nifH* gene to the 16S rRNA gene (~1 copies of *nifH* gene: 50 copies of 16S rRNA gene) were similar to ratios seen between the 16S rRNA gene and genes used in nitrogen cycling found in other studies [67,68].

Seasonal effects observed in this study may also be related to the crop management practices that occur throughout the year. Samples taken in March are from a relatively undisturbed soil with no plant cover. June samples may be affected by frequent cultivations for weed control, especially in the organic crop protection plots. This makes it difficult to separate soil temperature and moisture effects in this study from the effects of seasonal management practices. Stage of crop growth can also influence microbial community composition. Certain members of the soil bacterial community, particularly *Acidobacteria*, *Bacteroidetes* and *Alpha-*, *Beta-*, and *Gammaproteobacteria*, have previously been observed to be diminished in summer in crop land [49]. It has been demonstrated that growth stage and seasonal effects

significantly affect diversity in soil under potato and maize [69]. For example, when culture dependent and independent (cloning and DGGE) methods were used to assess bacterial diversity in bulk and rhizosphere soil in 3 species of potato, bacterial communities were observed to change as the plant developed. Higher diversity was observed around 25 days after planting, compared to growth 65 and 140 days after planting [70]. Similarly in maize, bacterial activity, as measured by PLFA and BIOLOG, changed as maize went through five leaf stage, flowering and maturity [71]. It is assumed that these observations reflect changes in the amount and quality of root exudates as the plant reaches maturity [47].

We found that management activity, temporal and seasonal factors appeared to exert no significant effect on the most abundant diazotrophs identified by sequencing the DGGE bands (Table S2). A follow up study is currently underway using pyrosequencing to more thoroughly resolve the taxonomic structure of the diazotrophic communities in these soils. Previous work looking at the impact of differing levels of nitrogen fertilization on the diazotrophic communities of soil showed that the predominant taxa were present in all soils regardless of the amounts of nitrogen fertilizer used [52,72]. It has been suggested that the predominant taxa remain unaffected by the level of N fertilization, whereas the minor members of this community are more sensitive to such changes [73]. In conclusion we found the dominant factors affecting the diversity and numbers of both the nitrogen fixing and the total bacterial community are temporal. The only exception was the impact of conventional crop protection protocols that seemed to reduce the number of

Table 5. Average copy numbers for *nifH* gene amplified from DNA and reverse transcribed RNA for each year of the trial.

	Copies of *nifH* RNA/g soil (mean±SE)		Copies of *nifH* DNA/g soil (mean±SE)			Copies of 16S rRNA gene/g soil (mean±SE)		
	2007	2009	2007	2008	2009	2007	2008	2009
Sample date (SD)								
March	$1.65\times10^6\pm$ 9.71×10^5b	$4.56\times10^4\pm$ 1.16×10^4b	$3.31\times10^5\pm$ 1.08×10^5a	$3.72\times10^6\pm$ 3.16×10^5a	$2.94\times10^5\pm$ 1.78×10^5a	$6.19\times10^7\pm$ 1.88×10^7a	$1.66\times10^7\pm$ 4.24×10^6a	$8.20\times10^8\pm$ 3.58×10^8b
June	$1.32\times10^6\pm$ 4.10×10^5b	$8.97\times10^3\pm$ 1.13×10^2a	$5.65\times10^5\pm$ 1.20×10^5ab	$3.54\times10^6\pm$ 3.46×10^5a	$1.59\times10^5\pm$ 6.35×10^4a	$7.21\times10^7\pm$ 1.54×10^7a	$1.52\times10^7\pm$ 3.09×10^7a	$4.20\times10^7\pm$ 1.20×10^7a
September	$2.49\times10^7\pm$ 7.15×10^6a	$6.28\times10^4\pm$ 5.95×10^3c	$8.10\times10^5\pm$ 1.00×10^5b	$4.39\times10^6\pm$ 6.31×10^5a	$7.70\times10^4\pm$ 2.47×10^4a	$1.08\times10^8\pm$ 2.50×10^7a	$1.26\times10^8\pm$ 6.65×10^7b	$2.62\times10^7\pm$ 4.49×10^6a
Crop protection (CP)								
ORG	$1.01\times10^7\pm$ 5.00×10^6a	$3.42\times10^4\pm$ 6.79×10^3a	$5.75\times10^5\pm$ 9.46×10^4a	$4.37\times10^6\pm$ 2.73×10^5a	$2.81\times10^5\pm$ 1.23×10^5a	$7.68\times10^7\pm$ 1.41×10^7a	$3.12\times10^7\pm$ 8.73×10^6a	$2.08\times10^8\pm$ 1.07×10^8a
CON	$8.50\times10^6\pm$ 2.86×10^6a	$4.41\times10^4\pm$ 8.33×10^3a	$5.62\times10^5\pm$ 1.00×10^5a	$3.40\times10^6\pm$ 4.30×10^5a	$7.26\times10^4\pm$ 1.94×10^4a	$8.43\times10^7\pm$ 1.90×10^7a	$7.39\times10^7\pm$ 4.53×10^7a	$3.83\times10^8\pm$ 2.36×10^8a
Fertility mgt (FM)								
ORG	$9.91\times10^6\pm$ 2.76×10^6a	$4.67\times10^4\pm$ 8.08×10^3a	$7.62\times10^5\pm$ 1.07×10^5a	$4.03\times10^6\pm$ 3.12×10^5a	$1.26\times10^5\pm$ 4.32×10^4a	$8.75\times10^7\pm$ 1.76×10^7a	$7.33\times10^7\pm$ 4.53×10^7a	$2.08\times10^8\pm$ 1.04×10^8a
CON	$8.62\times10^6\pm$ 4.929×10^6b	$3.16\times10^4\pm$ 6.75×10^3b	$3.76\times10^5\pm$ 6.44×10^4b	$3.74\times10^6\pm$a 4.25×10^5	$2.27\times10^5\pm$ 1.20×10^5a	$7.36\times10^7\pm$ 1.57×10^7a	$3.18\times10^7\pm$ 8.70×10^6a	$3.83\times10^8\pm$ 2.37×10^8a
ANOVA *P*-values								
SD	0.001	<0.001	0.050				0.011	0.009
FM	0.026	0.016	0.005					
SD*FM					0.036			
SD*CP*FM							0.048	

P-values are only shown for terms with P<0.05; data for *nifH* RNA 2007 and 2009, *nifH* DNA 2009 only, and 16S rRNA all years, were log-transformed before analysis. Means followed by the same letter for a given factor are not significantly different (P<0.05; Tukey's HSD test where there are more than two treatment levels).

diazotrophs within the soils but not their activity. Fertility management appeared to have little effect on the diversity of both the nitrogen fixing and the total bacterial community, although soil parameters, particularly pH and the concentrations of nitrate and ammonium, were significant factors in determining community structures. The combination of our study and the work of others suggests that rather than the bacterial communities being affected directly by the nature of the fertilizers applied they are more likely to respond to changes in carbon and nitrogen levels in the soil [10,43,44,74]. Although crop management practices were found to impact on the activity and function of soil bacteria, the overriding factor was consistently the year and date of sampling.

Supporting Information

Table S1 Summary of environmental conditions measured in the experimental field during the 14 days prior to each sample date.

Table S2 The closest matches for the 22 sequenced bands derived from the NCBI database.

Table S3 Significant explanatory variables for *nifH* and 16S rRNA gene activity (*q*PCR) and diversity (DGGE H') determined by stepwise regression.

Table S4 The impact of farm management and year of sampling on environmental variables measured in each soil

Author Contributions

Conceived and designed the experiments: CHO JMC SPC CL. Performed the experiments: CHO. Analyzed the data: CHO JMC SPC. Contributed reagents/materials/analysis tools: SPC CL JMC. Wrote the paper: SPC CHO JMC.

References

1. Kibblewhite MG, Ritz K, Swift MJ (2008) Soil health in agricultural systems. Phil Trans Roy Soc B-Biol Sci 363: 685–701.
2. Hsu SF, Buckley DH (2009) Evidence for the functional significance of diazotroph community structure in soil. ISME 3: 124–136.
3. Hayden HL, Drake J, Imhof M, Oxley APA, Norng S, et al. (2010) The abundance of nitrogen cycle genes amoA and nifH depends on land-uses and soil types in South-Eastern Australia. Soil Biol Biochem 42: 1774–1783.
4. Coelho MRR, Marriel IE, Jenkins SN, Lanyon CV, Seldin L, et al. (2009) Molecular detection and quantification of nifH gene sequences in the rhizosphere of sorghum (*Sorghum bicolor*) sown with two levels of nitrogen fertilizer. Appl Soil Ecol 42: 48–53.
5. Hauggaard-Nielsen H, Mundus S, Jensen E (2009) Nitrogen dynamics following grain legumes and subsequent catch crops and the effects on succeeding cereal crops. Nut Cycl Agroecosyst 84: 281–291.

6. Gamble M, Bagwell C, LaRocque J, Bergholz P, Lovell C (2010) Seasonal Variability of Diazotroph Assemblages Associated with the Rhizosphere of the Salt Marsh Cordgrass, *Spartina alterniflora*. Microb Ecol 59: 253–265.

7. Hartmann M, Fliessbach A, Oberholzer HR, Widmer F (2006) Ranking the magnitude of crop and farming system effects on soil microbial biomass and genetic structure of bacterial communities. FEMS Microbiol Ecol 57: 378–388.

8. Esperschütz J, Gattinger A, Mäder P, Schloter M, Fließbach A (2007) Response of soil microbial biomass and community structures to conventional and organic farming systems under identical crop rotations. FEMS Microbiol Ecol 61: 26–37.

9. Bossio DA, Scow KM, Gunapala N, Graham KJ (1998) Determinants of soil microbial communities: effects of agricultural management, season, and soil type on phospholipid fatty acid profiles. Microb Ecol 36: 1–12.

10. van Diepeningen AD, de Vos OJ, Korthals GW, van Bruggen AHC (2006) Effects of organic versus conventional management on chemical and biological parameters in agricultural soils. Appl Soil Ecol 31: 120–135.

11. Widmer F, Rasche F, Hartmann M, Fließbach A. (2006) Community structures and substrate utilization of bacteria in soils from organic and conventional farming systems of the DOK long-term field experiment. Appl Soil Ecol 33: 294–307.

12. Wessén E, Hallin S, Philippot L (2010) Differential responses of bacterial and archaeal groups at high taxonomical ranks to soil management. Soil Biol Biochem 42: 1759–1765.

13. Fernández-Calviño D, Bååth E (2010) Growth response of the bacterial community to pH in soils differing in pH. FEMS Microbiol Ecol 73:149–156.

14. Mäder P, Fließbach A, Dubois D, Gunst L, Fried P, et al. (2002) Soil Fertility and Biodiversity in Organic Farming. Science 296: 1694–1697.

15. Pimentel D, Hepperly P, Hanson J, Douds D, Seidel R (2005) Environmental, Energetic, and Economic Comparisons of Organic and Conventional Farming Systems. BioSci 55: 573–582.

16. Keeling AA, Cook JA, Wilcox A (1998) Effects of carbohydrate application on diazotroph populations and nitrogen availability in grass swards established in garden waste compost. Biores Technol 66: 89–97.

17. Hartley AE, Schlesinger WH (2002) Potential environmental controls on nitrogenase activity in biological crusts of the northern Chihuahuan Desert. J Arid Environ 52: 293–304.

18. Bürgmann H, Meier S, Bunge M, Widmer F, Zeyer J (2005) Effects of model root exudates on structure and activity of a soil diazotroph community. Environ Microbiol 7: 1711–1724.

19. Fierer N, Jackson RB (2006) The diversity and biogeography of soil bacterial communities. PNAS 103: 626–631.

20. Hallin S, Jones CM, Schloter M,Philippot L. (2009) Relationship between N-cycling communities and ecosystem functioning in a 50-year-old fertilization experiment. ISME J 3: 597–605.

21. Reed SC, Seastedt TR, Mann CM, Suding KN, Townsend AR, et al. (2007) Phosphorus fertilization stimulates nitrogen fixation and increases inorganic nitrogen concentrations in a restored prairie. Appl Soil Ecol 36: 238–242.

22. DeLuca TH, Drinkwater LE, Wiefling BA, DeNicola DM (1996) Free-living nitrogen-fixing bacteria in temperate cropping systems: Influence of nitrogen source. Biol Fert Soil 23: 140–144.

23. Orr CH, James A, Leifert C, Cooper JM, Cummings SP (2011) Diversity and Activity of Free-Living Nitrogen Fixing Bacteria and Total Bacteria in Organic and Conventionally Managed Soils. Appl Env Microbiol 77: 911–919.

24. Hussain S, Siddique T, Saleem M, Arshad M, Khalid A, et al. (2009) Impact of Pesticides on Soil Microbial Diversity, Enzymes, and Biochemical Reactions. In: Advances in Agronomy. New York. USA: Academic Press. 159–200.

25. Cooper J, Niggli U, Leifert C (2007) Handbook of organic food safety and quality. Cambridge, UK: Woodhead Publishing Limited. 521 p.

26. Bending GD, Rodríguez-Cruz MS, Lincoln SD (2007) Fungicide impacts on microbial communities in soils with contrasting management histories. Chemosphere 69: 82–88.

27. Cycoń M, Piotrowska-Seget Z (2009) Changes in bacterial diversity and community structure following pesticides addition to soil estimated by cultivation technique. Ecotoxicol 18: 632–642.

28. Spyrou I, Karpouzas D, Menkissoglu-Spiroudi U. (2009) Do botanical pesticides alter the structure of the soil microbial community? Microb Ecol 58: 715–727.

29. Omar SA, Abd-Alla MH (1992) Effect of pesticides on growth, respiration and nitrogenase activity of *Azotobacter* and *Azospirillum*. World J Microbiol Biotech 8: 326–328.

30. Fox JE, Gulledge J, Engelhaupt E, Burow ME, McLachlan JA (2007) Pesticides reduce symbiotic efficiency of nitrogen-fixing rhizobia and host plants. PNAS 104: 10282–10287.

31. Fox JE, Starcevic M, Jones PE, Burrow ME, McLachlan JA (2004) Phytoestrogen signalling and Symbiotic Gene Activation Are Disrupted by Endocrine-Disrupting Chemicals. Environ Health Persp 112: 672–677.

32. Soil Association (2005). Soil Association organic standards, Bristol, UK.

33. Wartiainen I, Eriksson T, Zheng W, Rasmussen U (2008) Variation in the active diazotrophic community in rice paddy–*nifH* PCR-DGGE analysis of rhizosphere and bulk soil. Appl Soil Ecol 39: 65–75.

34. Poly F, Monrozier LJ, Bally R (2001) Improvement in the RFLP procedure for studying the diversity of nifH genes in communities of nitrogen fixers in soil. Res Microbiol 152: 95–103.

35. Baxter J, Cummings SP (2008) The degradation of the herbicide bromoxynil and its impact on bacterial diversity in a top soil. J Appl Microbiol 104: 1605–1616.

36. Cummings SP, Gyaneshwar P, Vinuesa P, Farruggia FT, Andrews M, et al. (2009) Nodulation of *Sesbania* species by *Rhizobium (Agrobacterium)* strain IRBG74 and other rhizobia. Environ Microbiol 11: 2510–2525.

37. Karlen Y, McNair A, Perseguers S, Mazza C, Mermod N (2007) Statistical significance of quantitative PCR. BMC Bioinform 8: 131.

38. R Development Core Team (2006) R: A language and environment for statistical computing. Vienna, Austria: R Foundation for Statistical Computing.

39. Venables WN, Ripley BD (2002) Modern applied statistics with S, 4th ed. New York, USA: Springer.

40. Minitab (2006) Minitab Statistical Software. Release 15 for Windows. State College, Pennsylvania, Minitab® is a registered trademark of Minitab Inc.

41. Lindström ES, Bergström AK (2005) Community composition of bacterio-plankton and cell transport in lakes in two different drainage areas. Aquatic Sci 67: 210–219.

42. Prosser JI (2010) Replicate or lie. Environ Microbiol 12: 1806–1810.

43. Postma J, Schilder MT, Bloem J, van Leeuwen-Haagsma WK (2008) Soil suppressiveness and functional diversity of the soil microflora in organic farming systems. Soil Biol Biochem 40: 2394–2406.

44. Toljander JF, Santos-González JC, Tehler A, Finlay RD (2008) Community analysis of arbuscular mycorrhizal fungi and bacteria in the maize mycorrhizo-sphere in a long-term fertilization trial. FEMS Microbiol Ecol 65: 323–338.

45. Birkhofer K, Bezemer TM, Bloem J, Bonkowski M, Christensen S, et al. (2008) Long-term organic farming fosters below and aboveground biota: Implications for soil quality, biological control and productivity. Soil Biol Biochem 40: 2297–2308.

46. Tamm L, Thürig B, Bruns C, Fuchs J, Köpke U, et al. (2010) Soil type, management history, and soil amendments influence the development of soil-borne (*Rhizoctonia solani*, *Pythium ultimum*) and air-borne (*Phytophthora infestans*, *Hyaloperonospora parasitica*) diseases. Eur J Plant Path 127: 465–481.

47. Ngosong C, Jarosch M, Raupp J, Neumann E, Ruess L (2010) The impact of farming practice on soil microorganisms and arbuscular mycorrhizal fungi: Crop type versus long-term mineral and organic fertilization. Appl Soil Ecol 46: 134–142.

48. Lejon DPH, Chaussod R, Ranger J, Ranjard L. (2005) Microbial community structure and density under different tree species in an acid forest soil (Morvan, France). Microb Ecol 50: 614–625.

49. Jangid K, Williams MA, Franzluebbers AJ, Sanderlin JS, Reeves JH, et al. (2008) Relative impacts of land-use, management intensity and fertilization upon soil microbial community structure in agricultural systems. Soil Biol Biochem 40: 2843–2853.

50. Drenovsky RE, Vo D, Graham KJ, Scow KM (2004) Soil Water Content and Organic Carbon Availability Are Major Determinants of Soil Microbial Community Composition. Microb Ecol 48: 424–430.

51. Moreno B, Garcia-Rodriguez S, Cañizares R, Castro J, Benítez E (2009) Rainfed olive farming in south-eastern Spain: Long-term effect of soil management on biological indicators of soil quality. Agri Ecosys Environ 131: 333–339.

52. Coelho MRR, de Vos M, Carneiro NP, Marriel IE, Paiva E, et al. (2008) Diversity of *nifH* gene pools in the rhizosphere of two cultivars of sorghum *Sorghum bicolor* treated with contrasting levels of nitrogen fertilizer. FEMS Microbiol Letts 279: 15–22.

53. Knowles R, Denike D (1974) Effect of ammonium-nitrogen, nitrite-nitrogen and nitrate-nitrogen on anaerobic nitrogenase activity in soil. Soil Biol Biochem 6: 353–358.

54. Chapin DM, Bliss LC, Bledsoe LI (1991) Environmental-regulation of nitrogen-fixation in a high arctic lowland ecosystem. Can J Bot 69: 2744–2755.

55. Kitoh S, Shiomi N (1991) Effect of mineral nutrients and combined nitrogen-sources in the medium on growth and nitrogen-fixation of the azolla-anabaena association. Soil Sci Plant Nutr 37: 419–426.

56. Roper MM, Turpin JE, Thompson JP (1994) Nitrogenase activity (C2H2 reduction) by free-living bacteria in soil in a long term tillage and stubble management experiment on a vertisol. Soil Biol Biochem 26: 1087–1091.

57. Doneche B, Seguin G, Ribereau-Gayon P (1983) Mancozeb Effect on Soil Microorganisms and Its Degradation in Soils. Soil Sci 135: 361–366.

58. Cycoń M, Piotrowska-Seget Z (2007) Effect of selected pesticides on soil microflora involved in organic matter and nitrogen transformations: pot experiment. Pol J Ecol 55: 207–220.

59. Sturz AV, Kimpinski J (1999) Effects of fosthiazate and aldicarb on population of plant-growth-promoting bacteria, root lesion nematodes and bacteria-feeding nematodes in the root zome of potatoes. Plant Pathol 48: 26–32.

60. Miloševiа NA, Govedarica MM (2002) Effect of herbicides on microbiological properties of soil. Proceedings for Natural Sciences 102: 5–21.

61. Zablotowicz RM, Reddy KN (2007) Nitrogenase activity, nitrogen content, and yield responses to glyphosate in glyphosate-resistant soybean. Crop Prot 26: 370–376.

62. Bohm GMB, Alves B JR, Urquiaga S, Boddey RM, Xavier GR, et al (2009) Glyphosate- and imazethapyr-induced effects on yield, nodule mass and biological nitrogen fixation in field-grown glyphosate resistant soybean. Soil Biol Biochem 41: 420–422.

63. Reddy KN, Zablotowicz RM (2002) Glyphosate-resistant soybean response to various salts of glyphosate and glyphosate accumulation in soybean nodules. Weed Sci 51: 496–502.

64. Vieira R, Silva C, Silveira A (2007) Soil microbial biomass C and symbiotic processes associated with soybean after sulfentrazone herbicide application. Plant Soil 300: 95–103.
65. Beauchamp CJ, Lévesque G, Prévost D, Chalifour FP (2006) Isolation of free-living dinitrogen-fixing bacteria and their activity in compost containing de-inking paper sludge. Biores Technol 97: 1002–1011.
66. Eckford R, Cook FD, Saul D, Aislabie J, Foght J (2002) Free-living Heterotrophic Nitrogen-fixing Bacteria Isolated from Fuel-Contaminated Antarctic Soils. Appl Environ Microbiol 68: 5181–5185.
67. Kandeler E, Deiglmayr K, Tscherko D, Bru D, Philippot L (2006) Abundance of narG, nirS, nirK, and nosZ Genes of Denitrifying Bacteria during Primary Successions of a Glacier Foreland. Appl Environ Microbiol 72: 5957–5962.
68. Morales SE, Cosart T, Holben WE (2010) Bacterial gene abundances as indicators of greenhouse gas emission in soils. ISME J 4: 799–808.
69. Diallo S, Crépin A, Barbey C, Orange N, Burini JF, et al (2010) Mechanisms and recent advances in biological control mediated through the potato rhizosphere. FEMS Microbiol Ecol 75: 351–364.

70. van Overbeek L, van Elsas JD (2008) Effects of plant genotype and growth stage on the structure of bacterial communities associated with potato (*Solanum tuberosum* L.) FEMS Microbiol Ecol 64: 283–296.
71. Griffiths BS, Caul S, Thompson J, Birch ANE, Scrimgeour C, et al (2006) Soil Microbial and Faunal Community Responses to Maize and Insecticide in Two Soils. J Environ Qual 35: 734–741.
72. Ogilvie L, Hirsch P, Johnston A (2008) Bacterial Diversity of the Broadbalk 'Classical' Winter Wheat Experiment in Relation to Long-Term Fertilizer Inputs. Microbial Ecol 56: 525–537.
73. Knauth S, Hurek T, Brar D, Reinhold-Hurek B (2005) Influence of different *Oryza* cultivars on expression of *nif*H gene pools in roots of rice. Environ Microbiol 7: 1725–1733.
74. Campbell BJ, Polson SW, Hanson TE, Mack MC, Schuur EAG (2010) The effect of nutrient deposition on bacterial communities in Arctic tundra soil. Environ Microbiol 12: 1842–1854.

3

Estimation of Wheat Agronomic Parameters using New Spectral Indices

Xiu-liang Jin[1,2], Wan-ying Diao[3], Chun-hua Xiao[3], Fang-yong Wang[4], Bing Chen[4], Ke-ru Wang[1,3]*, Shao-kun Li[1,3]*

1 Institute of Crop Science, Chinese Academy of Agricultural Sciences/Key Laboratory of Crop Physiology and Production Ministry of Agriculture, Beijing, China, **2** Beijing Research Center for Information Technology in Agriculture, Beijing, China, **3** Key Laboratory of Oasis Ecology Agriculture of Xinjiang Construction Crops, Shihezi, China, **4** Institute of Cotton, Xinjiang Academy of Agricultural Reclamation Sciences, Shihezi, China

Abstract

Crop agronomic parameters (leaf area index (LAI), nitrogen (N) uptake, total chlorophyll (Chl) content) are very important for the prediction of crop growth. The objective of this experiment was to investigate whether the wheat LAI, N uptake, and total Chl content could be accurately predicted using spectral indices collected at different stages of wheat growth. Firstly, the product of the optimized soil-adjusted vegetation index and wheat biomass dry weight (OSAVI×BDW) were used to estimate LAI, N uptake, and total Chl content; secondly, BDW was replaced by spectral indices to establish new spectral indices (OSAVI×OSAVI, OSAVI×SIPI, OSAVI×$CI_{red\ edge}$, OSAVI×$CI_{green\ mode}$ and OSAVI×EVI2); finally, we used the new spectral indices for estimating LAI, N uptake, and total Chl content. The results showed that the new spectral indices could be used to accurately estimate LAI, N uptake, and total Chl content. The highest R^2 and the lowest RMSEs were 0.711 and 0.78 (OSAVI×EVI2), 0.785 and 3.98 g/m² (OSAVI×$CI_{red\ edge}$) and 0.846 and 0.65 g/m² (OSAVI×$CI_{red\ edge}$) for LAI, nitrogen uptake and total Chl content, respectively. The new spectral indices performed better than the OSAVI alone, and the problems of a lack of sensitivity at earlier growth stages and saturation at later growth stages, which are typically associated with the OSAVI, were improved. The overall results indicated that this new spectral indices provided the best approximation for the estimation of agronomic indices for all growth stages of wheat.

Editor: Ive De Smet, University of Nottingham, United Kingdom

Funding: The study is supported by the National Natural Science Foundation of China (grant numbers 31071371, 30760109, 41161068) and the Key Projects in the National Science & Technology Pillar Program during the Twelfth Five-Year Plan Period (grant number 2012BAH27B04). The National Natural Science Foundation of China had role in study design, data collection and analysis, the Key Projects in the National Science & Technology Pillar Program during the Twelfth Five-Year Plan Period had role in decision to publish,or preparation of the manuscript.

Competing Interests: The authors have declared that no competing interests exist.

* E-mail: wkeru01@163.com (KRW); lishk@mail.caas.net.cn (SKL)

Introduction

The development of remote sensing has provided opportunities to quantitatively describe agronomic parameter changes across all growth stages of crops. The application of remote sensing to agronomic problems has created new methods to effectively improve field crop management. Many authors have provided detailed information about the relationships between spectral indices and agronomic parameters, including the leaf area index (LAI), nitrogen (N) uptake, total chlorophyll (Chl) content, and so on.

The LAI is a key variable for the diagnosis and prediction of crop growth and yield. This makes the LAI critical for effective understanding of the biophysical processes of plant canopies and the prediction of plant growth and productivity [1–6]. Rouse et al. suggested that the most well-known and widely used vegetation index was the normalized difference vegetation index (NDVI), and found that NDVI was linearly correlated with the leaf area index (LAI) in crop fields [7]. However, the NDVI does possess certain limitations related to soil background brightness, in that the NDVI tends to be affected by different soil color and moisture conditions [8–10]. To overcome this problem, Rondeaux et al. proposed using an optimized soil-adjustment factor, and obtained an optimized soil-adjusted vegetation index (OSAVI), which mitigated the effects of soil and moisture conditions [11].

Nitrogen is a critically important element that is monitored in an effort to maintain crop health, while there is a good relationship between nitrogen content and chlorophyll content [12–13], therefore chlorophyll content is an important indicator for nitrogen fertilizer applications. Stone et al. detected and predicted N uptake in winter wheat using hand-held sensors [14–15]. Osborne et al. identified the important reflectance wavelengths for the prediction of N concentration changes at different growth stages [16]. Gitelson et al. suggested that the green NDVI (GNDVI) was more sensitive than the NDVI for wheat N uptake over 100 kg/ha [17], but Moges et al. indicated that the NDVI was more robust than the GNDVI for prediction of crop N uptake [18]. Certain researchers have proposed many N indicators for the assessment of crop N change according to the spectral features of chlorophyll in the visible and red-edge bands [19–20]. A good relationship between a combined index (the ratio of modified chlorophyll absorption ratio index and modified triangular vegetation index2, MCARI/MTVI2) and leaf nitrogen concentration was found by Eitel et al. [21]. Fitzgerald et al. reported that a spectral index, the canopy chlorophyll content index (CCCI), could predict canopy N, and specifically found a good relationship between the CCCI and canopy N [22]. Vigneaua et al. used field hyperspectral imaging as a non-destructive method to assess leaf nitrogen content in wheat [23]. And recently, remote sensing methods have been developed and applied for the prediction of Chl content for field crop management [24–30]. Gitelson et al. identified the best vegetation indices for the estimation of Chl content [30]. A good correlation between total Chl content and (R_{NIR} which is in the near infrared band reflectance/$R_{red edge}$ which is in the red edge position reflectance)–1, (R_{NIR}/R_{green} which is in the green band reflectance)–1 was found by Gitelson et al. [31], and Gitelson et al. applied the new vegetation index ($R_{NIR}/R_{red edge}$–1 and R_{NIR}/R_{green}–1) for the estimation of Chl content and indirectly estimated gross primary production (GPP) [32,33].

However, spectral indices still eixst insensitivity at earlier growth stages, and saturation at later growth stages. Because biomass dry weight (BDW) was accumulated gradually as growth stages progressed, and insensitivity at earlier growth stages and saturation at later growth stages was not observed. Consequently, the objectives of this study were to (1) combine OSAVI and BDW for estimating LAI, N uptake, and total Chl content in wheat; (2) replace BDW with spectral indices; (3) and attempt to propose new spectral indices, in an effort to improve insensitivity at earlier growth stages and saturation at later growth stages.

Materials and Methods

2.1. Design of Experiment

Field experiments were conducted in 2009 and 2010 at the ShiHezi University experiment site (44°20′N, 86°3′E), Xinjiang Province, China. The experiment site had representative soil types and crop management practices for Xinjiang Province, China. The soil was fine-loamy with a total N content of 42.6 mg kg^{-1}, Olsen P of 26.5 mg kg^{-1}, exchangeable K of 139.4 mg kg^{-1}, and organic matter content of 11.6 g kg^{-1} in the 0–30 cm layer. Three local wheat cultivars: Xinchun 6, Xinchun 17, and Xinchun 22, were planted on April 5th 2008 and April 8th 2009. Nitrogen fertilizer as urea was applied at four rates (0, 105, 225, and 345 kg N ha^{-1}) before planting. The N application was distributed at three stages in the growth process in the following percentages: 50% at seeding, 25% at jointing, and 25% at booting. For all treatments, 99 kg ha^{-1} P_2O_5 (as monocalcium phosphate [$Ca(H_2PO_4)_2$]) and 150 kg ha^{-1} K_2O (as KCl) was applied prior to seeding. The experiment was a 2-way factorial arrangement of treatments in a randomized complete block design with three replications for each treatment. Other management practices followed local standard wheat production practices.

2.2. Measurement of Canopy Reectance

Spectral measurements were carried out at the following stages (2009, 2010): tillering (8th May, 6th May), jointing (20th May, 25th May), heading (8th June, 10th June), anthesis (18th June, 20th June), and filling (27th May, 25th May), respectively. All canopy spectral measurements were mounted on the tripod boom and held in a nadir orientation 1.0 m above the canopy. Measurements were taken under clear sky conditions between 10:00 and 14:00 (Beijing local time) using an ASD Field Spec Pro Spectrometer (Analytical Spectral Devices, Boulder, CO, USA). This spectrometer is fitted with a 25° field of view fiber optics, operating in the 350–2500 nm spectral region with a sampling interval of 1.4 nm between 350 and 1050 nm, and 2 nm between 1050 and 2500 nm, and with spectral resolution of 3 nm at 700 nm, and 10 nm at 1400 nm. A 40 cm × 40 cm BaSO4 calibration panel was used for calculating the black and baseline

reectance. To reduce the possible effect of sky and field conditions, spectral measurements were taken at four sites in each plot and were averaged to represent the canopy reflectance of each plot. Vegetation radiance measurement was taken by averaging 16 scans at an optimized integration time, with a dark current correction at every spectral measurement. A panel radiance measurement was taken before and after the vegetation measurement by two scans each time.

2.3. Agronomic Parameters Measurement

Immediately following spectral measurements, the leaf area index (LAI) was measured using the LAI-2000 Plant Canopy Analyzer (LI-COR Inc., Lincoln, NE, USA) with spectrometric measurements at the same position and gained by destructive sampling. The wheat was cut at the ground level and wet weights were recorded in 0.24 m^2. Each plant was then dried at 70°C for 3 d and the dry weight was recorded. Dry plant material was then milled and analyzed in the laboratory. Total N content was estimated using a dry combustion method in a Dumas Elementary Analyser (Macro-N, Foss Heraeus, Hanau, Germany) [34].

The spectral measurements positions of the wheat leaves were collected using a hole puncher (diameter, 0.4 cm). Then, about 0.2 g wheat leaves of each sample was punched off in the laboratory. The selected samples were placed in 95% ethanol or acetone solution and allowed to stand for 24 hr in the dark. Following the 24-hr treatment, the leaves were white-green in color. Finally, leaf pigment density was measured using a colorimetric spectro-photometer. Absorbance of the supernatant was measured at 645, 652 and 663 nm, and chlorophyll a plus chlorophyll b content per unit leaf area was then calculated by the method of McKinney [35].

2.4. Selection of Spectral Indices

This study tested the spectral indices that were considered to be good candidates for estimation of plant chlorophyll content and LAI (Table 1) [11,31,36–37].

2.5. Statistical Analysis

Linear and nonlinear regression analysis was carried out using the biomass dry weight, OSAVI and OSAVI×biomass dry weight (the product of OSAVI and biomass dry weight) as independent variables, and the LAI, total Chl content, and N uptake (leaves yield N concentration of measured leaves) as dependent variables.

Statistically significant ($p<0.05$ or 0.01) coefficients of determination (R^2) between three crop parameters (LAI, total Chl content, N uptake) and new vegetation indices or biomass dry weight were analyzed using SPSS software (16.0, SPSS, Chicago, IIinois, USA). The R^2 and root mean square error (RMSE) were used as metrics for quantifying the amount of variation explained by the relationships developed, as well as their accuracy. The performance of the model was evaluated through R^2 and RMSE for the estimation of in-situ measured LAI, total Chl content and N uptake. Generally, the performance of the model was estimated by comparing the differences of the R^2 and RMSE between the measured value and predicted value. The higher the R^2 and the lower the RMSE were, the higher the precision and accuracy of model to predict agronomic parameters was considered to be.

Results

3.1. Leaf Area Index (LAI)

A significant relationship was found between the OSAVI and LAI for the selected data, ranging from all growth stages of wheat across 2009 (Figure 1a). As noted above, the OSAVI was

Table 1. Summary of spectral indices, wavebands and citations for LAI, nitrogen uptake and total chlorophyll content in this paper.

Index	Name	Formula	Developer(s)
OSAVI	Optimized soil-adjusted vegetation index	$1.16 \times (R_{800} - R_{670})/(R_{800} + R_{670} + 0.16)$	Rondeaux et al. (1996) [11]
SIPI	Structure insensitive pigment index	$(R_{800} - R_{445})/(R_{800} - R_{680})$	Penuelas et al. (1995) [36]
$CI_{red\ edge}$	Red edge model	$(R_{800}/R_{700}) - 1$	Gitelson et al. (2005) [31]
$CI_{green\ model}$	Green model	$(R_{800}/R_{550}) - 1$	Gitelson et al. (2005) [31]
EVI2	Enhanced vegetation index 2	$2.5 \times (R_{800} - R_{660})/(1 + R_{800} + 2.4 \times R_{660})$	Jiang et al. (2008) [37]

Note: Ri denotes reectance at band i (nanometer).

calculated and measured for each stage by averaging ASD-2500 sensor readings from taken at the vertical height of the canopy at 1.0 m. Steven showed that the LAI has a strong relationship with the OSAVI [38]. Figure 1a has shown that for the OSAVI index, there was a problem with saturation at later growth stages in wheat, determination coefficient (R^2) value was 0.536. It is mainly for this reason that the OSAVI was not sensitive to the LAI (ranges from 4 to 6) at later growth stages. Possibly, this was because the band combination only considered the visible and near-infrared bands and was thus affected by environmental conditions.

Biomass dry weight and LAI were highly correlated ($R^2 = 0.688$, $P<0.01$), independent of the area the wheat occupied. At earlier growth stages, the correlation between biomass dry weight and LAI (Figure 1b) was better than that between OSAVI and LAI (Figure 1a), R^2 value was 0.688. The improvement was even greater at later growth stages. This is important because it indicates that biomass dry weight can be used to estimate wheat LAI, and to compensate for the OSAVI's lack of sensitivity to the later growth stages changes in LAI. So, we used the OSAVI× biomass dry weight (BDW) index to estimate LAI changes using the data from all growth stages in wheat. The results showed that the OSAVI×BDW index was a good predictor of LAI, though the biomass dry weight alone was a more accurate predictor (Figure 1c). We performed non-linear regression on the OSA-

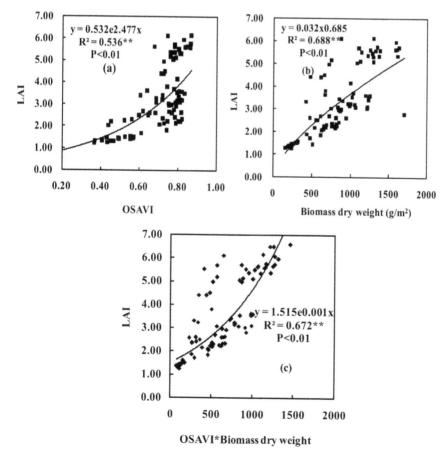

Figure 1. The coefficient of determination (R^2) between OSAVI, biomass dry weight (BDW), BDW×OSAVI and LAI under the different nitrogen treatments for wheat (n = 90). Note: Probability levels are indicated by n.s., * and ** for 'not significant', 0.05, and 0.01, respectively.

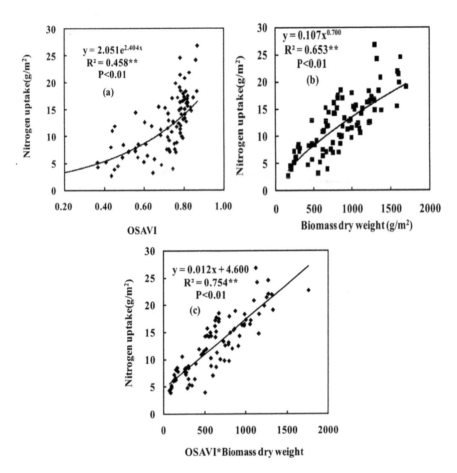

Figure 2. Relationship between OSAVI, biomass dry weight (BDW) and BDW×OSAVI and nitrogen uptake under the different nitrogen treatments for wheat (n = 90). Note: Same as above.

VI×BDW and LAI at all growth stages, and the resulting R^2 was 0.672. These results suggested that the OSAVI×biomass dry weight index was an improvement over the OSAVI index, because the OSAVI alone was not sensitive to LAI changes at later growth stages.

3.2. Nitrogen Uptake

The amount of nitrogen (N) taken up in wheat was correlated with the OSAVI with an R^2 value of 0.458 (Figure 2a). The results showed that the relationship between the OSAVI and N uptake was highly correlated ($P<0.01$) at all growth stages. However, at later growth stages when N uptake ranged from 15 to 27 g/m^2, saturation was observed. A main reason for this problem was that the OSAVI was affected by other environmental factors [11]. Specifically, in this paper, the OSAVI saturation was mainly influenced by the structure of the wheat canopy. For example, the plant height of wheat canopy vary from 80 cm to 120 cm at the later growth stages, the reflectance at the top of wheat canopy (0–40 cm) can be detected by spectrometer (direct light), but the reflectance at the bottom of canopy (40–120 cm) cannot be detected by spectrometer (diffused light). The high-density wheat affected the spectral measurement and plant height. This caused the proportion of the diffused light to increase, thereby increasing the OSAVI saturation. The above problems led to a lack of sensitivity to the changes in N uptake at later growth stages. This suggested that with the increase in wheat biomass, the OSAVI became slightly affected by the surrounding environmental

conditions, and the OSAVI and N uptake were correlated. The correlation between the OSAVI and N uptake could be explained by the ability of OSAVI to detect differences in red absorption. Moges et al. showed a similar finding, their research investigated the relationships of green, red, and near infrared bands to N uptake [16]. The results showed that the correlation between biomass dry weight and N uptake was better than that between OSAVI and N uptake, and the R^2 value was 0.653 (Figure 2b).

The OSAVI×biomass dry weight index (BDW) could be used to estimate N uptake changes, and was a good predictor of N uptake. It performed better than either biomass dry weight or OSAVI alone in the prediction of N uptake for wheat. The results showed that a linear regression performed between the OSAVI×BDW and N uptake at all growth stages resulted in an R^2 value of 0.754 (Figure 2c). This result suggested that the OSAVI×BDW index could be effectively used to improve the OSAVI's prediction accuracy for N uptake.

3.3. Total Chlorophyll Content

The total chlorophyll (Chl) content was correlated with OSAVI, with a coefficient of determination (R^2) of 0.632. The results suggested that the total Chl content in the wheat canopy could be estimated using OSAVI (Figure 3a). Possible explanation could be the red edge absorption waveband (670 nm) is very sensitive to total Chl content [23–24,26,29]. Previous research also showed a good relationship between total Chl content and biomass [39], and biomass changes influenced the spectral reflectance in the near-

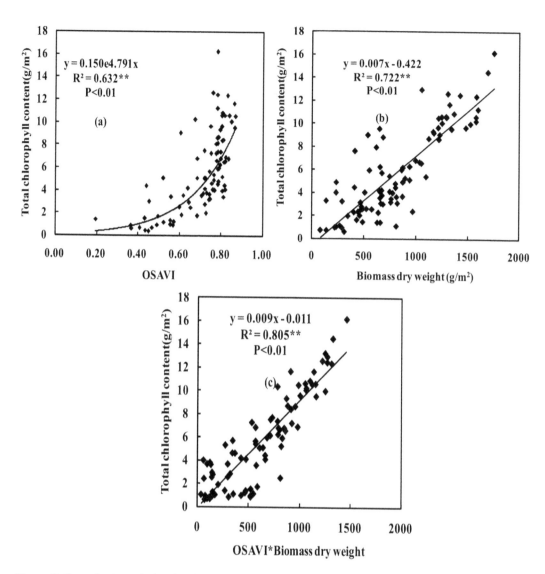

Figure 3. Quantitative relationship between OSAVI, biomass dry weight (BDW) and BDW×OSAVI and total chlorophyll content under the different nitrogen treatments for wheat (n = 90). Note: Same as above.

infrared waveband (800 nm). Both red edge absorption waveband and near-infrared waveband are involved in OSAVI calculation.

Table 2. Relationship between the new spectral indices and LAI under the different nitrogen treatments for wheat (n = 90).

Spectral indices	Simulated equations	Determination coefficient (R^2)
OSAVI×OSAVI	$y = 0.253\ln(x)+0.221$	0.634**
OSAVI×SIPI	$y = 0.174\ln(x)+0.507$	0.563**
OSAVI×CI$_{red\ edge}$	$y = 3.756\ln(x)+0.337$	0.692**
OSAVI×CI$_{green\ model}$	$y = 3.534\ln(x)+0.527$	0.706**
OSAVI×EVI2	$y = 0.269\ln(x)+0.151$	0.711**

Note: n = number of pairs of data. x represents spectral indices, y represents LAI. Probability levels are indicated by n.s.,
*and ** for 'not significant', 0.05, and 0.01, respectively.

Therefore, variations in Chl content changes were detected by OSAVI.

Figure 3b shows that biomass dry weight and total chlorophyll content were more highly correlated than were the OSAVI and total Chl content, with an R^2 value of 0.722. Linear regression was performed between biomass dry weight and total Chl content, and between biomass dry weight and total Chl content at the different growth stages. The results showed that the relationship between the OSAVI and total chlorophyll content exhibited similar results to that between OSAVI and N uptake vis-à-vis a lower correlation at earlier growth stages.

The OSAVI×biomass dry weight index could be used to estimate total Chl content changes, and it was a good indicator of total Chl content for wheat. It improved upon the results of the biomass dry weight and OSAVI alone for estimation of the total Chl content for wheat, and the R^2 was 0.805 (Figure 3c). This result indicated that the OSAVI×biomass dry weight index improved upon the prediction accuracy for total Chl content.

Taken together, these results indicated that the OSAVI×biomass dry weight index (BDW) could be used to improve the estimation accuracy of LAI, nitrogen (N) uptake, and total

Table 3. Relationship between the new spectral indices and nitrogen uptake under the different nitrogen treatments for wheat (n = 90).

Spectral indices	Simulated equations	Determination coefficient (R^2)
OSAVI×OSAVI	$y = 0.246\ln(x) - 0.095$	0.542**
OSAVI×SIPI	$y = 0.164\ln(x) + 0.302$	0.641**
OSAVI×CI$_{red\ edge}$	$y = 0.145x^{1.331}$	0.785**
OSAVI×CI$_{green\ model}$	$y = 0.203x^{1.200}$	0.776**
OSAVI×EVI2	$y = 0.274\ln(x) - 0.212$	0.783**

Note: n = number of pairs of data. x represents spectral indices, y represents nitrogen uptake.
Probability levels are indicated by n.s.,
*and ** for 'not significant', 0.05, and 0.01, respectively.

chlorophyll (Chl) content, respectively. The most accurate estimation was gained for total Chl content, the worst for LAI, and the median for N uptake, among three agronomic parameters.

The results showed that biomass dry weight and LAI were highly correlated ($R^2 = 0.688$, $P<0.01$) (Fig. 1b). Most of results indicated that the R^2 and RMSE of the curves fitting was better than the linear fitting [40–43]. But sometimes, the linear fitting was better than the curves fitting, it may be a high related to measurement data and crops physiological mechanism. In this paper, we try to find the best fitting lines by using the curves fitting or the linear fitting, the result indicated that the equations of curves fitting is more than the equations of linear fitting, our results was consistent with previous researches [40–43]. Therefore, we justify the curves fitting was better than the linear fitting according to the R^2 and RMSE of regression model. The results showed that biomass dry weight and LAI were highly correlated (Fig. 1b). It indicated that the size of biomass dry weight represented the size of LAI. It was mainly because LAI growth is synchronized with biomass dry weight according to the certain proportion. The biomass dry weight and nitrogen content were also highly correlated (Fig. 2b). It suggested that the appropriate nitrogen was used to increase the biomass dry weight accumulation. Because the nitrogen was applied for improving crops photosynthesis, and there was a good relationships between photosynthesis and biomass dry weight, thereby biomass dry weight accumulation

Table 4. Relationship between the new spectral indices and total chlorophyll content under the different nitrogen treatments for wheat (n = 90).

Spectral indices	Simulated equations	Determination coefficient (R^2)
OSAVI×OSAVI	$y = 0.264x^{0.400}$	0.652**
OSAVI×SIPI	$y = 0.504x^{0.279}$	0.683**
OSAVI×CI$_{red\ edge}$	$y = 0.137x^{0.769}$	0.846**
OSAVI×CI$_{green\ model}$	$y = 1.307x^{0.742}$	0.824**
OSAVI×EVI2	$y = 0.208x^{0.482}$	0.798**

Note: n = number of pairs of data. x represents spectral indices, y represents total chlorophyll content.
Probability levels are indicated by n.s.,
*and ** for 'not significant', 0.05, and 0.01, respectively.

was increased. Similarly, the appropriate total chlorophyll content could be used to increase crops photosynthesis and biomass dry weight accumulation, thus there was a high correlation between biomass dry weight and total chlorophyll content (Fig. 3b).

A straight linear relationships between biomass and LAI before heading stages, but this relationship between biomass and LAI have changed after heading stages. The rate of increase in the LAI is less than in the biomass (stem and spike weights increase more than leaf weight) before and after flowering. The LAI arrived at the highest values at flag leaf fully expanded, but the biomass dry weight (BDW) is still increased (it is mainly increased from spike weight). Therefore, a curvilinear relationship between wheat BDW and LAI should be better after heading stages. Metabolic physiology exists different before and after flowering. The wheat metabolic physiology was dominated by nitrogen and supplemented by carbon before flowering, canopy leaf nitrogen content change was relatively small, canopy leaf nitrogen content was accurately estimated by more accurately monitoring LAI. However, the wheat metabolic physiology was dominated by carbon and supplemented by nitrogen after flowering, the differences are significant in canopy leaf nitrogen content, nitrogen transfers from lower leaves to the upper leaves and spike in wheat. The nitrogen was not estimated accurately by monitoring LAI because of the lower leaves nitrogen was moved away by nitrogen transfer, resulting in a relative large error will be generated. If biomass factors are taken into account, thereby the transfer of nitrogen will also be included in the monitoring results, therefore it could be used to improve the nitrogen estimation accuracy. These results indicated that the biomass dry weight was highly related to LAI, nitrogen content and total chlorophyll content because biomass dry weight had a close relationship with LAI, nitrogen content and total chlorophyll content in crops physiological mechanism. These results also suggested that the biomass dry weight could be used to estimate the LAI, nitrogen content and total chlorophyll content for wheat. The results showed that a good relationships among the LAI, nitrogen content, total chlorophyll content and OSA-VI×BDW (Figs. 1c, 2c and 3c). It provided a basis for BDW was replaced by spectral indices to establish new spectral indices.

3.4. New Spectral Indices

The new spectral index (OSAVI×BDW) had a good relationship with LAI, N uptake and total Chl content, but BDW was gained by destructive sampling, and required significant time investment for data acquisition. Thus, we attempted to build new spectral indices by replacing the BDW with others spectral indices. Tables 2, 3 and 4 showed that the new spectral indices: OSAVI×OSAVI, OSAVI×SIPI, OSAVI×CI$_{red\ edge}$, OSAVI× CI$_{green\ model}$ and OSAVI×EVI2, could be used to improve LAI, N uptake and total Chl content estimation accuracy. For LAI, the lowest and highest determination coefficient (R^2) observed were OSAVI×SIPI and OSAVI×EVI2 with R^2 values of 0.563 and 0.711, respectively (Table 2); similarly, the lowest and highest R^2 were OSAVI×CI$_{green\ model}$ and OSAVI×OSAVI, for which R^2 were 0.542 and 0.785 for nitrogen uptake, respectively (Table 3); the lowest and highest R^2 were OSAVI×CI$_{red\ edge}$ and OSAVI×OSAVI, with R^2 values of 0.652 and 0.846 for total Chl content, respectively (Table 4).

3.5. Validation Model and Comparison

To validate the model accuracy, we compared the predicted values (the predicted values were gained by the LAI, N uptake and total Chl content of regression equations in 2009) with the actual values (the actual values were gained by the LAI, N uptake and total Chl content field measurement data in 2010). A good

Table 5. The relationships the measured value and predicted value for LAI, nitrogen uptake and total chlorophyll content under the different nitrogen treatments for wheat.

LAI		Nitrogen uptake		Leaf chlorophyll content	
	RMSE (g/m^2)		RMSE (g/m^2)		RMSE (g/m^2)
OSAVI	1.41	OSAVI	7.93	OSAVI	3.42
biomass dry weight	1.02	biomass dry weight	4.45	biomass dry weight	3.12
OSAVI×BDW	1.12	OSAVI×BDW	4.23	OSAVI×BDW	2.11
OSAVI×OSAVI	1.18	OSAVI×OSAVI	6.46	OSAVI×OSAVI	3.36
OSAVI×SIPI	1.36	OSAVI×SIPI	4.85	OSAVI×SIPI	3.02
OSAVI×CI$_{red\ edge}$	0.98	OSAVI×CI$_{red\ edge}$	3.98	OSAVI×CI$_{red\ edge}$	0.65
OSAVI×CI$_{green\ model}$	0.84	OSAVI×CI$_{green\ model}$	4.12	OSAVI×CI$_{green\ model}$	0.79
OSAVI×EVI2	0.78	OSAVI×EVI2	4.03	OSAVI×EVI2	1.02

correlation between the predicted values and the actual values was observed for the following indices: OSAVI, biomass dry weight, and OSAVI×biomass dry weight (BDW). The corresponding root mean square errors (RMSEs) were 1.42, 1.02, and 1.12 for LAI; 7.93 g/m^2, 4.45 g/m^2, and 4.23 g/m^2 for nitrogen uptake; 3.42 g/m^2, 1.85 g/m^2, and 2.23 g/m^2 for leaf chlorophyll content, respectively (Table 5). These data indicated that OSAVI×biomass dry weight could be used to improve the estimation accuracy of LAI, nitrogen uptake, and leaf chlorophyll content. The new spectral indices were proposed by replacing BDW with spectral indices, and then obtained OSAVI×OSAVI, OSAVI×SIPI, OSAVI×CI$_{red\ edge}$, OSAVI×CI$_{green\ model}$ and OSAVI×EVI2. The results showed that the new spectral indices were better than OSAVI alone for estimating LAI, nitrogen uptake and total chlorophyll content (Tables 2, 3, 4 and 5 and Figs. 1a, 2a and 3a). The results indicated that the products of spectral indices and OSAVI could be used to improve the LAI, N uptake and total Chl content estimation accuracy.

Discussion

Crop leaf area index (LAI), total chlorophyll (Chl) content and nitrogen (N) uptake was estimated using new spectral indices. The results showed that the new spectral indices could be used to improve the LAI, total Chl content and N uptake estimation accuracy. We used the OSAVI×biomass dry weight (BDW) index (the product of OSAVI and biomass dry weight) to improve the relationship between the LAI and spectral indices. The results indicated that the OSAVI×BDW index was better than the OSAVI at estimating wheat LAI. The OSAVI×BDW index was more sensitive to LAI due to the further decreased effects of soil at earlier growth stages. For example, the spectrometer was sensitive to the soil color and moisture, thus reducing the wheat canopy spectral information detection at earlier growth stages, and leading decreased sensitive to canopy spectral reflectance changes. The OSAVI×BDW ameliorated the saturation at later growth stages because of the addition of biomass dry weight, which was increased gradually with progression through the wheat growth stages. Overall, the saturation problem observed for the spectral indices was mainly influenced by the structure of the wheat canopy at later growth stages (see section 3.2). The OSAVI×BDW index was used to improve the estimation of nitrogen, and the results demonstrated that it was better than OSAVI alone for nitrogen assessment. The reason for this was similar to that explaining the reasonable estimation of LAI, the results were similar to those of

previous research [14–19]. We proposed that the OSAVI×BDW index could be used to effectively improve the estimation accuracy of total Chl content indirectly, and the results demonstrated that it could accomplish the proposed tasks. Again, the reasons for such are similar to those explaining the reasonable estimation of N uptake.

The new spectral indices in Tables 2, 3 and 4 are derived from BDW×OSAVI. Biomass dry weight (BDW) was gained by destructive sampling, but required a greater time investment for data acquisition. To quickly obtain data and estimate leaf area index (LAI), total chlorophyll (Chl) content and nitrogen (N) uptake, we attempted to build new spectral indices by replacing BDW with other spectral indices. We obtained the OSAVI×O-SAVI, OSAVI×SIPI, OSAVI×CI$_{red\ edge}$, OSAVI×CI$_{green\ model}$ and OSAVI×EVI2. The results suggested the new spectral indices were better than OSAVI alone for estimating LAI, total Chl content and N uptake (Tables 2, 3, 4 and 5 and Figs. 1a, 2a and 3a). The main reason is that the SIPI, CI$_{red\ edge}$, CI$_{green\ model}$ and EVI2 include the 800 nm band is sensitive to LAI changes. The products of OSAVI and SIPI, CI$_{red\ edge}$, CI$_{green\ model}$ and EVI2 will further improve the OSAVI sensitivity in LAI changes. Therefore, these new spectral indices could be used for improving the LAI. For N uptake, the products of OSAVI and SIPI, CI$_{red\ edge}$, CI$_{green\ model}$ and EVI2, making OSAVI increase the sensitive bands of the chlorophyll content changes (445 nm, 550 nm, 660 nm and 680 nm), these new spectral indices are sensitive to detect chlorophyll content changes. Thus, the N uptake estimation was improved by using the new spectral indices. For total Chl content, it was similar to those explained reasonable estimation of N uptake.

Taken together, the results indicated that it was feasible to use new methods to improve agronomic parameters (LAI, N uptake and total Chl content) assessment accuracy. This paper evaluated the estimation accuracy of agronomy parameters by multiplying the spectral indices in OSAVI, it showed the OSAVI is unnecessary and able to be replaced by others spectral indices. For example, if you want to better estimate total Chl content or N uptake, you could be selected to the product of two spectral indices are highly related to chlorophyll. Further, we used these indices to improve the accuracy of these predictions for all crop growth stages. In future studies, we will try to multiply two spectral indices that are highly related to LAI, total Chl content or N uptake to estimate agronomic parameters of different crops.

Conclusions

Building upon previous studies, wheat leaf area index (LAI), nitrogen uptake, and total chlorophyll content were predicted using the products spectral indices and spectral indices methods, and the main results and conclusions follow. The results suggested that the LAI, nitrogen uptake and total Chl content could be accurately predicted using the products of OSAVI and biomass dry weight (OSAVI×biomass dry weight (BDW) index), the corresponding determination coefficient (R^2) and root mean square errors (RMSEs) were 0.672 and 1.12, 0.754 and 4.23 g/m^2, 0.805 and 2.11 g/m^2, respectively. We obtained the new spectral indices by using spectral indices replace BDW. The relationships between LAI, nitrogen uptake and total Chl content and new spectral indices, OSAVI×EVI2, OSAVI×CI$_{red\ edge}$ and OSAVI×CI$_{red\ edge}$ had the highest R^2 and the lowest RMSEs for LAI, nitrogen uptake and total Chl content, respectively, R^2 and RMSEs were 0.711 and 0.78, 0.785 and 3.98 g/m^2, 0.846 and 0.65 g/m^2, respectively.

Author Contributions

Conceived and designed the experiments: XLJ CHX WYD FYW BC SKL. Performed the experiments: XLJ WYD KRW FYW. Analyzed the data: XLJ. Contributed reagents/materials/analysis tools: CHX KRW SKL. Wrote the paper: XLJ SKL KRW.

References

1. Daughtry CST, Gallo KP, Goward SN, Prince SD, Kustas WD (1992) Spectral estimates of absorbed radiation and phytomass production in corn and soybean canopies. Remote Sensing of Environment 39: 141–152.
2. Goetz SJ, Prince SD (1996) Remote sensing of net primary production in boreal forest stands. Agricultural and Forest Meteorology 78: 149–179.
3. Liu J, Chen J, Cihlar J, Park WM (1997) A process-based boreal ecosystem productivity simulator using remote sensing inputs. Remote Sensing of Environment 62: 158–175.
4. Moran MS, Inoue Y, Barnes EM (1997) Opportunities and limitations for image-based remote sensing in precision crop management. Remote Sensing of Environment 61: 319–346.
5. Moran MS, Maas SJ, Pinter JPJ (1995) Combining remote sensing and modeling for estimating surface evaporation and biomass production. Remote Sensing of Environment 12: 335–353.
6. Tucker CJ, Holben BN, Elgin JJH, McMurtrey JE (1980) Relationship of spectral data to grain yield variations. Photogrammetric Engineering & Remote Sensing 46: 657–666.
7. Rouse JW, Haas RH, Schell JA, Deering DW, Harlan JC (1974) Monitoring the vernal advancements and retrogradation of natural vegetation. In: NASA/GSFC, Final Report, Greenbelt, MD, USA, 1974, pp.1–137.
8. Bausch WC (1993) Soil background effects on reectance-based crop coefficients for corn. Remote Sensing of Environment 46: 213–222.
9. Elvidge CD, Lyon RJP (1985) Inuence of rock-soil spectral variation on the assessment of green biomass. Remote Sensing of Environment 17: 265–279.
10. Huete AR, Jackson RD, Post DF (1985) Spectral response of a plant canopy with different soil backgrounds. Remote Sensing of Environment 17: 37–53.
11. Rondeaux G, Steven M, Baret F (1996) Optimization of soil adjusted vegetation indices, Remote Sensing of Environment 55: 95–107.
12. Yang JC, Wang ZQ, Zhu QS (1996) Effects of nitrogen nutrition on rice yield and its physiology mechanism under the different status of soil moisture, Chinese Agriculture Science 29: 58–66 (in Chinese).
13. Rhykerd CL, Noller CH (1974). The role of nitrogen in forage production, In D.A. Mays (ed.) Forage fertilization, American Society of Agronomy, Madison, WI, 416–424.
14. Stone ML, Solie JB, Raun WR, Whitney RW, Taylor SL, et al. (1996a) Use of spectral radiance for correcting in-season fertilizer nitrogen deficiencies in winter wheat. Transactions of the American Society Agriculture Engineers 39: 1623–1631.
15. Stone ML, Solie JB, Whitney RW, Raun WR, Lees HL (1996b) Sensors for the detection of nitrogen in winter wheat. Tech. Paper Series No. 961757. SAE, Warrendale, PA.
16. Osborne SL, Schepers JS, Francis DD, Schlemmer MR (2002) Detection of phosphorous and nitrogen deficiencies in corn using spectral radiance measurement. Agronomy Journal 94: 1215–1221.
17. Gitelson AA, Merzlyak MN (1996) Signature analysis of leaf reflectance spectra: algorithm development for remote sensing of chlorophyll. Journal of Plant Physiology 148: 494–500.
18. Moges SM, Raun WR, Mullen RW, Freeman KW, Johnson GV, et al. (2004) Estimating green, red, and near infrared bands for predicting winter wheat biomass, nitrogen uptake, and final grain yield. Journal of Plant Nutrition 27: 1431–1441.
19. Reyniers M, Walvoort DJJ, De Baardemaaker J (2006) A linear model to predict with a multi-spectral radiometer the amount of nitrogen in winter wheat. International Journal of Remote Sensing 27: 4159–4179.
20. Zhu Y, Yao X, Tian YC, Liu XJ, Cao WX (2008) Analysis of common canopy vegetation indices for indicating leaf nitrogen accumulations in wheat and rice. International Journal of Appllied earth Obsevation and Geoinformation 10: 1–10.
21. Eitel JUH, Long DS, Gessler PE, Smith AMS (2007) Using in-situ measurements to evaluate the new RapidEye satellite series for prediction of wheat nitrogen status. International Journal of Remote Sensing 28: 4183–4190.
22. Fitzgerald G, Rodriguez D, O'Leary G (2010) Measuring and predicting canopy nitrogen nutrition in wheat using a spectral index-The canopy chlorophyll content index (CCCI). Field Crops Research 116: 318–324.
23. Vigneaua N, Ecarnotb M, Rabatel G, Roumetb P (2011) Potential of field hyperspectral imaging as a non destructive method to assess leaf nitrogen content in wheat. Field Crops Research 122: 25–31.
24. Buschmann C, Nagel E (1993) In vivo spectroscopy and internal optics of leaves as basis for remote sensing of vegetation. International Journal of Remote Sensing 14: 711–722.
25. Gitelson AA, Merzlyak MN (1994a) Quantitative estimation of chlorophyll a using reflectance spectra: Experiments with autumn chestnut and maple leaves. Journal of Photochemistry and Photobiology Biology 22: 247–252.
26. Gitelson AA, Merzlyak MN (1994b) Spectral reflectance changes associated with autumn senescence of Aesculus hippocastanum andAcer platanoides leaves. Spectral features and relation to chlorophyll estimation. Journal of Plant Physiology 143: 286–292.
27. Markwell J, Osterman JC, Mitchell JL (1995) Calibration of the Minolta SPAD-502 leaf chlorophyll meter. Photosynthesis Research 46: 467–472.
28. Gamon JA, Surfus JS (1999) Assessing leaf pigment content and activity with a reflectometer. New Phytologist 143: 105–117.
29. Gitelson AA, Merzlyak MN, Chivkunova OB (2001) Optical properties and non-destructive estimation of anthocyanin content in plant leaves, Journal of Photochemistry Photobiology 74: 38–45.
30. Gitelson AA, Zur Y, Chivkunova OB, Merzlyak MN (2002) Assessing carotenoid content in plant leaves with reectance spectroscopy. Journal of Photochemistry Photobiology 75: 272–281.
31. Gitelson AA, Vina A, Ciganda V, Rundquist DC, Arkebauer TJ (2005) Remote estimation of canopy chlorophyll content in crops. Geophysical Research Letters 32: L08403, doi:10.1029/2005 GL022688.
32. Gitelson AA, Vina A, Verma SB, Rundquist DC, Arkebauer TJ, et al. (2006) Relationship between gross primary production and chlorophyll content in crops: Implications for the synoptic monitoring of vegetation productivity. Geophysical Research Letters 111: D08S11, doi:10.1029/2006JD006017.
33. Peng Y, Gitelson AA, Keydan G, Rundquist DC, Moses W (2011) Remote estimation of gross primary production in maize and support for a new paradigm based on total crop chlorophyll content. Remote Sensing of Environment 115: 978–989.
34. Schepers JS, Francis DD, Thompson MT (1989) Simultaneous determination of total C, total N and 15N on soil and plant material. Communications in Soil Science and Plant Analysis 20: 949–959.
35. McKinney G (1941) Absorption of light by chlorophyll solutions. Journal of Biology Chemistry 140: 315–322.
36. Penuelas J, Baret F, Filella I (1995) Semiempirical indexes to assess carotenoids chlorophyll-a ratio from leaf spectral reectance. Photosynthetica 31: 221–230.
37. Jiang Z, Huete AR, Didan K, Miura T (2008) Development of a two-band enhanced vegetation index without a blue band. Remote Sensing of Environment 112: 3833–3845.
38. Steven MD (1998) The densitivity of the OSAVI vegetation Index to observational parameters, Remote Sensing of Environment 63: 49–60.
39. Cheng JP, Cao CG, Cai ML, Yuan BZ, Zhai J (2008) Effect of different nitrogen nutrition and soil water potential on the physiology parameters and yield of hybrid rice. Plant Nutrition and Fertilizer Science 14: 199–206 (in Chinese).
40. Gitelson AA, Gurlin D, Moses WJ, Barrow T (2009) A bio-optical algorithm for the remote estimation of the chlorophyll-a concentration in case 2 waters. Environment Research Letters 045003, doi:10.1088/1748-9326/4/4/045003.
41. Houborg R, Anderson MC, Daughtry CST, Kustas WP, Rodell M (2011) Using leaf chlorophyll to parameterize light-use-efficiency within a thermal-based carbon, water and energy exchange model. Remote Sensing of Environment 115: 1694–1705.
42. Sakamoto T, Gitelson AA, Wardlow BD, Arkebauer TJ, Verma SB, et al. (2012) Application of day and night digital photographs for estimating maize biophysical characteristics. Precision Agriculture 13: 285–301.
43. Gurlin D, Gitelson AA, Moses WJ (2011) Remote estimation of chl-a concentration in turbid productive waters-Return to a simple two-band NIR-red model? Remote Sensing of Environment 115: 3479–3490.

4

Bacterial Indicator of Agricultural Management for Soil under No-Till Crop Production

Eva L. M. Figuerola[1,9], Leandro D. Guerrero[1,9], Silvina M. Rosa[1], Leandro Simonetti[1], Matías E. Duval[2], Juan A. Galantini[2], José C. Bedano[3], Luis G. Wall[4], Leonardo Erijman[1,5]*

1 Instituto de Investigaciones en Ingeniería Genética y Biología Molecular (INGEBI-CONICET) Vuelta de Obligado 2490, Buenos Aires, Argentina, 2 CERZOS-CONICET Departamento de Agronomía, Universidad Nacional del Sur, Bahía Blanca, Argentina, 3 Departamento de Geología, Universidad Nacional de Río Cuarto, Río Cuarto, Córdoba, Argentina, 4 Departamento de Ciencia y Tecnología, Universidad Nacional de Quilmes, Roque Sáenz Peña 352, Bernal, Argentina, 5 Facultad de Ciencias Exactas y Naturales, Universidad de Buenos Aires, Ciudad Universitaria, Pabellón 2, Buenos Aires, Argentina

Abstract

The rise in the world demand for food poses a challenge to our ability to sustain soil fertility and sustainability. The increasing use of no-till agriculture, adopted in many areas of the world as an alternative to conventional farming, may contribute to reduce the erosion of soils and the increase in the soil carbon pool. However, the advantages of no-till agriculture are jeopardized when its use is linked to the expansion of crop monoculture. The aim of this study was to survey bacterial communities to find indicators of soil quality related to contrasting agriculture management in soils under no-till farming. Four sites in production agriculture, with different soil properties, situated across a west-east transect in the most productive region in the Argentinean pampas, were taken as the basis for replication. Working definitions of Good no-till Agricultural Practices (GAP) and Poor no-till Agricultural Practices (PAP) were adopted for two distinct scenarios in terms of crop rotation, fertilization, agrochemicals use and pest control. Non-cultivated soils nearby the agricultural sites were taken as additional control treatments. Tag-encoded pyrosequencing was used to deeply sample the 16S rRNA gene from bacteria residing in soils corresponding to the three treatments at the four locations. Although bacterial communities as a whole appeared to be structured chiefly by a marked biogeographic provincialism, the distribution of a few taxa was shaped as well by environmental conditions related to agricultural management practices. A statistically supported approach was used to define candidates for management-indicator organisms, subsequently validated using quantitative PCR. We suggest that the ratio between the normalized abundance of a selected group of bacteria within the GP1 group of the phylum Acidobacteria and the genus *Rubellimicrobium* of the Alphaproteobacteria may serve as a potential management-indicator to discriminate between sustainable *vs.* non-sustainable agricultural practices in the Pampa region.

Editor: Hauke Smidt, Wageningen University, The Netherlands

Funding: The authors are members of the BIOSPAS consortium (http://www.biospas.org/en). ELMF, JCB, LGW and LE are members of Consejo Nacional de Investigaciones Científicas y Técnicas (CONICET), Argentina. JAG is member of Comisión de Investigaciones Científicas of Buenos Aires Province, Argentina (CIC). MD and SR received fellowships from ANPCyT, Argentina. LDG and LS were fellows of CONICET. The funders had no role in study design, data collection and analysis, decision to publish, or preparation of the manuscript.

Competing Interests: The authors have declared that no competing interests exist.

* E-mail: erijman@dna.uba.ar

9 These authors contributed equally to this work.

Introduction

Sowing crop into no-till soil is a farming method that has initially been developed as an alternative to conventional tillage practices, with the aims of using less fossil fuels, reducing the erosion of soils, and increasing the soil carbon pool [1]. Soil structure can be significantly modified through reduced-till management practices [2]. Soil aggregates are less subjected to dry and wet cycles in no-tilled soil, compared to conventional-tilled soil, due to the protection exerted by surface residues. Therefore, it appears that reduced-till management reduces the risk of surface runoff, increase soil aggregation, and improve soil hydrological properties [3]. This is particularly true if no-till management is combined with diverse crop rotation [4].

Additional driving forces for no-till agriculture are the lower production costs, the higher yields and the incorporation of less fertile areas into crop production [4]. During the past several decades, no-till agriculture has been increasingly adopted in many areas of the world [5]. In Argentina, this practice has spread steadily in the last 30 years [6], covering presently almost 20 million hectare, which represents 70% of the total cultivated area [4]. Through the adoption of this novel agriculture management, farmers have been gradually incorporated novel technologies for weed, disease and fertilizer management through trial-and-error learning. The combination of these technologies with no-till management led to a farmers' definition of good agricultural practices on the basis of economic yield alongside soil conservation and gain in productivity. Yet this situation rapidly highlighted the need for new working hypotheses to aid in soil quality monitoring.

Driven by the influence of favorable market conditions, a substantial portion of that area is presently dedicated to soybean monoculture, often combined with minimal nutrient restoration. From the noticeable increase in soil born diseases caused by residue- and soil-inhabiting pathogens selected by the previous crop, questions arise about the ability to maintain soil fertility and sustainability if monoculture prevails over the crop rotation [7].

The use of soil quality indicators is important in order to guide land and resource management decisions. Traditionally, soil quality research has focused primarily on soil chemical and physical properties [8]. In general, assessment of soil quality will be influenced by management factors, and by climate and soil type as well. In view of that, different data sets of soil quality indicators have been proposed to discriminate between soil textural classes for different agricultural management systems and a variety of crops [9,10,11,12,13]. Besides the well known chemical and physical parameters used as soil quality indicators, such as soil organic matter and soil structure, there is still no consensus about biological soil indicators of sustainable agricultural systems. The massive adoption of no-till practices in extensive agriculture in Argentina gave rise to many situations, in which improvement of crop yield could not be associated to established quality indicators, but to the history of the soil management, suggesting that additional biological parameters might be necessary to describe changes in soil quality.

By driving crucial soil processes, such as decomposition of organic materials and nutrient cycling, soil bacteria are key players in ecosystem functioning. The structure of the microbial community in soil, the distribution of microbial biomass and enzyme activity may be affected by several factors, such as farming systems [14], plant species [15,16,17], tree species, soil pH [18], soil type [19], tillage and crop rotation [20,21,22,23,24]. This is why it is also important to take into consideration microbiological indicators when evaluating soil quality [25]. Yet, understanding about the influence of bacterial community structure on soil quality, and inversely, revealing the effect of soil characteristics on the structuring of bacterial communities is still scarce. In particular, to our knowledge, no previous study has addressed these issues in the framework of crop productivity, as assessed by farmers' records.

This work is a part of a larger effort to find microbiological indicators of sustainable agriculture in the framework of no-till farming. The project BIOSPAS (http://www.biospas.org/en) is a multidisciplinary research project, in which agricultural soil biology is approached by means of a polyphasic description [26]. We have considered three treatments, which were replicated as blocks in four agricultural sites located across a west-east transect in Argentine Central Pampas, having documented history of no-till management. Two treatments were related to contrasting agricultural management practices under no-till in terms of crop rotation, fertilization, pest management and agrochemical use, which in coincidence to farmers' records of crop yield can be regarded as "Good no-till Agricultural Practices (GAP)" and "Poor no-till Agricultural Practices (PAP)". The third treatment corresponded to non-cultivated soils nearby the agricultural sites, which were used as references for natural environments (NE).

Pyrosequencing of 16S RNA gene using barcoded sequence tags is a high-throughput technique that has the capability to provide sufficient coverage and sequence length to afford an extensive taxonomic description of soil biota, comparing multiple samples in a single run [27]. A highly variable region of the 16S rRNA gene is individually PCR-amplified using primers containing a barcoded sequence (pyrotag) that allows distinction between samples. Tagged amplicons are pooled at equimolar concentration and sequenced in a single reaction. Reads were later assigned to individual samples based on the barcode sequence. Subsequent comparison to databases allows the identification of bacterial taxa and their relative abundance within the community. Here, we used tag-encoded pyrosequencing to deeply sample the 16S rRNA

gene from bacteria residing in soils corresponding to the three treatments at the four sites with the objective of finding out potential candidate bacterial species as indicators of agricultural management. As we have considered soils with varied characteristics, in terms of texture and organic matter, the identification of statistically based soil management-associated taxa can provide useful diagnostic tools for agricultural soil quality across the surveyed region.

Materials and Methods

Study Sites

The management and sites for this study were selected after a thoughtfully discussion between scientists and farmers participants of the BIOSPAS Project (www.biospas.org/en).

Whereas the sites selected may not fulfilled a rigorous definition of replicates, due to slight differences in management (historical crop sequence, years on no-till agriculture, were not the same), the experimental design privileged the perspective of farmers in terms of the relation between soil management and crop productivity. We have therefore followed a working definition of soil management, according to a set of definitions of Certified Agriculture by the Argentine No Till Farmers Association (AAPRESID, www.ac.org.ar/descargas/PyC_eng.pdf) and the Food and Agricultural Organization of the United Nations (FAO, www.fao.org/prods/GAP/index_en.htm).

Three treatments were defined (Table 1): 1) "Good no-till Agricultural Practices" (GAP): Sustainable agricultural management under no-till, subjected to intensive crop rotation (basically wheat/other winter crop/soybean/maize and sometimes including the use of cover crops, such as vicia/triticale), nutrient replacement, minimized agrochemical use (herbicides, insecticides and fungicides) and showing higher yield compared to PAP (Table 1); 2) "Poor no-till Agricultural Practices" (PAP): Non-sustainable agricultural management under no-till with high crop monoculture (soybean), low nutrient replacement, high agrochemical use (herbicides, insecticides and fungicides) and showing lower yields compared to GAP (Table 1); 3) "Natural Environment" (NE): As reference, natural grassland was selected in an area of approximately 1 hectare, close to the cultivated plots (less than 5 km), where no cultivation was practiced for (at least) the last 30 years.

Treatments were replicated 4 times (blocks) in agricultural fields located across a west-east transect in the most productive region in the Argentinean Pampas. Sites of soil sampling were near the following locations in Argentina: Bengolea at Córdoba Province (33° 01′ 31″ S; 63° 37′ 53″ W); Monte Buey at Córdoba Province (32° 58′ 14″ S; 62° 27′ 06″ W); Pergamino at Buenos Aires Province (33° 56′ 36″ S; 60° 33′ 57″ W); Viale at Entre Ríos Province (31° 52′ 59,6″ S; 59° 40′ 07″ W). See Table 2 for a description of soil characteristics.

Sampling

Samples were taken in June 2009 (winter) as triplicate for each treatment-site in three 5 m^2 sampling points separated at least 50 m from each other, taking care not to follow the sowing line in the field. Three additional samplings in the exact same locations were performed in February 2010, September 2010 and February 2011. Samplings were performed at private productive fields, which belong to any of the funders of this work. None of the sampling areas belong to a protected area or land. Permissions were obtained directly from the owners or responsible persons. At Bengolea and Monte Buey locations sampling was allowed by Jorge Romagnoli, from La Lucía SA, at Pergamino sampling was

Table 1. Description of the agricultural management and crop yield, averaged over the five years before the first sampling date, in June 2009 (2005–2009).

	Bengolea		Monte Buey		Pergamino		Viale	
	GAP	PAP	GAP	PAP	GAP	PAP	GAP	PAP
% no-tillage	100	80	100	100	100	100	100	100
Soybean/maize ratio[a]	1.5	4	0.67	4	1.5	5	1,5	4
% Winter with wheat[b]	60	40	60	20	40	0	40	20
% Winter cover crops[c]	20	0	40	0	0	0	20	0
Herbicide (L) used[d]	27.7	43.8	25.2	38.9	29.3	46.5	34.5	43.1
Soybean yield (kg.ha^{-1})	3067	2775	3167	2675	2933	2825	3000	1805
Maize yield (kg.ha^{-1})	10500	2700	12550	8000	9500	–[e]	7030	3450

[a]Number of soybean cycles to number of maize cycles over the last 5years.
[b]Percentage of winters that wheat was planted as a winter crop.
[c]Percentage of winters that a cover crop (*Vicia* sp., *Melilotus alba* or *Lolium perenne*) was planted. Cover crops were chemically burned before summer crops are planted.
[d]Calculated as liters of low-toxicity herbicides plus liters of moderate-toxicity herbicides weighted by two. Toxicity was defined according to EPA Toxicity Categories. Unit: total liters over 5 years.
[e]No maize was planted in the last 5 years.

allowed by Gustavo Gonzalez Anta, from Rizobacter Argentina SA, sampling at Viale was allowed by Pedro Barbagelatta, member of Aapresid.

Each sample of the top 10 cm of mineral soil was collected as a composite of 16–20 randomly selected subsamples. Composite soil samples were homogenized in the field and transported to the laboratory at 4°C. Within 3 days after collection, samples were sieved through 2-mm mesh to remove roots and plant detritus. Soils were stored at −20°C until DNA extraction.

Chemical and Physical Soil Properties

Soils were classified according to Soil Taxonomy and INTA (Instituto Nacional de Tecnologia Agropecuaria, Argentina) soil map. The main chemical properties of soils were determined by standard methods on samples that were air-dried, crushed and passed through a 2-mm sieve after removal of plant residues. The pH was determined on mixtures of 1:2.5 sample:water. The contents of organic matter as total organic carbon were measured by dry combustion using a LECO CR12 Carbon analyzer (LECO, St. Joseph, MI, USA). The total nitrogen contents in whole soils were obtained by the Kjeldahl method. Extractable phosphorus was determined by the method of Bray and Kurtz. Data is summarized in Table 2.

DNA Extraction

Further homogenization was performed by careful grounding 10–15 g of each soil sample in a mortar before DNA extraction. DNA was extracted from 0.5 g of soil using FastDNA spin kit for soil extraction kit (Mpbio Inc), following the manufacturer's instructions. In order to reduce the presence of humic substances

Table 2. Soil characteristics according to site and agricultural management at the first sampling date, in June 2009.

	Bengolea			Monte Buey			Pergamino			Viale		
	NE	GAP	PAP	NE	GAP	PAP	NE	GAP	PAP	NE	GAP	PAP
Climate	Temperate Subhumid			Temperate Subhumid			Temperate Humid			Temperate Humid		
MAT[1] (°C)	17			17			16			18		
MAP[2] (mm yr^{-1})	870			910			1000			1160		
Altitude (m)	224	222	223	112	111	108	64	68	65	73	80	81
Slope (%)	0.5	0.75	0.5	0.01	0.5	0.2	0.25	0.5	0.5	0.75	0.75	0.2
Years of no-till		13	5		28	10		6	5		13	9
Soil classification	Entic Haplustoll			Typic Argiudoll			Typic Argiudoll			Argic Pelludert		
Texture	Sandy loam			Silt loam			Silt loam			Silty clay/Silty clay loam		
Carbon %	1.7	1.5	1.1	3.5	2.1	1.7	2.7	1.7	1.8	5	3.5	2.5
Nitrogen %	0.146	0.156	0.125	0.328	0.181	0.132	0.233	0.153	0.136	0.369	0.283	0.179
Extractable P (ppm)	44.3	53.1	17.8	296.5	126.5	20.6	10.5	18	11.9	20.2	40.4	41.8
pH	6.3	6.2	6.2	5.6	5.5	6.2	6.2	6	5.7	6.4	6.7	6.3
Moisture	10.58	7.96	6.32	25.47	21.87	18.03	22.83	22.03	12.73	17.8	25.3	18.2

[1]Mat: Mean annual temperature.
[2]MAP: mean annual precipitation.

that inhibited the subsequent PCR reaction, an additional purification step was performed on the DNA sample using polyvinyl polypyrrolidone (PVPP). Eluted DNA was stored at $-20°C$.

Pyrosequencing

Barcoded pyrosequencing analysis was run on samples from the first sampling date of the BIOSPAS project, in June 2009. DNA samples were diluted to 10 ng/μL and 1.5 μL DNA aliquots of each sample were used for 50 μL PCR reactions. A fragment of the 16S rRNA gene of approximately 525 bp in length was amplified using bacterial primer 27F and universal primer 518R, both containing a unique 10-bp barcode sequence per sample to facilitate sorting of sequences from a single pyrosequencing run. PCR was conducted with 0.3 mM of each forward and reverse barcoded primer, 1.5 μl template DNA, 2X buffer reaction, 0.2 mM of dNTPs, 1 mM $MgSO_4$ and 1U of Platinum Pfx DNA polymerase (Invitrogen). Samples were initially denatured at 94°C for 5 min, then amplified using 35 cycles of 94°C for 30s, 50°C for 30 s and 68°C for 30s. A final extension of 10 min at 68°C was added at the end of the program to ensure complete amplification of the target region. Each of the triplicate subsamples was amplified separately and later combined and used as a representative composite of each sample. Amplicons were gel purified using GFX PCR DNA and Gel Band Purification Kit (GE Healthcare), and sent to the Genome Project Division Macrogen Inc. Seoul, Republic of Korea to be run on a Roche Diagnostics (454 Life Science) GS-FLX instrument with Titanium chemistry.

Sequence Data Analysis

Data were processed using MOTHUR v.1.22.2 following the Schloss SOP [28]. Briefly, the 10-bp barcode was examined in order to assign sequences to samples. Sequencing errors were reduced by implementation of the AmpliconNoise algorithm and low-quality sequences were removed (minimum length 200 bp, allowing 1 mismatch to the barcode, 2 mismatches to the primer, and homopolymers no longer than 8 bp). Sequences with ambiguous bases were eliminated as well, as their presence appear to be a strong indication of defective sequences [29]. The choice of these parameters for filtering follows the recommendation of Schloss et al. to reduce the error rate [30].

Chimera were removed with 'chimera.uchime' Mothur command. Sequences were aligned and classified against the SILVA bacterial SSU reference database v 102 [31]. Following the OTU approach [28], sequences from forward primers were clustered according the furthest neighbor-clustering algorithm.

Pyrosequencing raw reads were deposited in the NCBI Short-Read Archive under accession SRA057382. Sequence profile of processed sequences is shown in table S1. Raw and filtered reads per sample are shown in table S2.

Shared tables were created indicating the number of times an OTU appears in each sample. Venn diagrams were constructed with package gplots in R 2.10.1 (http://www.R-project.org/) from shared tables at the 0.05 distance level. Names and sequences from shared phylotypes were retrieved with scripts written in Python.

An indicator value analysis was performed for detecting statistically significant associations between taxa and soil management [32]. The indicator value combines the abundance of the OTU in the target group compared to other groups (specificity), with its relative frequency of occurrence in that particular group (fidelity). The value of the IndVal index was calculated using function IndVal in [R] package 'labdsv' (http://ecology.msu.montana.edu/labdsv/R/labdsv). Data were previously ANOVA filtered to reduce the number of tests and therefore increase the

power to detect true differences. In order to perform multiple testing corrections, analysis of false-discovery rate (FDR) of 0.05 of significance were calculated for the complete set of p-values with qvalue.gui() in [R] package 'qvalue' (http://genomics.princeton.edu/storeylab/qvalue/). The FDR estimates the chance of reporting a false-positive result in all the significant results [33].

Real-time PCR Quantification

Bacterial taxa were quantified using the taxon-specific 16S rRNA primers designed in this work. All primers were designed using the PRIMROSE software [34].

For bacteria within the GP1 group of the phylum Acidobacteria, two specific primers were designed using the sequences available at the Ribosomal Database Project (RDP) database (http://rdp.cme.msu.edu/probematch/search.jsp) and the 100 sequences of the selected OTU.

Primers for the genus *Rubellimicrobium* Rub290F (GAGAGGAT-GATCAGCAAC) and Rub547R (CGCGCTTTACGCC-CAGTC) were designed using all the *Rubellimicrobium* sequences available in the RDP database. The specificity of the primers Gp1Ac650R (TTTCGCCACAGGTGTTCC) and SubGp1-143F (CGCATAACATCGCGAGGG) were initially checked by *in silico* analysis against RDP probe match (data set options: good and >1200 bp). The Gp1Ac650R primer matched with 92.4% of the sequences within the class Acidobacteria Gp1 (2268/2454), and with other 394 non-target bacteria. The SubGp1-143F primer matched with 99/100 sequences of the indicator group and only 3 sequences of non-target bacteria in the RDP database. More than 98% of the sequences in the RDP database (105/110) matched the primers combination Rub290F (GAGAGGATGATCAGCAAC) and Rub547R (CGCGCTTTACGCCCAGTC). Only 2 sequences of non-target bacteria in the RDP database matched with this primers combination.

To further test the specificity of the qPCR assays, clone libraries were constructed for each primer set using DNA extracted from GAP soil samples of Pergamino and Monte Buey for Acidobacteria GP1, and PAP soil samples of Bengolea and Monte Buey for *Rubellimicrobium*. Twenty four positive clones of each library (n = 48 for each primer set) were sent to Macrogen Inc. for complete sequencing. Sequences were assigned to taxonomic groups using the RDP classifier program. 100% of the cloned amplicons that could be identified belonged to the correct target groups.

Quantification was based on the increasing fluorescence intensity of the SYBR Green dye during amplification. The qPCR assay was carried out in a 20μL reaction volume containing the SYBR green PCR Master Mix (Applied Biosystems, UK), 0.5 μM of each primer, 0.25 μg/μL of BSA and 10 ng of soil DNA. Primer annealing temperature was optimized for PCR specificity in temperature-gradient PCR assays, utilizing the DNA Engine Opticon 2 System (MJ Research, USA). Optimal conditions for PCR were defined as 10 minutes at 94°C, 35 cycles of 94°C for 30 seconds, 59°C for 20 seconds and 72°C for 30 seconds, for both sets of primers. Standard curves were obtained using at least five ten-fold serial dilutions of a known amount of PCR amplicon mixtures as templates, purified through QIAquick PCR purification columns. Controls with no DNA templates gave null or negligible values.

Statistical Analysis

Patterns of similarity between samples were investigated using Correspondence Analysis (CA) on the relative abundances of $OTUs_{0.05}$. Due to the fact that the number of sequences obtained for NE of Monte Buey was markedly lower than those obtained for all other samples, this sample was excluded from this type of analysis.

Correlation between relative abundances of all significant indicator $OTUs_{0.05}$ and soil environmental gradients was assessed using Canonical Correspondence Analysis (CCA). The model used to explain variability included moisture content, total nitrogen content, total carbon content, ratio of total carbon to total nitrogen content and pH. Abiotic variables were standardized by subtracting the mean and dividing by the standard deviation (z-score standardization), making quantitative variables dimensionless. Significance was assessed using permutation tests. Multivariate analyses were performed in [R] package 'vegan'.

The effect of site and management on the number of copies of 16S rRNA genes for GP1A (the indicator group within the GP1 of the Acidobacteria) and *Rubellimicrobium*, was determined for each sampling date by mixed models. Treatment and seasons (summer, winter) were considered as fixed factors, whereas year, site, and subsample were included in the random structure. For this analysis, data were log transformed to achieve normal distribution of residues. The mean number of 16S rRNA gene copies of each taxon was compared for the effect of treatment by orthogonal contrasts.

Management indicator value was defined as the logarithm of the ratio of the normalized abundance (i.e. fold-change relative to a non-cultivated soil) of the GP1A and the normalized abundance of *Rubellimicrobium* template. Two-tailed one-sample *t*-tests were performed on mean management indicator values (n = 48 for each type of management) to test the null hypothesis that the mean was equal to 0, and 95% confidence intervals were calculated. Statistical analysis was carried out with InfoStat Plus version 2011 (http://www.infostat.com.ar).

Results

Soil Quality and Productivity According to Sites and Agricultural Practices

The information on the agricultural management and crop yields of the different sites under study are summarized in Table 1. Before the first sampling, all sites had been managed under no-till for at least the preceding five years, with the exception of a single year (2004/2005) in Bengolea, where the PAP site was chisel-plowed. In the four localities, GAP had in average a 62% higher proportion of maize in the crop rotation than the PAP. GAP had in the last five years 50% of the winters with crop, whereas PAP sites had only 20%. In addition, cover crops had been implanted in winter in three of the four GAP localities. Management also differed in terms of the amount of herbicides used, as soils under PAP had used 36% more herbicides than GAP during the previous five years. Soybean yield had been in average 24.7% higher in GAP than in PAP, whereas maize yield had been 149.9% higher in GAP.

Soil chemical and physical properties of the studied sites are presented in Table 2. There is a difference in soil texture among localities, with increasing clay and decreasing sand content from West (Bengolea) to East (Viale). Values of soil organic matter follow the relation NE>GAP>PAP at the different localities, except in Pergamino where the Good no-till Agricultural Practices (GAP) and the Poor no-till Agricultural Practices (PAP) showed similar values. Soil N content also followed the same pattern, with the exception of Bengolea, where GAP had higher values than NE. No clear association was observed between values of extractable P and soil type or management. The pH, which ranged from 5.5 to 6.7, did not appear to correlate with soil type or soil management. A more detailed analysis, comparing physical and chemical soil properties of the different agricultural management under study, exceeds the purpose of this paper and will be presented elsewhere (Duval et al, unpublished).

Bacterial Community Structure

The structure of bacterial communities related to the agricultural management practices was obtained from the massive parallel sequencing data of the 12 samples, i.e. three management scenarios over the four locations. A total of 210579 sequences with an average read length of 284 bp, were obtained after trimming, sorting, and quality control of the pyrosequencing data (table S2). 80% of these sequences were classified to a known phylum in the domain Bacteria (Fig. S1).

We examined OTU distribution across the pyrosequencing-based data sets. Using 3% sequence variation criterion, the patterns of the rarefaction curves were roughly comparable in all samples and none of the curves reached a plateau (Fig. S2). Therefore it was not possible to establish a trend in the differences of richness as a function of either geographical location or soil management. Since, despite quality filtering, pyrosequencing has a large intrinsic error that could lead to overestimation of rare phylotypes, further estimation of bacterial richness was not attempted.

Taxa Overlap between Soil Samples

Correspondence Analysis (CA) was applied to the data set of relative abundances for taxa defined at 0.05 distance (Fig. 1). Axes 1 and 2 account for 28.8% of the total inertia (16.4% and 12.4%, for axes I and II, respectively). Fitting of environmental factors to ordination indicate that samples were distributed according geographical location (site) with p = 0.001.

The majority of OTUs were unique to the samples in which they were found. The Venn diagrams in Fig. 2 show the number of shared OTUs among the different sample types. When analyzed by soil management, 254 OTUs were common to GAP and PAP samples across all sites (Table S3). Considering only sequences that were found in one type of management, but absent in the other, GAP and PAP samples had respectively 142 and 200 OTUs in common among the four sites, which corresponded to around 1.0% and 1.4% of the total number of sequences (Tables S4 and S5).

Considered by geographical location, the number of common OTUs was around 11% of the total OTUs identified in each location. In these cases, the overlap for the three samples in each sampling location was similar to the numbers of OTUs shared by the pair GAP and NE and the pair GAP and PAP, which in turn were consistently higher than the overlap of OTUs shared by NE and PAP (Fig. S3). We did not detect any OTU that was common to NE and PAP, but absent in GAP, even when the sample of NE from Monte Buey, which had less sequences, was excluded from the analysis. This finding disputes the possibility that the overlaps between groups of samples were due to chance.

Indicator Taxa of Agricultural Management

Table 3 shows the result of the IndVal analysis for indicators containing more than 20 sequences, which identified four significant indicators of GAP and five significant indicators of PAP.

Canonical Correspondence analysis (CCA) was applied on the bacterial taxa identified as indicators using IndVal to study the association of physico-chemical soil properties to sites and taxa (Fig. 3). Bacterial taxa clustered into three well-separated groups associated with the different soil management practices despite the different geographic origin of the data. The first ordination axis was strongly correlated to total nitrogen content (0.96, p<0.05),

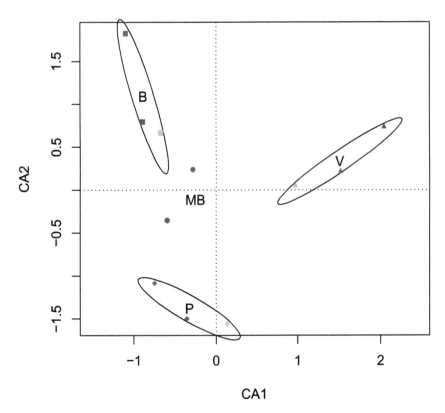

Figure 1. Ordination diagram from the Correspondence Analysis of the relative abundances for taxa defined at 0.05 distance. The 2-D CA diagram account for 29% of inertia. Locations of soils are indicated by squares (Bengolea, B), circles (Monte Buey, MB), diamond (Pergamino, P) and triangles (Viale, V). Colors indicate soil management type: Poor no-till Agricultural Practices in red, Good no-till Agricultural Practices in blue and Natural Environment in green. Standard error ellipses show 95% confidence areas.

meaning the bacterial indicators of natural environments were associated with higher than average nitrogen content. The second canonical axis correlated in descending order to the pH (0.49), moisture (−0.47), carbon to nitrogen ratio (0.45) and total carbon (0.22). Separation between management indicators was influenced by the second canonical axis. Indicators of PAP are located in the positive quadrant. i.e. they occur at sites with higher than average pH, and carbon to nitrogen ratio, and lower than average values of moisture. Inversely, GAP indicators were associated with higher moisture content, lower pH and lower carbon to nitrogen ratio.

Although significant after the application of the false discovery rate, most of the indicators had low abundance across the samples. To obtain meaningful quantitative results, we analyzed only significant indicator containing at least 75 sequences. As a result, we selected single the list's top indicators for GAP and PAP samples respectively, belonging to a taxa within the Acidobacteria Group 1 (GP1A), and to *Rubellimicrobium*, a genus of the order *Rhodobacterales* of the class Alphaproteobacteria (Table 3). Although this threshold can be considered somewhat arbitrary, it was selected on the basis of the fact that those taxa were represented more evenly across all studied locations.

PCR primers were designed to target the sequences detected in the pyrosequence data set of these selected groups. Cloning and sequencing of the PCR products derived from the primer specificity tests confirmed the specificity of these primers (see M&M). Using these newly designed primers, quantitative PCR was conducted to validate the results of the sequence analysis. Primer sets of both Acidobacteria GP1A and the genus *Rubellimicrobium* were calibrated using known concentrations of clones of PCR amplified 16S rRNA genes from the respective

controls. The quantification of a set of Acidobacteria GP1A and of the genus *Rubellimicrobium*, performed over samples from two successive winter-summer seasons (June 2009 to February 2011) are shown respectively in Fig. 4 A and B. Quantitative PCR data revealed that the abundance of both taxa were significantly different among managements (mixed models, p<0.0098 for GP1A and p<0.0001 for *Rubellimicrobium*). Post-hoc contrasts indicated that the number of copies of 16S rRNA genes targeted with GP1A-specific primers were statistically higher in samples of soils defined as GAP (p = 0.005), compared to poorly managed soils. In opposition, the number of copies of 16S rRNA genes targeted with *Rubellimicrobium*-specific set of primers were statistically higher in samples of soils defined as PAP (p = 0.004).

The observation that the GP1A-specific primers target sequences that increase in GAP samples and that the *Rubellimicrobium*-specific primers target sequences that increase in PAP samples, prompted us to evaluate their combined use as potential indicator of agricultural management under no-till regime in the Pampa Region. For that purpose, we calculated the ratio of the normalized abundance (i.e. fold-change relative to a non-cultivated soil in the same geographical location) of the GP1A and the normalized abundance of *Rubellimicrobium* template. By definition, this value is equal to one for NE samples. Log transformation was applied to achieve normal distribution. The resulting indicator value will take therefore a value of zero for NE samples. For all other samples, the sign of the indicator will depend on whether the GP1A and *Rubellimicrobium* abundances increase or decrease relative to NE. The results are indicated in Fig. 5. Despite the high variability of the data, likely due in part to the heterogeneous distribution of bacteria in subsamples within

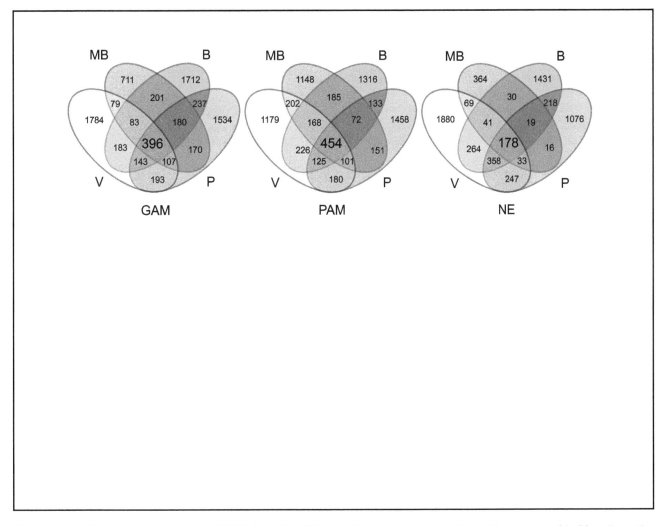

Figure 2. Venn diagram of the overlap of OTUs from the different soil management practices at four geographical locations. The numbers of overlapping tag sequences are indicated in the graph. Management practices are indicated at the bottom of each diagram: GAP: Good no-till Agricultural Practices, PAP: Poor no-till Agricultural Practices; NE: Natural Environment. Location labels are indicated with B (Bengolea), MB (Monte Buey), P (Pergamino) and V (Viale).

each given sample, the mean of the value calculated for all GAP samples across the four sites and four sampling dates (n = 48) was significantly higher than 0 (p = 0.0018), whereas the mean of the value in all PAP samples was significantly lower than 0 (p<0.001).

Discussion

The main hypothesis tested in this study was that the relative abundance of selected soil bacterial taxa could be used as indicator of the impact of agronomic management at a regional scale. Based on massive parallel sequencing and quantitative PCR, we have found that the combined use of the abundance of two bacterial taxa could potentially fulfill this task. The bacteria, belonging to Acidobacteria Group 1, and to the genus *Rubellimicrobium* of the Alphaproteobacteria, were augmented in soils under no-till crop production, managed with sustainable and non-sustainable practices, respectively. What makes our finding more compelling is that the taxa that appeared to be specific of soil management, were present in soils with different physical properties (Table 1) with various crop sequences (Table 2), suggesting that the physiology of these group of bacteria might be affected by nutrient and carbon shifts, and probably different soil microstructure, produced by the

different crop rotation practices: intense crop rotation vs. monoculture practice, the most important characteristic differentiating soil managements, with consistent different yields.

Agricultural soil activities should sustain crop productivity while preserving soil environmental quality. After several years of no-till agriculture and the widespread practice of monoculture, farmers have realized the impact of management on soil quality, which ultimately impacted on crop productivity. This has led to a working definition of "Good no-till Agricultural Practices" (GAP) and "Poor no-till Agricultural Practices" (PAP), according to criteria based on yield, crop rotation, fertilization, pest management and agrochemical use (http:// www.ac.org.ar/index_e.asp). In this context, indicators of soil quality are essential tools to evaluate the impact of management on the soil ecosystem. Physical properties, such as soil structure, water storage capacity and soil aeration, as well as soil chemical characteristics are currently used as indicators of soil health. In addition, microbial properties are increasingly regarded as more sensitive and consistent indicators than biochemical parameters for monitoring the effect of management on soil quality [35]. This is because bacteria are in intimate contact with the soil

Table 3. Results of indicator species analysis.

	Size	IndVal	Freq	p value	q value	Phylogenetic affiliation
GAP	100	0.86	8	0.028	0.041	Acidobacteria_Gp1
PAP	76	0.78	9	0.032	0.041	*Rubellimicrobium*
GAP	55	0.91	7	0.050	0.041	Alphaproteobacteria
PAP	34	0.85	8	0.037	0.041	*Micromonosporaceae*
PAP	28	0.75	8	0.043	0.041	Acidobacteria_Gp16
GAP	26	0.85	7	0.014	0.041	Unclassified bacteria
GAP	23	0.83	6	0.044	0.041	Unclassified bacteria
PAP	20	1.00	4	0.009	0.041	Unclassified bacteria
PAP	20	0.80	7	0.038	0.041	Actinomycetales

For each of the taxa, we indicate the total number of sequences corresponding to the OTU that represents the specific groups of samples (size), the Indicator Value index (IndVal), the number of samples that contain the taxon (Freq), the statistical significance of the association (p-value), the chance of reporting a false-positive result (q-value), and the lowest taxonomic rank assigned with a bootstrap confidence greater than 80%. Agricultural managements GAP and PAP are defined in the main text. Results were sorted according to Size. Only OTUs containing 20 or more sequences are shown in this table. See Table S6 for a complete list of significant indicators with IndVal values ≥0.75.

availability of substrates, and therefore the energy available for growth [36].

Previous studies have related the effect of different management practices on the diversity and stability of microbial communities and the abundance of individual taxa, in carefully designed experimental plots, using PLFA profiling [37,38], phenotypic fingerprinting [23], ribosomal fingerprinting [39], and pyrosequencing [20,24]. However, we are not aware of a study that looked for indicators of soil management in the large spatial and temporal scale of agricultural practices tested in productive fields. This task is particularly challenging, as most bacteria present discernible biogeographical patterns, even within a given habitat type [40]. Accordingly, in a large-scale investigation on the relative importance of various soil factors and land-use regimes on soilborne microbial community composition, it was found that the main differences in the bacterial communities were related to soil factors [41].

In our samples we have observed that bacterial communities as a whole appeared indeed to be structured chiefly by geographical proximity, meaning that differences in composition are due mainly to soil characteristics at the landscape scale [42]. Nevertheless, it was particularly interesting that the distribution of certain bacterial populations was clearly shaped by factors determined by soil management as well, opening the window to find bacterial indicators of soil status across a broad spatial scale [24]. The numbers of OTUs, which were found to be common to the four soil locations subjected to similar management practice, was relatively large. We deem unlikely that this overlap was the outcome of chance alone, as for each type of management the

microenvironment. Microorganisms can be directly affected by some toxic effect, or indirectly, e.g. by changes in the

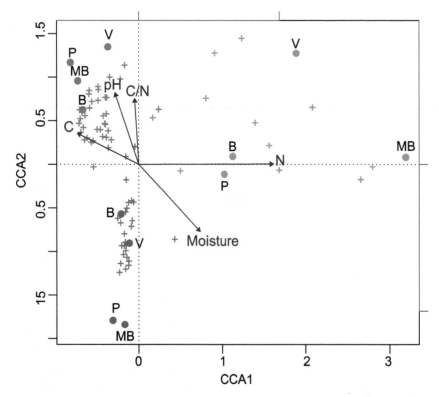

Figure 3. Ordination diagram from Canonical Correspondence Analysis of bacterial taxa identified as indicators using IndVal. Only OTUs identified with IndVal values higher than 0.75 were used in this analysis (Table S6). The 2-D ordination diagram CCA accounts for 66% of inertia. Samples are indicated by circles and site labels. OTUs are indicated by crosses, names are omitted. Arrows for quantitative variables show the direction of increase of each variable, and the length of the arrow indicates the degree of correlation with the ordination axes. Colors indicate soil management type: Poor no-till Agricultural Practices in red, Good no-till Agricultural Practices in blue; Natural Environment in green. Location labels are indicated with: B (Bengolea), MB (Monte Buey), P (Pergamino) and V (Viale).

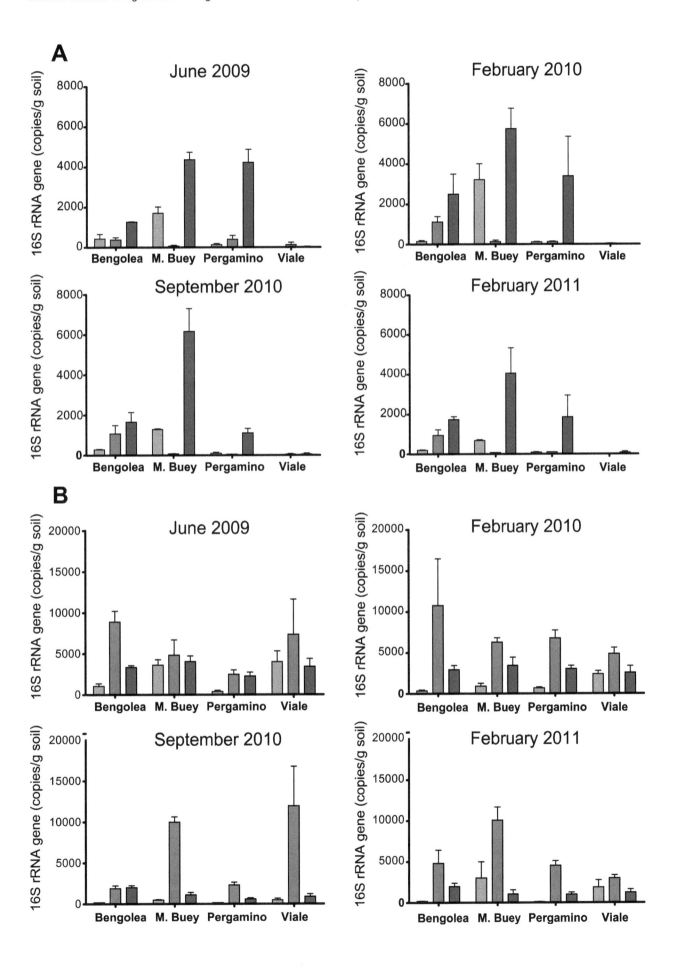

Figure 4. Quantitative phylogenetic group abundance of the OTUs targeted with a set of primers specific for Acidobacteria GP1A (panel A) and *Rubellimicrobium* genus (panel B). Each soil sample subjected to the indicated management in the four geographical locations was sampled at the date showed in the boxes. Bars correspond to the average qPCR data of three independent samples. Colors indicate soil management type: Poor no-till Agricultural Practices in red, Good no-till Agricultural Practices in blue, and Natural Environment in green. Error bars are standard error.

number of OTUs shared by any three of the four geographical locations was lower than the number of OTUs common to the four soil samples (Fig. 2). Even more striking is the observation of bacterial groups that can be associated with soil management in agricultural soils with dissimilar characteristics across a relatively wide regional scale. These data are consistent with both genomic and environmental perspectives suggesting the existence of ecological coherence of bacterial at different taxonomic ranks [43].

The set of indicator taxa were used to evaluate the correlation of their abundances with soil characteristics across sites and management. The ordination illustrated how indicator taxa were responsive to soil management practices. GAP indicators were associated with higher moisture content, and lower carbon to nitrogen ratio and lower pH. Slight changes in pH might have been caused by acidifying reactions (e.g. nitrification). Inversely, the occurrence of PAP indicators at sites was associated with higher than average carbon to nitrogen ratio, i.e. under conditions in which nitrogen becomes a limiting factor.

The phylum Acidobacteria ranked third in abundance in each of the twelve soil samples examined in this study. Acidobacteria constituted an average of 20% of soil bacterial taxa in 16S rRNA gene libraries, according to a published meta-analysis [44] and more recent analysis of agricultural soils indicated that three subgroups (GP4, GP6 and GP1) situate among the five most abundant genera in soils [24,45]. Although the phylum Acidobacteria it has been frequently associated with low nutrient availability [46], its wide global distribution and high diversity led to the proposition that its members are involved in a broad range of metabolic pathways [47]. Several findings point to the fact that

not all subdivisions within the phylum Acidobacteria share the same traits. Examples of these are the occurrence of numerically dominant as well as metabolically active Acidobacteria in rhizospheric soil [45], the lineage-dependent variations in relative abundance within a clay fraction of soil versus bulk soil [48] and the differences in the pH preferences for growth [44]. Interestingly, Mummey et al found that Acidobacteria were poorly represented in the inner fraction of aggregates [49], but were more abundant in soil macroaggregates and the outer fractions of microaggregates, i.e. in coarse pores, where they are supposed to have high turnover rates because of the effect of predation and desiccation events, and due to the transiently high oxygen and nutrient availability [50].

It is therefore not entirely surprising that a subgroup of the Acidobacteria group 1 emerges as a potential bacterial candidate for agronomic practices in soils managed under no till regime, in which carbon conservation and stability of macroaggregates are enhanced (Morras et al, personal communication). Unraveling specific details about the ecology of this particular lineage of Acidobacteria through cultivation [51] and genomic studies [52] are needed to gain a better understanding of its involvement in soil processes.

Neither is the natural habitat of Rubellimicrobia currently well characterized. To date, four species of the genus *Rubellimicrobium* had been described. One thermophilic species, *R. thermophilum*, which was isolated from slime deposits on paper machines and a pulp dryer [53] and three mesophilic species, two of which have been isolated from soils: *R. mesophilum* [54] and *R. roseum* [55], and *R. aerolatum*, which was isolated from air samples [56]. It is worth noting that fatty acids profiles of the same soil samples analyzed in this work, show in all PAP treatments that fatty acid $C_{18:1}\omega 7c$ is significantly augmented (Ferrari and Wall, unpublished). This is relevant because $C_{18:1}$ $\omega 7c$ is one of the major membrane fatty acids in most of the isolates belonging to the genus *Rubellimicrobium* [54,55,56]. Given the limited physiological and ecological information available on the genus *Rubellimicrobium*, it would be too speculative to suggest for it any indicator function in soils at the present time. Nevertheless, it is interesting to note that this genus appeared to respond to the use of the soil in a recent study of the impact of long-term agriculture on desert soil, in which it was shown that *Rubellimicrobium* was among the extremophilic bacterial groups that disappeared from soil after agricultural use [57]. Efforts to isolate *Rubellimicrobium* strains from the soils surveyed in the present study are currently under way in our laboratory, in order to perform a thorough physiological characterization.

The results of this study provide relevant information about the distribution of several groups of numerically abundant taxa in agricultural soils. It was also demonstrated that different taxa of bacteria respond differentially to geographical constraints and contemporary disturbances in no-till agriculture systems, highlighting the potential of high-resolution molecular tools to identify bacterial groups that may serve as potential indicator that might be used to assess the sustainability of agricultural soil management and to monitor trends in soil condition over time.

We note that the selection of indicator species based solely on the frequency of occurrence does not permit conclusions about the processes in which they are involved. In this regard, knowledge on

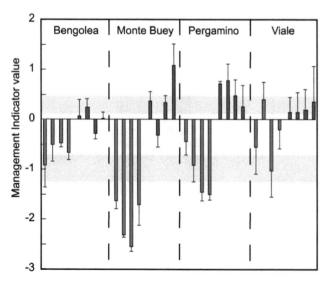

Figure 5. Indicator values for samples of soils under different agricultural management. The geographic sites are indicated in the box. In each site bars are ordered from left to right to the successive sampling dates: June 2009, February 2010, September 2010 and February 2011. PAP, Poor no-till Agricultural Practices (red) and GAP, Good no-till Agricultural Practices (blue). Shadow areas are 95% confidence intervals of indicator of GAP (0.24, 0.71) and PAP (−1.31, −0.41). Error bars are standard error.

their habitat specialization would be important, as this factor is not likely to be influenced by natural variations in environmental conditions [58]. However, considering the scarcity of data regarding the habitat preferences, physiology and in situ activity of Acidobacteria GP1 and *Rubellimicrobium*, a mechanistic link between the factors driving the relative distribution of these taxa and the different soil management is currently not feasible. Thus, although we have initially developed the indicator on a purely phenomenological basis, the understanding of the underlying ecological selection for both groups of taxa depending on the soil management remains a crucial goal for future studies.

Meanwhile, the proposed marker appears to fulfill several of the criteria required for appropriate ecological indicators. It is easily measured, it is sensitive to soil management actions and is integrative, i.e. it provides adequate coverage across a relatively wide range of ecological variables, e.g. soils types, climate, crop sequence, etc. [59]. Based on the data presented here, appropriate tests for simple monitoring can be elaborated to further validate if the proposed candidate biological indicator can be integrated into a minimum dataset, to allow measuring the impact of management practices under no-till at the regional scale.

Supporting Information

Figure S1 Complete set of sequences classified at phylum level against SILVA bacterial SSU refer- ence database v.102 by Bayesian method, with a confidence cutoff of 80% using classify.seqs command in Mothur.

Figure S2 Rarefaction analysis of pyrosequencing tags of the 16S rRNA gene in soils subjected at different agricultural practoces in the four geographic locations. Blue: Good no-till agricultural practices. Red: Poor no-till agricultural practices, Green: Natural environments.

Figure S3 Venn diagram of the overlap of OTUs from the different geographical locations subjected to different soil management practices. The numbers of overlapping tag sequences are indicated in the graph. Locations are indicated at the bottom of each diagram. Management practices are indicated with GAP: Good no-till Agricultural Practices, PAP: Poor no-till Agricultural Practices; NE: Natural Environment.

Table S1 Summary of processed 454-sequencing reads.

Table S2 Filtered and raw (in parenthesis) reads of 454 Pyrosequencing per sample.

Table S3 List of OTUs common to good no-till agricultural practices (GAP) and poor no-till agricultural practices (PAP) in the four locations. Sequences were assigned to taxonomic groups using the RDP classifier (http://rdp.cme.msu.edu/classifier/classifier.jsp). OTUs were sorted by the total number of sequences in the complete data set.

Table S4 List of OTUs only common to good no-till agricultural practices (GAP) in the four locations. Sequences were assigned to taxonomic groups using the RDP classifier (http://rdp.cme.msu.edu/classifier/classifier.jsp). OTUs were sorted by the total number of sequences in the complete data set.

Table S5 List of OTUs common only to poor no-till agricultural practices (PAP) in the four locations. Sequences were assigned to taxonomic groups using the RDP classifier (http://rdp.cme.msu.edu/classifier/classifier.jsp). OTUs were sorted by the total number of sequences in the complete data set.

Table S6 Results of indicator species analysis. For each of the taxa, we indicate the Indicator Value index (IndVal), the number of samples that contain the taxon (Freq), the statistical significance of the association (p-value), the total number of sequences corresponding to the OTU (size), and the lowest taxonomic rank assigned with a bootstrap confidence indicated in parenthesis. Agricultural managements GAP and PAP an NE are defined in the main text. Only OTUs with IndVal higher than 0.75 are shown.

Author Contributions

Conceived and designed the experiments: ELMF LDG LGW LE. Performed the experiments: ELMF LDG. Analyzed the data: ELMF LDG JCB SR LE. Contributed reagents/materials/analysis tools: LS MD JAG. Wrote the paper: ELMF LGW LE. Commented on manuscript: LDG JCB.

References

1. Hobbs PR, Sayre K, Gupta R (2008) The role of conservation agriculture in sustainable agriculture. Phil Trans Royal Soc B: Biol Sci 363: 543–555.
2. Bronick CJ, Lal R (2005) Soil structure and management: a review. Geoderma 124: 3–22.
3. Abid M, Lal R (2009) Tillage and drainage impact on soil quality: II. Tensile strength of aggregates, moisture retention and water infiltration. Soil Till Res 103: 364–372.
4. Derpsch R, Friedrich T, Kassam A, Li H (2010) Current status of adoption of no-till farming in the world and some of its main benefits. Intl J Agric Biol Eng 3: 1–25.
5. Montgomery DR (2007) Soil erosion and agricultural sustainability. Proc Natl Acad Sci USA 104: 13268–13272.
6. Viglizzo EF, Frank FC, Carreño LV, Jobbágy EG, Pereyra H, et al. (2011) Ecological and environmental footprint of 50 years of agricultural expansion in Argentina Global Change Biol 17: 959–973.
7. Cook RJ (2006) Toward cropping systems that enhance productivity and sustainability. Proc Natl Acad Sci USA 103: 18389–18394.
8. Bockstaller C, Guichard L, Makowski D, Aveline A, Girardin P, et al. (2008) Agri-environmental indicators to assess cropping and farming systems. A review. Agron Sustain Dev 28: 139–149.
9. Govaerts B, Sayre KD, Deckers J (2006) A minimum data set for soil quality assessment of wheat and maize cropping in the highlands of Mexico. Soil Tillage Res 87: 163–174.
10. Imaz M, Virto IP, Bescansa P, Enrique A, Fernandez-Ugalde O, et al. (2010) Soil quality indicator response to tillage and residue management on semi-arid Mediterranean cropland. Soil Tillage Res 107: 17–25.
11. Rodrigues de Lima AC, Hoogmoed W, Brussaard L (2008) Soil quality assessment in rice production systems: establishing a minimum data set. J Environ Qual 37: 623–630.
12. Shukla MK, Lal R, Ebinger M (2006) Determining soil quality indicators by factor analysis. Soil Tillage Res 87: 194–204.
13. Yemefack M, Jetten VG, Rossiter DG (2006) Developing a minimum data ser for characterizing soil dynamics in shifting cultivation Systems. Soil Tillage Res 86: 84–98.
14. Hartmann M, Fliessbach A, Oberholzer HR, Widmer F (2006) Ranking the magnitude of crop and farming system effects on soil microbial biomass and genetic structure of bacterial communities. FEMS Microbiol Ecol 57: 378–388.
15. Grayston SJ, Wang S, Campbell CD, Edwards AC (1998) Selective influence of plant species on microbial diversity in the rhizosphere. Soil Biol Biochem 30: 369–378.
16. Marschner P, Crowley D, Yang CH (2004) Development of specific rhizosphere bacterial communities in relation to plant species, nutrition and soil type. Plant and Soil 261: 199–208.
17. Smalla K, Wieland G, Buchner A, Zock A, Parzy J, et al. (2001) Bulk and rhizosphere soil bacterial communities studied by denaturing gradient gel

electrophoresis: plant-dependent enrichment and seasonal shifts revealed. Appl Environ Microbiol 67: 4742–4751.

18. Nacke H, Thürmer A, Wollherr A, Will C, Hodac L, et al. (2011) Pyrosequencing-based assessment of bacterial community structure along different management types in german forest and grassland soils. PloS One 6: 1387–1390.

19. Girvan MS, Bullimore J, Pretty JN, Osborn AM, Ball AS (2003) Soil type is the primary determinant of the composition of the total and active bacterial communities in arable soils. Appl Environ Microbiol 69: 1800–1809.

20. Acosta-Martinez V, Dowd S, Sun Y, Allen V (2008) Tag-encoded pyrosequencing analysis of bacterial diversity in a single soil type as affected by management and land use. Soil Biol Biochem 40: 2762–2770.

21. Lagomarsino A, Moscatelli MC, Di Tizio A, Mancinelli R, Grego S, et al. (2009) Soil biochemical indicators as a tool to assess the short-term impact of agricultural management on changes in organic C in a Mediterranean environment. Ecol Indic 9: 518–527.

22. Laudicina VA, Badalucco L, Palazzolo L (2011) Effects of compost input and tillage intensity on soil microbial biomass and activity under Mediterranean conditions. Biol Fertil Soils 47: 63–70.

23. Lupwayi NZ, Rice WA, Clayton GW (1998) Soil microbial diversity and community structure under wheat as influenced by tillage and crop rotation. Soil Biol Biochem 30: 1733–1741.

24. Yin C, Jones KL, Peterson DE, Garrett KA, Hulbert SH, et al. (2010) Members of soil bacterial communities sensitive to tillage and crop rotation. Soil Biol Biochem 42: 2111–2118.

25. Schloter M, Dilly O, Munch JC (2003) Indicators for evaluating soil quality. Agric Ecosyst Environ 98: 255–262.

26. Wall LG (2011) The BIOSPAS Consortium: Soil Biology and Agricultural Production. In: Bruijn FJd, editor. Handbook of Molecular Microbial Ecology I: Metagenomics and Complementary Approaches. Hoboken, NJ, USA: John Wiley & Sons, Inc. pp. ch. 34.

27. Parameswaran P, Jalili R, Tao L, Shokralla S, Gharizadeh B, et al. (2007) A pyrosequencing-tailored nucleotide barcode design unveils opportunities for large-scale sample multiplexing. Nucl Acids Res 35: e130.

28. Schloss PD, Westcott SL, Ryabin T, Hall JR, Hartmann M, et al. (2009) Introducing mothur: open-source, platform-independent, community-supported software for describing and comparing microbial communities. Appl Environ Microbiol 75: 7537–7741.

29. Huse SM, Huber JA, Morrison HG, Sogin ML, Welch DM (2007) Accuracy and quality of massively parallel DNA pyrosequencing. Genome Biol 8: R143.

30. Schloss PD, Gevers D, Westcott SL (2011) Reducing the effects of PCR amplification and sequencing artifacts on 16S rRNA-based studies. PLoS One 6: e27310.

31. Pruesse E, Quast C, Knittel K, Fuchs BM, Ludwig W, et al. (2007) SILVA: a comprehensive online resource for quality checked and aligned ribosomal RNA sequence data compatible with ARB. Nucl Acids Res 35: 7188–7196.

32. Dufrene M, Legendre P (1997) Species assemblages and indicator species: the need for a flexible asymmetrical approach. Ecol Monogr 67: 345–366.

33. Storey J, Tibshirani R (2003) Statistical significance for genomewide studies. Proc Natl Acad Sci USA 100: 9440–9445.

34. Ashelford KE, Weightman AJ, Fry JC (2002) PRIMROSE: a computer program for generating and estimating the phylogenetic range of 16S rRNA oligonucleotide probes and primers in conjunction with the RDP-II database. Nucleic Acids Res 30: 3481–3489.

35. Garbisu C, Alkorta I, Epelde L (2011) Assessment of soil quality using microbial properties and attributes of ecological relevance. Appl Soil Ecol 49: 1–4.

36. Bending GD, Turner MK, Rayns F, Marx MC, Wood M (2004) Microbial and biochemical soil quality indicators and their potential for differentiating areas under contrasting agricultural management regimes. Soil Biol Biochem 36: 1785–1792.

37. Romaniuk R, Giuffré L, Costantini A, Nannipieri P (2011) Assessment of soil microbial diversity measurements as indicators of soil functioning in organic and conventional horticulture systems. Ecol Indic 11: 1345–1353.

38. Zelles L (1999) Fatty acid patterns of phospholipids and lipopolysaccharides in the characterisation of microbial communities in soil: a review. Biol Fertil Soils 29: 111–129.

39. Wu T, Chellemi DO, Graham JH, Martin KJ, Rosskopf EN (2008) Comparison of soil bacterial communities under diverse agricultural land management and crop production practices. Microb Ecol 55: 293–310.

40. Martiny JB, Bohannan BJ, Brown JH, Colwell RK, Fuhrman JA, et al. (2006) Microbial biogeography: putting microorganisms on the map. Nat Rev Microbiol 4: 102–112.

41. Kuramae E, Gamper H, van Veen J, Kowalchuk G (2011) Soil and plant factors driving the community of soil-borne microorganisms across chronosequences of secondary succession of chalk grasslands with a neutral pH. FEMS Microbiol Ecol 77: 285–294.

42. Ge Y, He J, Zhu Y, Zhang J, Xu Z, et al. (2008) Differences in soil bacterial diversity: driven by contemporary disturbances or historical contingencies? ISME J 2: 254–264.

43. Philippot L, Andersson SG, Battin TJ, Prosser JI, Schimel JP, et al. (2010) The ecological coherence of high bacterial taxonomic ranks. Nat Rev Microbiol 8: 523–529.

44. Sait M, Davis KER, Janssen PH (2006) Effect of pH on isolation and distribution of members of subdivision 1 of the phylum Acidobacteria occurring in soil. Appl Environ Microbiol 72: 1852–1857.

45. Lee SH, Ka JO, Cho JC (2008) Members of the phylum Acidobacteria are dominant and metabolically active in rhizosphere soil. FEMS Microbiol Lett 285: 263–269.

46. Fierer N, Bradford MA, Jackson RB (2007) Toward an ecological classification of soil bacteria. Ecology 88: 1354–1364.

47. Ganzert L, Lipski A, Hubberten HW, Wagner D (2011) The impact of different soil parameters on the community structure of dominant bacteria from nine different soils located on Livingston Island, South Shetland Archipelago, Antarctica. FEMS Microbiol Ecol 76: 476–491.

48. Liles MR, Turkmen O, Manske BF, Zhang M, Rouillard JM, et al. (2010) A phylogenetic microarray targeting 16S rRNA genes from the bacterial division Acidobacteria reveals a lineage-specific distribution in a soil clay fraction. Soil Biol Biochem 42: 739–747.

49. Mummey D, Holben W, Six J, Stahl P (2006) Spatial stratification of soil bacterial populations in aggregates of diverse soils. Microb Ecol 51: 404–411.

50. Chenu C, Hassink J, Bloem J (2001) Short-term changes in the spatial distribution of microorganisms in soil aggregates as affected by glucose addition. Biol Fertil Soils 34: 349–356.

51. Davis KE, Sangwan P, Janssen PH (2011) Acidobacteria, Rubrobacteridae and Chloroflexi are abundant among very slow-growing and mini-colony-forming soil bacteria. Environ Microbiol 13: 798–805.

52. Ward NL, Challacombe JF, Janssen PH, Henrissat B, Coutinho PM, et al. (2009) Three genomes from the phylum Acidobacteria provide insight into the lifestyles of these microorganisms in soils. Appl Environ Microbiol 75: 2046–2056.

53. Denner EB, Kolari M, Hoornstra D, Tsitko I, Kampfer P, et al. (2006) *Rubellimicrobium thermophilum* gen. nov., sp. nov., a red-pigmented, moderately thermophilic bacterium isolated from coloured slime deposits in paper machines. Int J Syst Evol Microbiol 56: 1355–1362.

54. Dastager SG, Lee JC, Ju YJ, Park DJ, Kim CJ (2008) *Rubellimicrobium mesophilum* sp. nov., a mesophilic, pigmented bacterium isolated from soil. Int J Syst Evol Microbiol 58: 1797–1800.

55. Cao YR, Jiang Y, Wang Q, Tang SK, He WX, et al. (2010) *Rubellimicrobium roseum* sp. nov., a Gram-negative bacterium isolated from the forest soil sample. Antonie Van Leeuwenhoek 98: 389–394.

56. Weon HY, Son JA, Yoo SH, Hong SB, Jeon YA, et al. (2009) *Rubellimicrobium aerolatum* sp. nov., isolated from an air sample in Korea. Int J Syst Evol Microbiol 59: 406–410.

57. Koberl M, Muller H, Ramadan EM, Berg G (2011) Desert farming benefits from microbial potential in arid soils and promotes diversity and plant health. PLoS One 6: e24452.

58. Carignan V, Villard MA (2002) Selecting indicator species to monitor ecological integrity: A review. Environ Monit Assess 78: 45–61.

59. Dale VH, Beyeler SC (2001) Challenges in the development and use of ecological indicators. Ecol indic 1: 3–10.

5

Soil Microbial Substrate Properties and Microbial Community Responses under Irrigated Organic and Reduced-Tillage Crop and Forage Production Systems

Rajan Ghimire[1], Jay B. Norton[1]*, Peter D. Stahl[1], Urszula Norton[2]

1 Department of Ecosystem Science and Management, University of Wyoming, Laramie, Wyoming, United States of America, **2** Department of Plant Sciences, University of Wyoming, Laramie, Wyoming, United States of America

Abstract

Changes in soil microbiotic properties such as microbial biomass and community structure in response to alternative management systems are driven by microbial substrate quality and substrate utilization. We evaluated irrigated crop and forage production in two separate four-year experiments for differences in microbial substrate quality, microbial biomass and community structure, and microbial substrate utilization under conventional, organic, and reduced-tillage management systems. The six different management systems were imposed on fields previously under long-term, intensively tilled maize production. Soils under crop and forage production responded to conversion from monocropping to crop rotation, as well as to the three different management systems, but in different ways. Under crop production, four years of organic management resulted in the highest soil organic C (SOC) and microbial biomass concentrations, while under forage production, reduced-tillage management most effectively increased SOC and microbial biomass. There were significant increases in relative abundance of bacteria, fungi, and protozoa, with two- to 36-fold increases in biomarker phospholipid fatty acids (PLFAs). Under crop production, dissolved organic C (DOC) content was higher under organic management than under reduced-tillage and conventional management. Perennial legume crops and organic soil amendments in the organic crop rotation system apparently favored greater soil microbial substrate availability, as well as more microbial biomass compared with other management systems that had fewer legume crops in rotation and synthetic fertilizer applications. Among the forage production management systems with equivalent crop rotations, reduced-tillage management had higher microbial substrate availability and greater microbial biomass than other management systems. Combined crop rotation, tillage management, soil amendments, and legume crops in rotations considerably influenced soil microbiotic properties. More research will expand our understanding of combined effects of these alternatives on feedbacks between soil microbiotic properties and SOC accrual.

Editor: Jose Luis Balcazar, Catalan Institute for Water Research (ICRA), Spain

Funding: Agricultural Prosperity for Small and Medium-Sized Farms Competitive Grant no. 2009-55618-05097 from the USDA National Institute of Food and Agriculture. The funders had no role in study design, data collection and analysis, decision to publish, or preparation of the manuscript.

Competing Interests: The authors have declared that no competing interests exist.

* Email: jnorton4@uwyo.edu

Introduction

Many changes in soil properties after conversion from one agricultural management system to another result from changes in soil microbiotic properties, defined here as the quality of microbial substrate and its effects on soil microbial communities [1,2]. It is well known that management practices such as reduced-tillage, cover crops, and crop diversification increase soil microbial activity in general, and microbial biomass and diversity in particular [2]. Similarly, practices used in certified-organic food and feed production, including amendments and legume crops in rotations, support increased microbial biomass [1,4], arbuscular mycorrhizal fungi (AMF) [5], and soil fauna [4]. It is not as clear how beneficial these management practices are under marginally productive conditions of cold, semiarid agroecosystems. In the study reported here, we evaluated whole-system effects on soil microbiotic properties after conversion from irrigated maize monoculture to conventional, reduced-tillage, and organic crop rotation systems in the central High Plains region of North

America. Each management system combines a different suite of practices, including cultivation methods, crop rotations, and soil amendments.

Soil microbiotic properties are influenced by soil amendments, crop rotations, and tillage practices by different mechanisms. Organic amendments contribute diverse microbial substrates as heterogeneous organic materials in different states of decomposition [1,4], while crop rotations diversify the supply of plant residues, including fine roots, root exudates, sloughed off tissues, and rhizodeposited materials, which drive diversification of soil microbial communities [2,6,7]. Intensive tillage drives pulses of microbial activity that mineralize labile soil organic matter (SOM) and shift microbiotic properties toward C-limited conditions that favor bacteria and reduce SOM concentrations, while reduced-tillage conserves labile substrates and creates a more consistent soil environment for microbial activity [2,3,8,9]. In reduced-tillage systems, plant- and root-derived residues provide nucleation sites for fungal and bacterial growth, which further colonize soil particles to form aggregates and increase aggregate-protected, labile SOM and efficiency of substrate utilization (less C respired per unit of microbial biomass) [10,11]. Perennial legume and non-legume forage crops in rotations further reduce soil disturbance compared with annual crops, and stimulate SOM accrual and microbial activity through increases in root biomass and residues [11–13]. Combinations of organic amendments and perennial legumes in rotations, which are common practices in certified organic crop and forage production, support more efficient soil N utilization than conventional, synthetic fertilizer-based management [4,14] and can shift microbiotic properties toward N-limited conditions that favor fungi and accrue SOM.

Such management systems may be especially important in the central High Plains agroecosystem, where the semiarid environment, with inherently low SOM, cold winters, hot, dry summers, and irrigation-driven wetting-drying cycles, exacerbate mineralizing microbiotic conditions that drive losses of SOM [15–18]. Improved understanding of how reduced-tillage and organic crop and forage production systems affect soil microbial substrate quality, microbial biomass, community structure, substrate utilization, and soil organic C (SOC) sequestration in this cold and dry agroecosystem will help to design more sustainable systems during a time of uncertainty due to the changing climate, increasing operation costs, and changing markets [24–26].

The aim of this study was to evaluate SOC, DOC, C:N ratios of microbial substrates, soil microbial biomass and community structure, and substrate utilization after transition from mono-cropped corn to crop rotations under conventional, organic, and reduced-tillage crop and forage production. The experiments were set up on inherently low fertility, irrigated soils in the dry and cold central High Plains agroecosystem. We hypothesized that crop rotations developed in the previously monocropped field would increase microbial biomass and microbial community diversity by increasing the quantity and changing the quality of microbial substrates. In addition, organic and reduced-tillage management systems would favor greater increases in soil microbial biomass and more diverse microbial communities with higher substrate utilization efficiency compared with conventional management.

Materials and Methods

Experimental Site

The four-year study was established in 2009 at the University of Wyoming Sustainable Agriculture Research and Extension Center (SAREC) near Lingle, Wyoming (42°7′15.03″N; 104°23′13.46″W). The study area has cool temperatures and a short growing season with an average frost-free period of about 125 days and 60-year average maximum and minimum temperature of 17.8°C and 0.06°C, respectively, and precipitation of 332 mm [19]. In addition, maximum and minimum air temperature and precipitation were monitored at the SAREC weather station within 1 km of the research plots during the study period. Monthly average maximum and minimum air temperature and monthly total precipitation throughout the study period are presented in Figure S1. Soil at the study site is mapped as Mitchell loam (loamy, mixed, active, mesic Ustic Torriorthent) with low SOM content (<1%), and slightly alkaline soil pH [20]. Soil texture of the study site was loamy with sand, silt and clay content of 41.0 (13.5)%, 41.4 (10.5)% and 17.6 (4.0)%, respectively (standard deviation in parentheses; n = 24).

Experimental Design and Treatments

The study was designed as two independent randomized complete block experiments (row-crop production and forage production) laid out on a 15-ha half-circle under an irrigation

pivot (305-m radius) that was divided into four wedge-shaped blocks (replications) (Figure S2). Each block was further separated into six plots consisting of three 0.405-ha crop production plots (outer three circles) and three 0.81-ha forage production plots (inner three circles). The three management-system treatments (conventional, certified organic, and reduced-tillage) were then randomly assigned to the crop and forage production plots. Before establishment of the experiment the entire area was under conventionally managed corn for at least six years.

All treatments were managed under four-year rotations starting in 2009. Table 1 shows the specific rotations, which were determined by a project advisory committee consisting of local producers and the SAREC management team. Under the conventional system inputs are applied as needed to maximize production, namely commercial synthetic fertilizer based on soil-test recommendations to supply nutrients, and chemical pesticides to control weeds, insects, and diseases. Specific management details are provided in Table S1. Conventional plots were moldboard ploughed, disked, and harrowed, which typically incorporates crop residues into soils leaving <15% of the soil surface covered by residues. The reduced-tillage system used conservation tillage that does not invert surface soil and leaves > 15% residue cover on the soil surface. In the organic system, tillage was done as in conventional plots, and pest control and nutrient management were based on practices allowed by the USDA National Organic Program standards (http://www.ams.usda.gov/AMSv1.0/nop). Conventional and reduced-tillage systems had chemical weed and pest control.

For soil fertility management, conventional and reduced-tillage plots received chemical fertilizer based on soil-test recommendations (Table S1). Organic management received composted cattle manure (dry matter 78% and C:N:P:S = 24.6:0.88:0.22:0.25%) in both crop and forage system in 2010. Because of the limited availability of composted cattle manure in 2011 and 2012, the organic crop system received raw manure (dry matter 29.2% and C:N:P:S = 21.3:1.42:0.35:0.40%) and the organic forage system received composted manure.

In the crop production experiment, management systems had different crops in rotation (Table 1). In forage system plots, a legume-grasses mixture was planted at 22 kg ha^{-1} in all plots in 2009, and included 50% alfalfa (*Medicago sativa* L.), 30% orchard grass (*Dactylis glomerata* L.), 10% meadow brome (*Bromus riparius* Rehmann), and 10% oat (*Avena sativa* L.) by weight. The forage production system plots were winter grazed for three months during 2011–2012 at stocking density of 1.6 fall-weaned calves ha^{-1}.

Soil Sampling

Soil samples were collected during spring, early summer, late summer, and fall seasons of the first (Year 1; 2009) and the fourth year (Year 4; 2012) from each of the 24 plots. During each sampling event, soil cores (3.2-cm diameter) were collected from 0–15 cm at 16 sampling points along a 50-m transect set in each plot, composited, thoroughly homogenized, subsampled (~500 g), and placed on ice for transport to the laboratory. The 0–15 cm depth was considered to be sufficient because the focus was on near-surface microbial properties. Sampling transects were mapped using GPS (Trimble GeoXT, Sunnyvale, CA) to locate transects for subsequent sampling. In the laboratory, soil samples were stored at −20°C for PLFA analysis and at 4°C for DOC, TDN, and potential soil respiration. Phospholipid fatty acid contents in soil were analyzed within two weeks of soil sample collection. Soil bulk density was measured in a separate set of 2.1×15 cm cores collected from 8 sampling points along the 50-m

Table 1. Crop rotations and management practices under different conventional (CV), organic (OR), and reduced-tillage (RT) management systems for crop and forage production (see Table S2 for detailed dates and management activities).

System		year	Crop in rotation	Management practices
Crop	CV	2009	Pinto bean	Tillage with moldboard plow and disk (5–7 passes each year), use of chemical fertilizers based on soil test recommendation for each crop, pesticides application as needed, and no livestock grazing.
		2010	Corn	
		2011	Sugar beet	
		2012	Corn	
	OR	2009	Alfalfa	Tillage with moldboard plow and disk (5–7 passes each year) and use of USDA-NOP certified practices for soil fertility (organic manure application) and pest management (e.g., cultivation), and no livestock grazing.
		2010	Alfalfa	
		2011	Corn	
		2012	Pinto bean	
	RT	2009	Pinto bean	Reduced-tillage (1–2 tillage passes each year that leave >15% crop residue on surface), use of chemical fertilizers based on soil test recommendation, pesticides application as needed, and no livestock grazing.
		2010	Corn	
		2011	Sugar beet	
		2012	Corn	
Forage	CV	2009	Alfalfa/grasses	Conventional tillage (5–7 passes in year 1 and 4), use of chemical fertilizers based on soil test recommendation, pesticides application as needed, and grazing with fall weaned calves during winter 2011/12.
		2010	Alfalfa/grasses	
		2011	Alfalfa/grasses	
		2012	Corn	
	OR	2009	Alfalfa/grasses	Conventional tillage and use of USDA certified practices for soil fertility (compost application) and pest management (no pesticides), and grazing with fall weaned calves during winter 2011/12.
		2010	Alfalfa/grasses	
		2011	Alfalfa/grasses	
		2012	Corn	
	RT	2009	Alfalfa/grasses	Reduced-tillage in the first year and no-tillage after, use of chemical fertilizers based on soil test recommendation, pesticides application as needed, and grazing with fall weaned calves during winter 2011/12.
		2010	Alfalfa/grasses	
		2011	Alfalfa/grasses	
		2012	Corn	

transects. Soil samples from the first and the last sampling dates were analyzed for other soil properties described below.

Laboratory Analysis

Total soil C and N were analyzed by dry combustion (EA1100 Soil C/N analyzer, Carlo Erba Instruments, Milan, Italy), inorganic C by modified pressure-calcimeter [21], and soil moisture by the gravimetric method [22]. Soil organic C was determined by subtracting inorganic C from total soil C. Soil pH was measured in a 1:1 soil:water mixture using an electrode [23]. Soil texture was determined by the hydrometer method [24]. Microbial substrate quality was determined as the ratio of DOC to total dissolved N (TDN) present in soils expressed as the C:N ratio of microbial substrate. For this, 10 g of field-moist soil was extracted with 50 ml of 0.5 M K_2SO_4 and amounts of DOC and TDN were determined by 720°C combustion catalytic oxidation/chemiluminescence with a Schimadzu TOC Analyzer (TOC-VCPH with TNM-1, Schimadzu Scientific Instruments, Inc.) coupled with TOC-Control V Ver.2 analysis software. Dissolved inorganic C was removed by automatic acidification and sparging

within the instrument. Potential soil respiration was determined as the amount of CO_2-C mineralized during a two-week incubation period [25]. Soil bulk density was determined by the core method [26] and water filled soil pore space was calculated from bulk density and gravimetric moisture content [27].

Microbial biomass and community structure was analyzed by the Blight and Dyre [28] method of fatty acid methyl ester (FAME) analysis as modified by Frostegård et al. [29] and Buyer et al. [30]. Fatty acids were directly extracted from soil samples using a 1:2:0.8 chloroform:methanol:phosphate buffer mixture (0.15 M, pH 4.0), and PLFAs were separated from neutral and glycolipid fatty acids in a solid-phase extraction column (Agilent Technologies Inc.). The PLFAs were methylated using a mild methanoic KOH, and the FAMEs were analyzed using an Agilent 6890 gas chromatograph with autosampler, split-splitless injector (7683B series), and flame ionization detector (Agilent Technologies Inc.). The system was controlled with Agilent Chemstation and MIDI Sherlock software, and the fatty acid peaks were identified using the MIDI peak identification software (MIDI, Inc., Newark, DE, USA). All solvents and chemicals used were of analytical grade, and all glassware used was rinsed 10 times with deionized water, and sterilized overnight in 450°C in a Blue M lab heat box type muffle furnace (Blue M Electric, Richardson, TX). The PLFA signatures of 16 different fatty acids, which were quantified in almost all the field plots, were used to study soil microbial community structure and these fatty acids were grouped into gram positive, gram negative and other bacteria, AMF and other fungi, and protozoa (Table S3). In addition, the Shannon diversity index [31] was calculated as an index of soil microbial diversity as influenced by management systems in crop and forage production. The ratio of potential soil respiration to total PLFA microbial biomass was also calculated as an index of microbial substrate utilization.

Statistical Analysis

Crop and forage production experiments were each analyzed as separate randomized complete block designs (RCBD) with three management-system treatments (conventional, organic, reduced-tillage) and four replicates. The analysis of soil properties that were measured at the beginning and end of the study, such as SOC, STN, pH and EC, were analyzed as split plot in time analysis of variance set in an RCBD for each system (p = 0.05). This analysis considered year as a repeated observation and replication as a random term in the model. Soil properties measured four times each year, such as soil microbial PLFA contents, DOC, C:N ratio of microbial substrate, potential soil respiration, water filled pore space, and soil bulk density, were analyzed as a split plot in time analysis of variance that considered season and year as repeated observation terms in the model. Statistical computations for both designs were facilitated by the mixed model (Proc Mixed) procedure of the Statistical Analysis System (SAS, ver. 9.3, SAS Institute, Cary, NC). Means were separated using the PDIFF test in the LSMEANS procedure (p = 0.05) unless otherwise stated. There were no significant season×management system interactions for either system in the three way split plot in time analysis of variance, therefore, results are reported as average of all four seasons within a year. In addition, PLFA data for individual microbial groups were normalized to the total microbial PLFAs and the data (mole percent of total PLFAs) were reanalyzed through a multivariate method (principal component analysis) to compare shifts in microbial community structure. Relationships between soil microbial substrate properties, microbial biomass and community structure, and substrate utilization were analyzed using Pearson correlation. Principle component and Pearson correlation analyses were performed using a Minitab V.16.0 (Minitab Inc., State College, PA, USA) and the first two principal components are graphed to summarize the results.

Results

Monthly average maximum and minimum temperatures during growing seasons (May to September) of 2009–2012 varied from year to year (Figure S1). The average minimum temperature was lowest in December 2009 and February 2011. Average precipitation was the lowest in 2012, followed by 2009 and 2010, compared to that in 2011. The amount of irrigation water depended on crop demand and the amount of precipitation received, and more water was applied to meet the crop water requirement in 2012 than in 2009–2011. All plots were irrigated to 60% of field capacity. Water filled porosity was consistent across management systems, seasons, and study years in both production systems (data not presented).

Soil pH was consistent across management systems and study years (range 7.3–7.8) under both crop and forage production, as was SOC concentration (Table 2). Soil organic C concentrations

were, however, significantly influenced by a management system× year interaction. Soils under reduced-tillage (p = 0.034) and organic (p = 0.004) crop production had significantly more SOC than soils under conventional crop production. In addition, soils under organic crop production in the fourth year had significantly more SOC than in the first year (p = 0.02). Soils under reduced-tillage forage production had significantly more SOC than those under conventional forage production (p<0.01). Soil total N concentrations were not significantly influenced by management systems, years, or management system×year interactions under either crop or forage production. Soil bulk density was not significantly influenced by management system, season, or year, but was significantly influenced by a management system×year interaction under crop production. Specifically, soil bulk density was significantly higher under reduced-tillage than organic crop production in the fourth year (p = 0.007).

Soil microbial biomass concentrations were significantly influenced by a management system×year interaction, but not by season. Specifically, in the fourth year under crop production there was significantly more soil microbial biomass under organic than conventional and (p<0.001) and reduced-tillage management (p = 0.01) (Fig. 1a). In the fourth year under forage production there was significantly more microbial biomass in soils under reduced-tillage than conventional (p = 0.002) and organic management (p = 0.047) (Fig. 1b). Under crop production, the greatest increase in total microbial biomass over the four-year period was observed in soils under organic management (353%) followed by conventional (262%) and reduced-tillage (202%) (based on year-one values). Under forage production, the increase in total microbial biomass concentrations were statistically similar at 396, 378 and 361% higher in the fourth year than in the first year in soils under conventional, organic, and reduced-tillage management systems, respectively.

We also observed increases in soil bacterial PLFAs, fungal biomarker PLFAs, DOC, and TDN across all treatments (only DOC data presented in Table 2), but the increases differed in magnitude. Dissolved organic C per unit SOC was 0.28–0.56% in the first year and 1.49–2.04% in the fourth year, with highest amount of DOC per unit SOC under conventional management. These changes corresponded with significantly higher fungal to bacterial ratios (F:B ratios) (Figure 2) and C:N ratios of microbial substrates (Figure 3) in the fourth year than in the first year. In addition, C:N ratios of microbial substrates were greater in soils under organic forage production than under conventional and reduced-tillage forage production. Similarly, microbial substrate utilization (potential soil respiration per unit PLFA) was consistent across management systems (Figure 4), but was 85–90% lower under crop production and 61–77% lower under forage production in the fourth year than in the first year.

Principal component analysis of microbial community structure revealed that the first two principle components explained 64.9% and 25.4% of the total sample variance under the crop production, and 74.3% and 21.0% of total sample variance under the forage production (Figure 5). The soil samples collected in the first year clustered on the left side of the figures 5.a1 and b1, and those collected in the fourth year clustered on the right side, corresponding to the increase in microbial substrate quality and decrease in potential soil respiration. There was greater variance in microbial community data collected in the first year than in the fourth year under both crop and forage production. In addition, loading scores for management systems separated more clearly along PC2 in the fourth year than in the first year. Among microbial groups, protozoa, other bacteria, and other fungi had positive loadings, while AMF and gram-negative bacteria had

Table 2. Soil properties as influenced by conventional (CV), organic (OR), and reduced-tillage (RT) management systems for crop and forage production one and four years after transition from continuous conventional corn.

System		SOC[†‡]			STN			DOC			Db		
		Y1	Y4	Δ%	Y1	Y4	Δ%	Y1	Y4	Δ%	Y1	Y4	Δ%
		g kg⁻¹ soil			g kg⁻¹ soil			mg kg⁻¹ soil			g cm⁻³		
Crop	CV	6.70aA(1.31)	5.83bA(0.98)	−13.0	0.65(0.08)	0.73(0.12)	+12.3	19.5aB(2.08)	119cA(4.46)	+510	1.46aA(0.02)	1.43abA(0.02)	−2.05
	OR	6.52aB(0.16)	8.95aA(0.37)	+37.3	0.79(0.05)	0.90(0.08)	+13.3	36.4aB(5.45)	154aA(4.43)	+323	1.48aA(0.02)	1.40bA(0.02)	−5.41
	RT	7.31aA(0.86)	8.00aA(0.52)	+9.44	0.72(0.04)	0.80(0.05)	+11.8	20.6aB(3.78)	137bA(8.49)	+565	1.42aA(0.02)	1.48aA(0.03)	+4.23
Forage	CV	8.96aA(1.36)	7.23bA(0.46)	−19.3	0.77(0.09)	0.81(0.08)	+5.19	38.6aB(3.85)	139aA(13.3)	+360	1.47aA(0.03)	1.44aA(0.02)	−2.04
	OR	9.26aA(0.76)	8.50abA(0.35)	−8.31	0.77(0.02)	0.89(0.07)	+15.5	26.5aB(5.73)	147aA(7.36)	+455	1.47aA(0.02)	1.42aA(0.03)	−3.40
	RT	8.13aA(0.70)	10.2aA(1.04)	+25.5	0.86(0.14)	1.04(0.09)	+20.9	27.7aB(3.31)	152aA(2.67)	+449	1.39bA(0.02)	1.42aA(0.02)	+2.16

[†]Number in parenthesis indicates standard error.
[‡]SOC = soil organic carbon, STN = soil total nitrogen, DOC = dissolved organic carbon (g kg⁻¹ soil) and Db = soil bulk density (g cm⁻³). Different lowercase letters within a column indicate significant difference among management systems within a year and different uppercase letters indicate significant difference among years within a management system in crop as well as forage production (p = 0.05). No letter within a column or row indicates no significant management system×year difference.

negative loadings along the PC1 axis (Figure 5.a2 and b2). Along the PC2 axis, gram-positive bacteria under crop production, and both gram-positive bacteria and AMF under forage production, had positive loadings. Across management systems and crop and forage production, gram-positive bacteria, gram-negative bacteria, and AMF together constituted of 93% of total soil microbial biomass in the first year and 76–78% in the fourth year (Table 3). After four years under alternative management systems, biomarker PLFAs for these three microbial groups had increased 2–4 fold, while biomarker PLFAs for other bacteria, fungi, and protozoa had increased up to 26, 9, and 36 fold, respectively, corresponding with significant shifts in microbial community structure. Changes in microbial community structure over the four-year study period are also indicated by Shannon's diversity index in Table 3, which was significantly higher in the fourth year than in the first year across both crop and forage production and all three management systems.

Increases in microbial biomass and F:B ratios, along with other changes in microbial community structure, along PC1 were strongly positively correlated with substrate availability (DOC concentrations) and quality (C:N ratio of microbial substrate) (Table 4) under both crop and forage production. Microbial substrate utilization decreased significantly with increasing substrate availability, increasing C:N ratios of microbial substrates, and increasing microbial biomass. Similarly, microbial community changes along PC2 were not related with substrate properties and substrate utilization.

Discussion

Our results support our hypotheses and indicate that conversion from continuous corn to crop rotations positively impacted soil microbiotic properties across all three management systems, with higher substrate availability, substrate C:N ratios, and soil microbial biomass contents, but lower substrate utilization in the fourth year than in the first year following transition (Tables 2 and 4; Figures 1a, 3a, and 4a). Both reduced-tillage and organic management systems added to these effects.

Under organic crop production, combined effects of perennial legumes in the rotations, which eliminated tillage for two of the four years, with additions of manure and compost, apparently offset losses of microbial substrates due to heavy tillage during the annual crop phases and supported the highest year-four microbial biomass concentrations (Table 2). Inclusion of legumes in rotations and organic amendments typically favor microbial growth, SOC and N accumulation, and diversification of microbial substrates [6,7,33]. Under reduced-tillage crop production, a more consistent soil environment apparently facilitated higher soil microbial biomass concentrations and diversity, as well as higher fungal productivity than under conventional management (Table 3; Figure 1a). Similar effects of reduced disturbance have been noted [2,9,34] in which fungal hyphae improve soil aggregation, which protects labile SOM components and regulates microbial substrate utilization [3,12,13,33].

Under forage production, reduced-tillage management had the highest year-four microbial biomass concentrations of the three systems, probably due to lack of plowing with conversion from perennial forage to corn. Organic management, with applications of composted manure, created significantly higher C:N ratios of microbial substrates in the fourth year than under conventional and reduced-tillage with chemical fertilizer application (Figure 3b). The fact that this difference in substrate quality did not occur under the crop production experiment (Figure 3a) suggests that it resulted from a combined effect of the alfalfa-grasses mixture and

a.

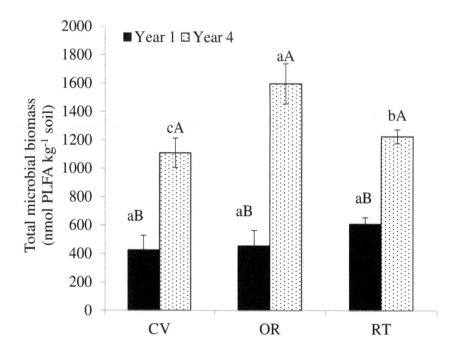

b.

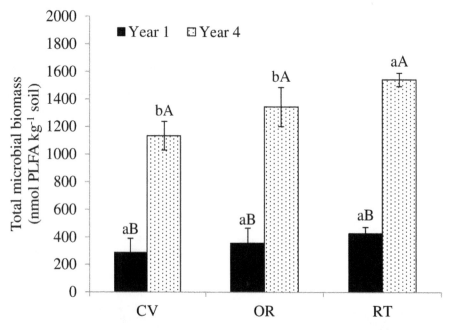

Figure 1. Total soil microbial biomass as influenced by conventional (CV), organic (OR), and reduced-tillage (RT) management systems for crop (a) and forage (b) production in the first and fourth years. Different lowercase letters indicate significant differences among management systems within a year and different uppercase letters indicate significant difference among years within a management system ($p = 0.05$).

compost applications. In the conventional and reduced-input forage systems, N from chemical fertilizer and alfalfa may have contributed to lower substrate C:N ratios than under the organic forage system.

Changes in soil microbiotic properties we observed in response to alternative management systems are consistent with results of previous studies, but greater in magnitude. Previous studies have reported two- to three-fold increases in microbial biomass with

a.

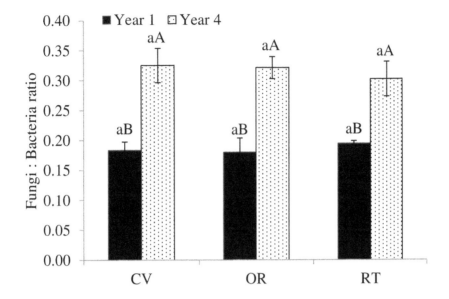

b.

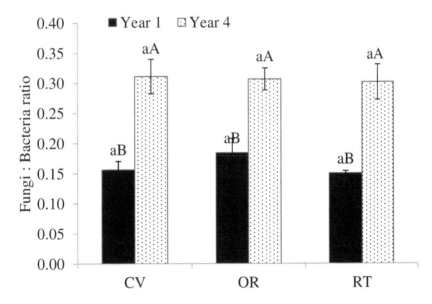

Figure 2. Fungal-to-Bacterial ratio as influenced by conventional (CV), organic (OR), and reduced-tillage (RT) management systems for crop (a) and forage (b) production in the first and fourth years. Different lowercase letters indicate significant differences among management systems within a year and different uppercase letters indicate significant difference among years within a management system (p = 0.05).

diversified rotations or reduced disturbance in place for several years [2,8,32,35]. The greater magnitudes we observed probably resulted from combined effects of transition from continuous corn to rotations with legumes, manure applications, and reduced-tillage on depleted, inherently-low-fertility soils. While variable precipitation during the seasons of the study may have influenced comparisons between years 1 and 4, air temperatures (Figure S1) and soil water filled pore space were consistent among management systems across years under both crop and forage production. Therefore, we believe that the changes in management, rather

than annual climatic variability, drove the observed changes in soil microbiotic properties.

The changes in quantity and quality of microbial substrates during the study period drove notable shifts in microbial community structure, including greater increases in fungal relative to bacterial PLFAs (Table 3; Figure 2), in gram-positive relative to gram-negative bacterial PLFAs, and in saprophytic fungi and protozoa relative to other groups. Higher DOC and C:N ratios of microbial substrates at the end of the study drove greater increases in saprophytic fungi, which rely on carbonaceous substrates, than AMF, which are often associated with more mineral-rich, low C:N

a.

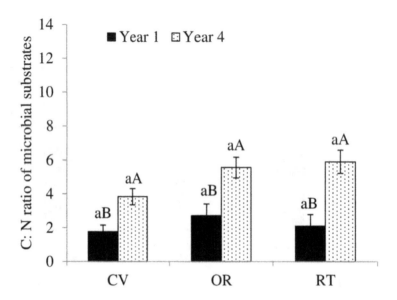

b.

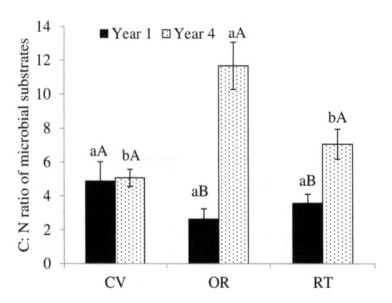

Figure 3. Carbon-to-nitrogen ratio of microbial substrate as influenced by conventional (CV), organic (OR), and reduced-tillage (RT) management systems for crop (a) and forage (b) production in the first and fourth years. Different lowercase letters indicate significant differences among management systems within a year and different uppercase letters indicate significant difference among years within a management system (p = 0.05).

substrates [36–38]. The observed increases in the non-mycorrhizal fungi strongly correlated with increased substrate availability, indicating changes in C:N ratios favored more fungal growth (saprotrophic fungi) than bacterial growth.

Increases in gram-positive relative to gram-negative bacteria are also often associated with increases in diversity of C sources in soils and decreases in mechanical soil disturbance, [36–39]. Gram-positive bacteria are associated with low substrate availability (high C:N) environments, while gram-negative bacteria dominate soils with more easily decomposable substrates [38]. Decreases in relative abundance of gram-negative bacteria may be beneficial

because many plant pathogenic microorganisms such as *Pseudomonas* and *Xanthomonas* species are gram negative [40,41]. Large increases in protozoa parallel the increases soil bacteria, which are their food source [42].

Greater amounts of higher C:N-ratio microbial substrates, more diverse communities of microorganisms with higher F:B ratios, and reduced potential soil respiration in the fourth year across all three management systems in general, and under reduced-tillage forage and organic crop systems in particular, suggests that minimum soil disturbance, application of organic amendments, and more legume crops in rotation increase soil microbial biomass,

a.

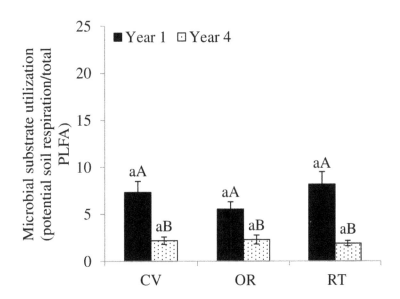

b.

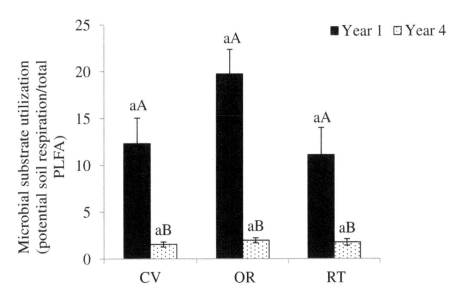

Figure 4. Microbial substrate utilization as influenced by conventional (CV), organic (OR), and reduced-tillage (RT) management systems for crop (a) and forage (b) production in the first and fourth years. Different lowercase letters indicate significant differences among management systems within a year and different uppercase letters indicate significant difference among years within a management system ($p = 0.05$).

alter community structure, and thereby influence SOC accrual (Figure 6). Although mechanisms of SOC regulation by specific groups of microorganisms are not well defined, given similar site characteristics, higher SOC sequestration potential is typically observed in soils with higher F:B ratios [43,44]. Higher SOC in fungal-dominated systems is mainly attributed to higher biomass C production per unit of C metabolized by fungus than by soil bacteria [44]. The changes in microbiotic properties we observed

indicate that substrate C was mainly transformed into microbial biomass or less labile SOM components, which may be reflected in year-four SOC contents.

Under crop production, both our organic system, with two years of alfalfa, and reduced-tillage system involved considerably less soil disturbance than our conventional system, and both had more year-4 SOC than the conventional system (Table 1, S1). Under forage production, with the same rotation across the three

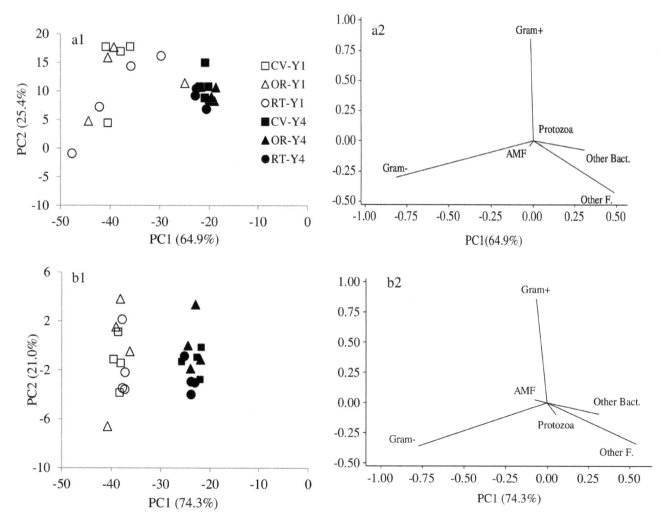

Figure 5. Score plots of the first two principle components (1) and loading of different microbial groups (2) as influenced by conventional (CV), organic (OR), and reduced-tillage (RT) management systems for crop (a) and forage (b) production. Gram+ = gram-positive bacteria, Gram− = gram-negative bacteria, AMF = arbuscular mycorrhizal Fungus, Other Bact. = other bacteria and Other F. = other fungus.

management systems, reduced tillage had significantly more year-4 SOC than the other two system, indicating that organic amendments combined with intensive tillage did not increase SOC. Taken together, these results suggest that reduced-tillage combined with legumes in rotation had the largest impacts on SOC accrual. Repeated tillage to plow down the grasses and alfalfa and establish corn in year 4 might have caused significant loss of SOC accrued during three years under forages in both organic and conventional systems.

Overall, increases in microbial substrate availability and microbial biomass over the 4-year study represent a small fraction of SOC reservoirs, even in this low-SOM environment. Therefore, longer-term evaluation of the effects of tillage, crop rotations, soil amendments, and legume integration in crop and forage production may further our understanding of the influence of microbiotic properties on SOC sequestration. Results of this cropping systems study bundle effects of reduced tillage, crop rotation, legumes in rotation, and soil fertility options into three management systems for crop and forage production. While overall effects are crucial to understanding how management alternatives affect system sustainability, evaluating individual

components will complement these results and contribute to design of best management practices for irrigated agriculture in cold, semiarid agroecosystems like the central High Plains.

Conclusions

In this study, management systems that included reduced tillage, perennial legumes, and organic amendments improved soil microbiotic properties that support SOC accrual. The greatest changes occurred with transition from continuous corn to crop rotation. The changes were enhanced under organic management in cash-crop production and reduced-tillage management under forage production. Under the different rotations of our crop production systems, more legume crops in rotation had greater influence on soil microbiotic properties than fewer legume crops. Under the same rotations of our forage production systems, reduced-tillage management had the greatest influences on soil microbiotic properties. These effects were driven by interactions between soil microbial community structure and microbial substrate quantity and quality that resulted in increases in fungal biomass that support SOC accrual. The results indicate that

Table 3. Soil microbial communities as influenced by conventional (CV), organic (OR), and reduced-tillage (RT) management systems for crop and forage production one and four years after transition from continuous conventional corn.

System		Gram positive bacteria[†] (nmol kg⁻¹ soil)			Gram negative bacteria (nmol kg⁻¹ soil)			Other Bacteria (nmol kg⁻¹ soil)			AMF (nmol kg⁻¹ soil)			Other Fungi (nmol kg⁻¹ soil)			Protozoa (nmol kg⁻¹ soil)			Shannon's diversity index		
		Y1	Y4	Δ%	Y1	Y4	Δ%	Y1	Y4	Δ%	Y1	Y4	Δ%	Y1	Y4	Δ%	Y1	Y4	Δ%	Y1	Y4	Δ%
Crop	CV‡	200aB (53.3)	414bA (43.7)	+107	160bB (33.8)	391cA (60.4)	+144	12.2aB (5.64)	60.3bA (11.7)	+394	28.7aB (6.21)	76.5bA (7.24)	+167	20.0bB (5.01)	152bA (15.8)	+665	3.30 (1.81)	14.4bA (2.70)	+336	3.50aB (0.25)	5.53aA (0.23)	+58.2
	OR	213aB (51.6)	555aA (48.8)	+161	166bB (34.4)	559aA (84.4)	+237	12.0aB (4.26)	105aA (18.9)	+775	28.8aB (7.00)	118aA (13.9)	+310	29.1aB (6.70)	230aA (16.1)	+201	3.02 (1.63)	28.7aA (2.17)	+850	3.55aB (0.29)	5.89aA (0.20)	+66.0
	RT	270aB (60.7)	467bA (54.1)	+72.9	236aB (45.9)	479bA (78.2)	+106	22.4aB (6.97)	70.9bA (13.5)	+217	44.3aB (8.88)	89.6bA (10.5)	+116	31.7aB (7.17)	173bA (23.0)	+147	4.38 (2.36)	21.4abA (3.28)	+754	3.80aB (0.27)	5.71aA (0.27)	+50.2
Forage	CV	124aB (21.5)	391bA (23.7)	+215	117aB (17.5)	408bA (55.1)	+249	2.30aB (1.61)	51.5aA (10.6)	+2139	24.9aB (3.50)	86.7bA (7.99)	+248	18.2aB (3.59)	176B (15.8)	+867	0.00	21.1 (2.31)	-	3.85aB (0.22)	5.69aA (0.20)	+47.9
	OR	157aB (22.3)	474abA (55.3)	+197	142aB (18.8)	490bA (93.0)	+245	3.69aB (2.26)	70.7aA (11.5)	+1575	29.5aB (3.64)	111aA (22.7)	+276	23.0aB (3.77)	177aB (16.1)	+670	0.00	22.8 (3.87)	-	4.00aB (0.16)	5.64aA (0.22)	+41.0
	RT	180aB (24.2)	526aA (41.6)	+192	177aB (19.5)	566aA (72.1)	+219	3.11aB (2.19)	80.4aA (19.4)	+2485	36.9aB (4.22)	116aA (12.3)	+214	29.1aB (3.30)	218aB (23.0)	+649	1.13 (1.13)	37.1 (2.94)	+3597	4.19aB (0.04)	5.71aA (0.22)	+36.1

[†]Number in parenthesis indicates standard error.
[‡]AMF = Arbuscular Mycorrhizal Fungi. Different lowercase letters within a column indicate significant difference among management systems within a year and different uppercase letters indicate significant difference among years within a management system in crop as well as forage production (p = 0.05). No letter within a column or row indicates no significant management system×year difference.

Table 4. Significant correlation coefficients (r) between soil microbial substrate properties, microbial community and substrate utilization in crop and forage production systems.

	DOC[†‡]	C:N MAS	Microbial biomass	F:B ratio	PC1[§]
Crop system					
Substrate C:N	0.69(<0.001)	-			
Microbial biomass	0.83(<0.001)	0.86(<0.001)	-		
F:B ratio	0.83(<0.001)	0.81(<0.001)	0.91(<0.001)	-	
PC1	0.78(<0.001)	0.63(0.001)	0.83(<0.001)	0.91(<0.001)	-
Soil Resp.	−0.64(0.001)	−0.47(0.02)	−0.70(<0.001)	−0.74(<0.001)	−0.84(<0.001)
Forage system					
Substrate C:N	0.65(0.001)	-			
Microbial biomass	0.85(<0.001)	0.56(0.006)	-		
F:B ratio	0.91(<0.001)	0.64(0.001)	0.83(<0.001)	-	
PC1	0.93(<0.001)	0.67(0.001)	0.91(<0.001)	0.97(<0.001)	-
Soil Resp.	−0.79(<0.001)	−0.40(<0.001)	−0.87(<0.001)	−0.73(<0.001)	−0.84(<0.001)

[†]Number in parenthesis indicates Pearson correlation p values.
[‡]DOC = dissolved organic carbon, F:B ratio = fungi to bacteria ratio, PC1 = First principal component and Soil Resp. = potential soil respiration.
[§]PC1 explains shift in soil microbial community structure from the first to the fourth year.

reducing disturbance, including legumes, and applying organic amendments positively impact soil processes in ways that enhance sustainable productivity of inherently low-fertility soils in a cold, semiarid environment, even over a relatively short time period. Further research may confirm the combined effects of crop rotations and alternative management systems on SOC accrual

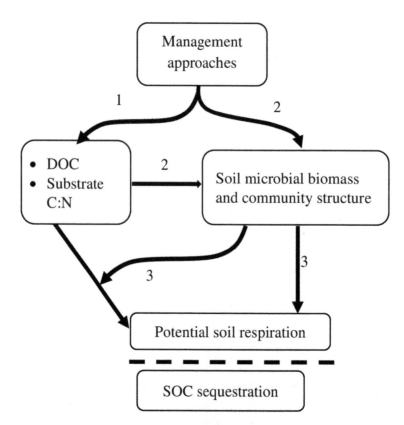

Figure 6. Conceptual framework illustrating influence of management system on microbial substrate properties, microbial communities, and SOC sequestration. Alternative management systems influence (1) microbial substrate availability and quality, (2) soil microbial biomass and community structure, and (3) soil respiration and SOC sequestration. DOC = dissolved organic carbon (microbial substrate).

and ecosystem services as influenced by soil microbial biomass and fungal productivity.

Supporting Information

Figure S1 Monthly average maximum and minimum temperature, and total precipitation at the SAREC weather station, Lingle, Wyoming (Apr. 2009–Sept. 2012).

Figure S2 Experimental design of crop and forage production plots under alternative management systems (C, conventional; O, organic; R, reduced tillage). Tiers A, B, and C are 1-acre cash-crop plots; tiers D, E, and F are 2-acre forage plots.

Table S1 Crop rotations under conventional (CV), organic (OR) and reduced-tillage (RT) management systems in crop and forage production.

Table S2 Timing of management practices by under conventional (CV), organic (OR), and reduce-tillage (RT) management systems on crop and forage production (2009–2012).

Table S3 Biomarker phospholipid fatty acids used for identifying taxonomic microbial groups.

Acknowledgments

We thank Jenna Meeks and Caley Kristian Gasch for field and laboratory assistance, Professor David Legg for assistance with statistical analyses, and Prakriti Bista and Professor Elise Pendall for their helpful comments to improve earlier drafts of this manuscript.

Author Contributions

Conceived and designed the experiments: JBN RG UN PDS. Performed the experiments: RG JBN. Analyzed the data: RG JBN. Contributed reagents/materials/analysis tools: JBN PDS UN. Wrote the paper: RG JBN. Reviewed the manuscript: UN PDS.

References

1. Berthrong ST, Buckley DH, Drinkwater LE (2013) Agricultural management and labile carbon additions affect soil microbial community structure and interact with carbon and nitrogen cycling. Microbial Ecol 66: 158–170.
2. Acosta-Martínez V, Mikha MM, Vigil MF (2007) Microbial communities and enzyme activities in soils under alternative crop rotations compared to wheat-fallow for the Central Great Plains. Appl Soil Ecol 37: 41–52.
3. Stahl PD, Parkin TB, Christensen M (1999) Fungal presence in paired cultivated and undisturbed soils in central Iowa. Biol Fertil Soils 29: 92–97.
4. Drinkwater LE, Letourneau DK, Workneh F, Vanbruggen AHC, Shennan C (1995) Fundamental differences between conventional and organic tomato agroecosystems in California. Ecol Appl 5: 1098–1112.
5. Oehl F, Sieverding E, Ineichen K, Mader P, Boller T, et al. (2003) Impact of land use intensity on the species diversity of arbuscular mycorrhizal fungi in agroecosystems of Central Europe. Appl Environ Microbiol 69: 2816–2824.
6. Kandeler E, Tscherko D, Spiegel H (1999) Long-term monitoring of microbial biomass, N mineralisation and enzyme activities of a Chernozem under different tillage management. Biol Fertil Soils 28: 343–351.
7. Shi Y, Lalande R, Ziadi N, Sheng M, Hu Z (2012) An assessment of the soil microbial status after 17 years of tillage and mineral P fertilization management. Appl Soil Ecol 62: 14–23.
8. Acosta-Martínez V, Dowd SE, Bell CW, Lascano R, Booker JD, et al. (2010) Microbial community structure as affected by dryland cropping systems and tillage in a semiarid sandy soil. Diversity 2: 910–931.
9. Halvorson AD, Wienhold BJ, Black AL (2002) Tillage, nitrogen, and cropping system effects on soil carbon sequestration. Soil Sci Soc Am J 66: 906–912.
10. Blanco-Canqui H, Lal R (2007) Regional assessment of soil compaction and structural properties under no-tillage farming. Soil Sci Soc Am J 71: 1770–1778.
11. Ghimire R, Norton J, Pendall E (2014) Alfalfa-grass biomass, soil organic carbon, and total nitrogen under different management systems in an irrigated agroecosystem. Plant Soil: 374: 173–184.
12. Six J, Feller C, Denef K, Ogle SM, Sa MJC, et al. (2002) Soil organic matter, biota and aggregation in temperate and tropical soils: Effects of no-tillage. Agronomie 22: 755–775.
13. Liebig M, Carpenter-Boggs L, Johnson JMF, Wright S, Barbour N (2006) Cropping system effects on soil biological characteristics in the Great Plains. Renew Agric Food Syst 21: 36–48.
14. Jenkinson DS, Parry LC (1989) The nitrogen-cycle in the broadbalk wheat experiment - a model for the turnover of nitrogen through the soil microbial biomass. Soil Biol Biochem 21: 535–541.
15. Ghimire R, Norton JB, Norton U, Ritten JP, Stahl PD, et al. (2013) Long-term farming systems research in the central High Plains. Renew Agric Food Syst 28: 183–193.
16. Krall JM, Delaney RH, Taylor DT (1991) Survey of nonirrigated crop production practices and attitudes of Wyoming producers. J Agron Educ 20: 120–122.
17. Norton JB, Mukhwana EJ, Norton U (2012) Loss and recovery of soil organic carbon and nitrogen in a semiarid agroecosystem. Soil Sci Soc Am J 76: 505–514.
18. Frey SD, Elliott ET, Paustian K (1999) Bacterial and fungal abundance and biomass in conventional and no-tillage agroecosystems along two climatic gradients. Soil Biol Biochem 31: 573–585.
19. Western Regional Climate Center (2013) Historical climate information. Reno, NV: Desert Research Institute. http://www.wrcc.dri.edu/. Accessed: 06-15, 2013.
20. Soil Survey Staff (2013) Web Soil Survey. Natural Resources Conservation Service, United States Department of Agriculture. http://websoilsurvey.sc.egov.usda.gov/ Accessed: 06-15, 2013.
21. Sherrod LA, Dunn G, Peterson GA, Kolberg RL (2002) Inorganic carbon analysis by modified pressure-calcimeter method. Soil Sci Soc Am J 66: 299–305.
22. Gardner WH (1986) Water content. In: Klute A, editor. Methods of Soil Analysis Part 1: Physical and Mineralogical Methods. 2nd ed. Madison, WI: Agronomy Monograph 9. ASA, SSSA pp. 493–541.
23. Thomas GW (1996) Soil pH and soil acidity. In: Sparks DL, editor. Methods of Soil Analysis, part 3: Chemical Methods. Madison, WI: Agronomy Monograph 9. ASA, SSSA pp. 475–490.
24. Gee GW, Bauder JW (1986) Particle-size analysis. In: Klute A, editor. Methods of Soil Analysis Part 1: Physical and Mineralogical Methods. 2nd ed. Madison, WI: Agronomy Monograph 9. ASA, SSSA pp. 383–411.
25. Nie M, Pendall E, Bell C, Gasch CK, Raut S, et al. (2012) Positive climate feedbacks of soil microbial communities in a semi-arid grassland. Ecol Letters 16: 234–241.
26. Blake GR, Hartge KH (1986) Bulk density. In: Klute A, editor. Methods of Soil Analysis, part 1: Physical and Mineralogical Methods. 2nd ed Madison, WI. ASA, SSSA 363–375.
27. Linn DM, Doran JW (1984) Effect of water-filled pore-space on carbon-dioxide and nitrous-oxide production in tilled and nontilled soils. Soil Sci Soc Am J 48: 1267–1272.
28. Blight EG, Dyre WJ (1959) A rapid method of total lipid extraction and purification. Can J Biochem Physiol 37: 911–917.
29. Frostergard A, Tunlid A, Baath E (1991) Microbial biomass measured as total lipid phosphate in soils of different organic content. J Microbiol Methods 14: 151–163.
30. Buyer JS, Roberts DP, Russek-Cohen E (2002) Soil and plant effects on microbial community structure. Can J Microbiol 48: 955–964.
31. Shannon CE, Weaver W (1949) The Mathematical Theory of Communication. Urbana, IL: Univ. of Illinois Press. 132 p.
32. Minoshima H, Jackson LE, Cavagnaro TR, Sanchez Moreno S, Ferris H, et al. (2007) Soil food webs and carbon dynamics in response to conservation tillage in California. Soil Sci Soc Am J 71: 952–963.
33. Delate K, Cambardella CA (2004) Agroecosystem performance during transition to certified organic grain production. Agron J 96: 1288–1298.
34. Ngosong C, Jarosch M, Raupp J, Neumann E, Ruess L (2010) The impact of farming practice on soil microorganisms and arbuscular mycorrhizal fungi: Crop type versus long-term mineral and organic fertilization. Appl Soil Ecol 46: 134–142.
35. Reganold JP, Andrews PK, Reeve JR, Carpenter-Boggs L, Schadt CW, et al. (2010) Fruit and soil quality of organic and conventional strawberry agroecosystems. Plos One 5: e12346.
36. Fliessbach A, Oberholzer HR, Gunst L, Maeder P (2007) Soil organic matter and biological soil quality indicators after 21 years of organic and conventional farming. Agric Ecosyst Environ 118: 273–284.
37. Carpenter-Boggs L, Stahl PD, Lindstrom MJ, Schumacher TE, Barbour NW (2003) Soil microbial properties under permanent grass, conventional tillage, and no-till management in South Dakota. Soil Till Res 71: 15–23.
38. Fierer N, Schimel JP, Holden PA (2003) Variations in microbial community composition through two soil depth profiles. Soil Biol Biochem 167–176.
39. Schaad NW, Jones JB, Chun W, editors (2001) Laboratory guide for identification of plant pathogenic bacteria: APS press, Minnesota. 398 p.

40. De Vos P, Goor M, Gills M, De Ley J (1985) Ribosomal ribonucleic acid Cistron similarities of phytopathogenic Pseudomonas species. Int J Syst Evol Microbiol 35: 169–184.

41. Stout JD (1980) The role of Protozoa in nutrient cycling and energy flow. Adv Microbial Ecol 4: 1–50.

42. Jastrow JD, Amonette JE, Bailey VL (2007) Mechanisms controlling soil carbon turnover and their potential application for enhancing carbon sequestration. Clim Change 80: 5–23.

43. Bailey VL, Smith JL, Bolton Jr H (2002) Fungal:bacterial ratios in soils investigated for enhanced C sequestration. Soil Biol Biochem 34: 997–1007.

44. Strickland MS, Rousk J (2010) Considering fungal: bacterial dominance in soils - Methods, controls, and ecosystem implications. Soil Biol and Biochem 42: 1385–1395.

6

Estimating Seasonal Nitrogen Removal and Biomass Yield by Annuals with the Extended Logistic Model

Richard V. Scholtz III*, Allen R. Overman

Agricultural & Biological Engineering Department, University of Florida, Gainesville, Florida, United States of America

Abstract

The Extended Logistic Model (ELM) has been previously shown to adequately describe seasonal biomass production and N removal with respect to applied N for several types of annuals and perennials. In this analysis, data from a corn (*Zea mays* L.) study with variable applied N were analyzed to test hypotheses that certain parameters in the ELM are invariant with respect to site specific attributes, like environmental conditions and soil type. Invariance to environmental conditions suggests such parameters may be functions of the crop characteristics and certain other management practices alone (like plant population, planting date, harvest date). The first parameter analyzed was Δb, the difference between the N uptake shifting parameter and the biomass shifting parameter. The second parameter tested was N_{cm}, the maximum N concentration. Both parameters were shown to be statistically invariant, despite soil and site differences. This was determined using analysis of variance with normalized nonlinear regression of the ELM on the data from the study. This analysis lends further evidence that there are common parameters involved in the ELM that do not rely on site-specific or situation-specific factors. More insight into the derivation of, definition of, and logic behind the various parameters involved in the model are also given in this paper.

Editor: Manuel Reigosa, University of Vigo, Spain

Funding: The funding of this analysis was provided by the Florida Experiment Station. The funders had no role in study design, data collection and analysis, decision to publish, or preparation of the manuscript.

Competing Interests: The authors have declared that no competing interests exist.

* E-mail: rscholtz@ufl.edu

Introduction

Effective water and nutrient management plays an essential role in future attempts at sustainable agricultural production. As the world's population continues to grow, the potable water supply is limited and must be guarded from unnecessary withdrawals and contamination from excessive nutrient loads. Strict monitoring and exercises in groundwater modeling of all agricultural operations is cost prohibitive. It therefore becomes necessary to investigate crop nutrient removal from their environment, and to adopt management procedures and rules that are based on a sound scientific foundation.

Overman et al. first proposed the logistic model as a nutrient management tool to describe seasonal biomass yield dependence of forage grasses on applied N [1]. The original application of the logistic model to plant biomass production was based on inductive reasoning, a process where inferences are made from "real world" observations [2]. While inductive reasoning is innately a more empirical method of model development, all models, no matter the complexity, have some element of empiricism [3]. It is because of the application to the "real world," that engineering and the applied sciences are, for practical reasons, inherently more empirical. The logistic model was extended to include seasonal plant N uptake (removal from the environment) dependence on applied N by forage grasses [4] and then for annuals, like corn [5]. The ELM is a five parameter, non-linear, parametric model that is capable of describing the seasonal biomass yields, N uptake, and N concentration with respect to applied N. Work conducted over the years has indicated that the ELM can effectively describe both

annual [6–11] and perennial [8,12–20] crops, for a wide range of nutrient inputs.

The ELM, begins with the simple logistic expression that relates N uptake, N_u, to the N applied, N, which is given by the following relationship:

$$N_u = \frac{A_N}{1 + \exp(b_N - cN)} \tag{1}$$

where A_N is the relative maximum N uptake in kg ha^{-1}, b_N is the dimensionless N uptake shifting parameter, and c is the applied N response parameter given in ha kg^{-1}. The phase relationship between biomass production, Y, and N uptake, N_u, is given by

$$Y = \frac{Y_m N_u}{k_N + N_u} \tag{2}$$

where Ym is the maximum potential biomass production in Mg ha^{-1}, and k_N is the N uptake response parameter in kg ha^{-1}. The transformation of Eq. (1) by using Eq. (2) yields the following logistic expression:

$$Y = \frac{A}{1 + \exp(b - cN)} \tag{3}$$

that relates biomass production to applied N, where A is the relative maximum biomass production in Mg ha^{-1}, b is the biomass yield shifting parameter, and c is the same applied N response parameter that applies to the N uptake logistic equation. The parameters A_N, A, b_N, b, and c are currently key parameters used with the ELM, and are the easiest to determine from regression analysis.

From the transformation of Eq. (1) into Eq. (3), the relative maximum biomass yield parameter can be written in terms of the maximum potential biomass yield parameter, the relative maximum N uptake parameter, and the N uptake response parameter.

$$A = \frac{Y_m A_N}{k_N + A_N} \quad (4)$$

The biomass shifting parameter can be written in terms of the N uptake shifting parameter, the relative maximum N uptake parameter, and the N uptake response coefficient.

$$b = b_N + \ln\left(\frac{k_N}{k_N + A_N}\right) \quad (5)$$

The difference between the shifting parameter for N uptake and the biomass yield can be written as the following:

$$\Delta b = b_N - b = -\ln\left(\frac{k_N}{k_N + A_N}\right) \quad (6)$$

N concentration is simply defined as the ratio of N uptake to biomass production. This leads to the following relationship:

$$N_c = \frac{N_u}{Y} = N_{cm} \frac{1 + \exp(b - cN)}{1 + \exp(b_N - cN)} \quad (7)$$

The N concentration model suggests that as N applied is increased to exorbitant levels, there is a maximum limit to the N concentration, N_{cm}. This maximum limit is simply the ratio of the relative maximum N uptake with respect to applied N, A_N, to the relative maximum biomass production with respect to applied N, A.

$$\lim_{N \to +\infty} N_c = \frac{A_N}{A} = N_{cm} \quad (8)$$

It seems logical to suggest that as the background amount of N present in the soil is decreased, the N concentration would be reduced to some prescribed lower limit, as there would be lower limit on the percent of proteins present in a given crop to sustain any growth. Mathematically, the model suggests that this lower limit of N concentration, N_{cl}, is a function of the maximum concentration and the difference between the N uptake shifting parameter and the biomass shifting parameter.

$$\lim_{N \to -\infty} N_c = N_{cm} \exp(-\Delta b) = N_{cl} \quad (9)$$

Also, this lower limit of N concentration, N_{cl}, can be found by taking the ratio of the N uptake response coefficient to the maximum potential biomass production parameter.

$$N_{cl} = \frac{k_N}{Y_m} \quad (10)$$

From the phase relationship between biomass production and N uptake, Eq. (2), N concentration can be found from the following equation with respect to N uptake:

$$N_c = \frac{k_N}{Y_m} + \frac{1}{Y_m} N_u \quad (11)$$

This predicts a linear relationship between N concentration, N_c, and N uptake, N_u. The line should have a slope equal to the inverse of the maximum potential biomass production and an intercept that equals the ratio of the N uptake response parameter to maximum potential biomass production. As this is a phase relationship, this is a functional segment that is bounded between N uptake values from 0 to the peak of A_N; and between N concentration values between N_{cl} and N_{cm}.

From the earlier work of Overman et al. [5], it has been shown that for a given site the applied N response parameter, c, and the N uptake intercept parameter, b_N, and the biomass intercept parameter, b, are not unique to the ELM when applied to grain or the whole plant. Meaning that the harvest index is constant for a given site. Their analysis showed that the only differences between the grain and the whole plant appear in the relative maximum N uptake parameter, A_N, and the relative maximum biomass production parameter, A [5]. Because the b_N, b, and c parameters were shown to be constant for both grain fraction and total biomass production, all the differences in both grain N uptake and biomass production, and the differences in the total plant N uptake and biomass production can be estimated with seven model parameters and a value for the seasonal amount of N applied. This is a comparative reduction of three parameters when b_N, b, and c are not held constant between grain and total plant biomass production.

The goal of this work is to continue to elevate the Extended Logistic Model (ELM) beyond the empiricism of it nascent beginnings and achieve a balance between what can be measured and what should be modeled, as called for by Montieth [21]. The intent is to shed new light on the significance of parameters used in the ELM and to contribute to the search for commonality among parameters. Normalized non-linear regression and analysis of variance (ANOVA) were used to show the invariance of two model parameters with respect to environmental differences, namely soil type and water availability.

Methods

Data Set

This analysis uses data collected by Eugene Kamprath from a corn (Pioneer 3320) N-rate field study that was conducted at three regional research stations in North Carolina from 1981 to 1984. A detailed explanation of the field experiment has been previously reported [22]. Supplemental irrigation was provided at the Clayton experiment station for the well-drained Dothan loamy fine sand (fine-loamy, siliceous, thermic Plinthic Kandiudults), at a rate of 10 to 12 cm a season, except for 1982 when no additional water was supplied. No irrigation was provided at the Kinston station for the well-drained Goldsboro sandy loam (fine-loamy, siliceous, thermic Aquic Paleudults). At the Plymouth experiment station, no irrigation was provided for the poorly-drained Portsmouth very fine sandy

Table 1. Analysis of variance for model parameters for corn grain and total plant biomass production and for corn grain and total plant N uptake, grown on three different soils.

Mode	Parameters Estimated	Degrees of Freedom	Normalized Residual Sum of Squares	Normalized Mean Sum of Squares	F Value
I	21	39	0.0247	0.000633	----
II, Common Δb.	19	41	0.0259	0.000631	----
II–I	--	2	0.00119	0.000595	0.940
III, Common Δb, & N_{cm}.	15	45	0.0267	0.000594	----
III–I	--	6	0.00202	0.000337	0.533

loam *(fine-loamy over sandy or sandy-skeletal mixed, thermic Typic Umbraquults)*. The experiments at each station were set up as a RCB design, with four replications. Both total plant and grain fraction biomass were sampled, and every year the experiment was conducted at a new location within the same soil type at each station. This was to limit the impact on the experiment of any residual N in the soil from the previous year. The fertilizer treatments were applied in the form of NH_4NO_3 at rates of 0, 56, 112, 168, and 224 kg ha^{-1} of N. Average values over the four year period were combined for each of the different treatments and the model parameters were evaluated based on those combined averages.

Normalization

Parameters A_N and A for grain, A_N and A for total plant, and b_N, b and c for each site are determined simultaneously, using Newton-Raphson non-linear regression. A detailed description of Newton-Raphson non-linear regression of logistic equation can be found in Overman and Scholtz [8]. The attempt of this methodology is to consistently distribute the standard error amongst all those parameters for further analysis. Because of the unit and an order of magnitude difference between biomass and N uptake parameter values, as well as a subsequent order of magnitude difference between grain and total plant parameter values, a normalization routine is also employed. The error sum of squares for each individual site is initially written as

$$ESS_{NORM} =$$

$$\underbrace{\sum\left(\frac{Y_g}{A_g} - \frac{1}{1+\exp(b-cN)}\right)^2 + \sum\left(\frac{N_{ug}}{A_{Ng}} - \frac{1}{1+\exp(b_N-cN)}\right)^2}_{grain\ fraction}$$

$$+ \underbrace{\sum\left(\frac{Y_t}{A_t} - \frac{1}{1+\exp(b-cN)}\right)^2 + \sum\left(\frac{N_{ut}}{A_{Nt}} - \frac{1}{1+\exp(b_N-cN)}\right)^2}_{total\ plant}$$

$$(12)$$

where the total normalized error is resultant from the sum of the normalized error from the three sites. For this study the initial Hessian matrix is 21 by 21 elements and paired with a 21 element Jacobian vector. As a result of the normalization procedure, performing the Newton-Raphson procedure can diverge more readily than a non-normalized procedure. Because of this, it is important to establish a reasonable initial guess for each parameter. First, individual logistic response are evaluated for each data set from all sites, for grain biomass, grain N uptake, for total plant biomass, and total plant N uptake. This results in six A values, six A_N values, six b values, six b_N values, and 12 values for the c parameter. Using the fact that it can be shown for two straight lines

$$y_1 = m_1 x + b_1 \qquad (13)$$

and

$$y_2 = m_2 x + b_2, \qquad (14)$$

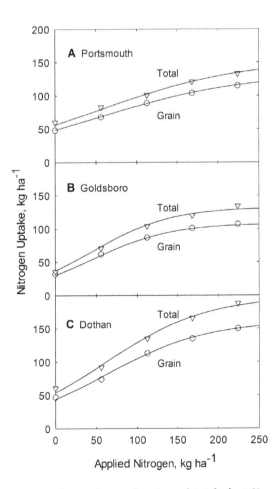

Figure 1. Dependence of grain and total plant N uptake on applied N for corn grown at the Plymouth (A), Kinston (B), and Clayton (C) experiment stations in North Carolina. Curves are constructed from Eq. 1 and from parameters listed in Table 2.

Table 2. Standard logistic model parameters invariant to corn grain and total plant biomass production and for corn grain and total plant N uptake, grown on three different soils.

Soil	Component	Parameter	Estimate
Dothan	Biomass	b	0.319
	N Uptake	b_N	0.978
	Both	c, ha kg^{-1}	0.0161
Goldsboro	Biomass	b	0.282
	N Uptake	b_N	0.941
	Both	c, ha kg^{-1}	0.0209
Portsmouth	Biomass	b	−0.105
	N Uptake	b_N	0.555
	Both	c, ha kg^{-1}	0.0111

who share the same sampling of the independent variable, the best fit single slope shared between them is given by

$$\hat{m} = \frac{m_1 + m_2}{2} \tag{15}$$

and the corresponding intercepts become

$$\hat{b}_1 = b_1 + \frac{1}{2}(m_1 - m_2)\bar{x} \tag{16}$$

and

$$\hat{b}_2 = b_2 + \frac{1}{2}(m_2 - m_1)\bar{x}, \tag{17}$$

the initial guess for the c parameter is the average of all 12 values, and the initial guess for each value of bN can be found from

$$\hat{b}_N = b_N + (c - \bar{c})\bar{N} \tag{18}$$

and for each value of b can be found from

$$\hat{b} = b + (c - \bar{c})\bar{N}. \tag{19}$$

The problem is bounded between the maximum and minimum values of the c parameter and each value of b_N is bounded between

$$\hat{b}_{Nmin} = b_N + (c - c_{max})\bar{N} \tag{20}$$

and

$$\hat{b}_{Nmax} = b_N + (c - c_{min})\bar{N} \tag{21}$$

and each value of b is bounded between

$$\hat{b}_{min} = b + (c - c_{max})\bar{N} \tag{22}$$

and

$$\hat{b}_{max} = b + (c - c_{min})\bar{N}. \tag{23}$$

Analysis

The first hypothesis of this analysis is that the difference between N uptake intercept parameter and the biomass intercept parameter, Δb, is invariant with respect to the differences in soil type and water availability for a given variety of an annual crop. Note that there is no attempt in this work to identify the effects of water availability or site characteristics on the ELM parameters, but to determine which are invariant to those characteristics. For this analysis, the same genetic line of corn is propagated by seeding and harvested at the same relative age. The second hypothesis is that maximum N concentration, N_{cm}, is also invariant with respect to the differences in soil type and water availability. A consequence of both hypotheses being affirmed is that the lower limit to the N concentration, N_{cl}, in the same annual crop is also invariant with respect to soil type and water availability. Parameters were estimated by minimization of the normalized error sum of squares, and analysis of variance (ANOVA) was used to determine the validity of both hypotheses.

For the analysis of variance, three scenarios or modes were used, each with a targeted reduction in the number of parameters used in the ELM to describe the corn data in the Kamprath study. Mode I had 21 separate parameters that were estimated by minimization of the normalized error sum of squares. In Mode I, there are individual values for A, and A_N, for both grain and for total plant, and corresponding values for b, b_N, and c at each of the three sites. For Mode II, the number of parameters estimated dropped to 19, because the Δb parameter was held constant across the three sites. For Mode III, the Δb and the N_{cm} parameters were both held constant across the three sites, reducing the number of estimated parameters to 15.

Nonlinear Coefficients of Determination [23] (Nash-Sutcliffe Model Efficiency Coefficient [24]) will be provided for grain and total plant N uptake and for grain and total plant biomass production just as a relative comparison of fit.

Results

Table 1 contains the summary of the analysis of variance test. The comparison between Modes I and II leads to an increase the *degrees of freedom* to 41 and results in a variance ratio of 0.940. Because the critical F(2,39,95%) value is 3.24, it is concluded that there is no significant difference between the two modes. Thus in

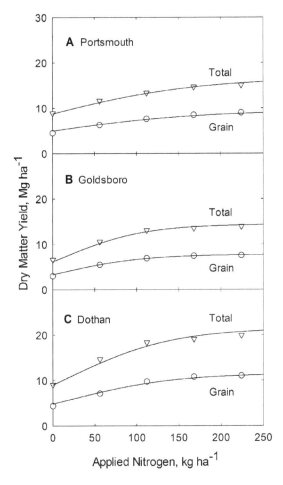

Figure 2. Dependence of grain and total plant biomass production on applied N for corn grown at the Plymouth (A), Kinston (B), and Clayton (C) experiment stations in North Carolina. Curves are constructed from Eq. 3 and from parameters listed in Table 2.

this study the Δb parameter is invariant to all soil and site differences.

Also from Table 1, the comparison between Modes I and III results in an increase the *degrees of freedom* to 45, and in a variance ratio of 0.533. With a critical F(6,39,95%) value of 2.34, not only is the Δb parameter is invariant, but so are the total plant and grain N_{cm} parameters. Thus, the soil, the field conditions, the environmental constraints, and even water availability play no role in either is the Δb or the two N_{cm} parameters. This leads to an invariance in the total plant and grain N_{cl} parameters, by virtue of Eq. (9).

The dependence of grain and whole plant N uptake on applied N at harvest is represented by Figure 1 for the three soil types. In general there is good agreement between the model line and the data. The resulting N uptake model lines (depicted in Figure 1), are generated from Eq. (1), using parameter values for b_N and c from Table 2 and values for A_{Ng} and A_{Nt} found in Table 3. Equation specific Non-linear Coefficient of Determination values and Error Sum of Squares are provided in Table 4.

Grain and whole plant biomass production versus applied N is shown in Figure 2 for all three soil types. In general there is good agreement between the model line and the data. The resulting biomass model lines (depicted in Figure 2), are generated from Eq. (3), using parameter values for b and c from Table 2 and values for A_g and A_t found in both Table 3. Equation specific Non-linear Coefficient of Determination values and Error Sum of Squares are provided in Table 4.

N concentration dependence on applied N is shown in Figure 3 for all three soil types. The resulting N concentration model lines (depicted in Figure 3), are generated from Eq. (7), using parameter values for b, b_N and c from Table 2 and values for $N_{cm\,g}$ and $N_{cm\,t}$ from Table 6.

The phase relationship between biomass production and N uptake for the corn grain and the whole plant is represented by Figure 4 for each of the three soils. The resulting biomass – N uptake phase model lines (depicted in Figure 4), are generated from Eq. (2), using parameter values for k_{Ng}, k_{Nt}, Y_{mg} and Y_{mt} found in Table 5.

The phase relationship between N Concentration and N uptake for the grain and the whole plant is represented by Figure 4 for each of the three soils. The resulting between N Concentration –

Table 3. Standard logistic model parameters specific to corn grain and total plant biomass production and for corn grain and total plant N uptake, grown on three different soils.

Soil	Component	Parameter	Plant Fraction	Estimate
Dothan	Biomass	A_g, Mg ha^{-1}	Grain	11.5
		A_t, Mg ha^{-1}	Total Plant	21.5
	N Uptake	A_{Ng}, kg ha^{-1}	Grain	161
		A_{Nt}, kg ha^{-1}	Total Plant	198
Goldsboro	Biomass	A_g, Mg ha^{-1}	Grain	7.76
		A_t, Mg ha^{-1}	Total Plant	14.4
	N Uptake	A_{Ng}, kg ha^{-1}	Grain	108
		A_{Nt}, kg ha^{-1}	Total Plant	132
Portsmouth	Biomass	A_g, Mg ha^{-1}	Grain	9.50
		A_t, Mg ha^{-1}	Total Plant	16.7
	N Uptake	A_{Ng}, kg ha^{-1}	Grain	133
		A_{Nt}, kg ha^{-1}	Total Plant	154

Table 4. Standard statistical measures of fit, based on specific component and plant fraction data.

Soil	Component	Plant Fraction	Coefficient of Determination	Specific Error Sum of Squares
Dothan	Biomass	Grain	0.984	0.523
		Total Plant	0.969	2.361
	N Uptake	Grain	0.997	24.5
		Total Plant	0.993	83.9
Goldsboro	Biomass	Grain	0.992	0.145
		Total Plant	0.983	0.468
	N Uptake	Grain	0.998	7.81
		Total Plant	0.993	35.2
Portsmouth	Biomass	Grain	0.977	0.169
		Total Plant	0.985	0.592
	N Uptake	Grain	0.999	5.01
		Total Plant	0.991	19.4

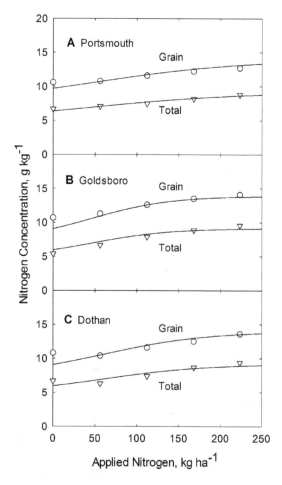

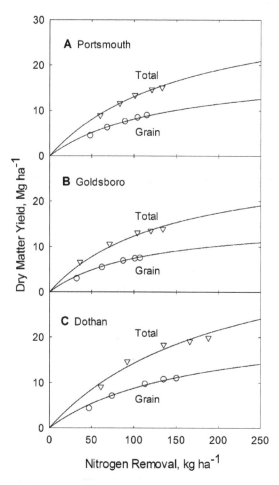

Figure 3. Dependence of grain and total plant N concentration on applied N for corn grown at the Plymouth (A), Kinston (B), and Clayton (C) experiment stations in North Carolina. Curves are constructed from Eq. 7 and from parameters listed in Table 2.

Figure 4. Dependence of grain and total plant biomass production on N uptake for corn grown at the Plymouth (A), Kinston (B), and Clayton (C) experiment stations in North Carolina. Curves are constructed from Eq. 2 and from parameters listed in Table 3.

Table 5. Standard phase model parameters for corn grain and total plant biomass production and for corn grain and total plant N uptake.

Soil	Parameter	Plant Fraction	Estimate
Dothan	k_{Ng}, kg ha^{-1}	Grain	157
	k_{Nt}, kg ha^{-1}	Total Plant	193
	Y_{mg}, Mg ha^{-1}	Grain	22.7
	Y_{mt}, Mg ha^{-1}	Total Plant	42.6
Goldsboro	k_{Ng}, kg ha^{-1}	Grain	104
	k_{Nt}, kg ha^{-1}	Total Plant	125
	Y_{mg}, Mg ha^{-1}	Grain	15.1
	Y_{mt}, Mg ha^{-1}	Total Plant	27.6
Portsmouth	k_{Ng}, kg ha^{-1}	Grain	122
	k_{Nt}, kg ha^{-1}	Total Plant	142
	Y_{mg}, Mg ha^{-1}	Grain	17.7
	Y_{mt}, Mg ha^{-1}	Total Plant	31.4

N uptake phase model lines (depicted in Figure 5), are generated from Eq. (11), using parameter values for k_{Ng}, k_{Nt}, Y_{mg} and Y_{mt} found in Table 5.

Discussion

From this analysis it is concluded that there are aspects of the ELM that are invariant with respect to both soil type and water availability for a given variety of annual crop propagated by seeding and harvested at the same relative age. This analysis has shown, for the Kamprath N-rate study conducted on corn in North Carolina [22] that both the difference between N uptake intercept parameter and the biomass intercept parameter, Δb, and the maximum N concentration, N_{cm}, are in fact invariant with respect to the crop's surrounding environmental conditions. From the model, these facts lead to the conclusion that both the upper limit N concentration, N_{cm}, and the lower limit N concentration, N_{cl}, are both invariant with respect to soil type and water availability in the study analyzed. This further suggests that the N_{cm} and N_{cl} parameters are of more importance to the model. While other parameters, such as

$$A = \frac{A_N}{N_{cm}} \tag{24}$$

$$\Delta b = \ln\left(\frac{N_{cm}}{N_{cl}}\right) \tag{25}$$

$$b = b_N - \Delta b = b_N - \ln\left(\frac{N_{cm}}{N_{cl}}\right) \tag{26}$$

$$Y_m = \frac{A_N}{N_{cm} - N_{cl}} \tag{27}$$

and

$$k_N = N_{cl} Y_m = \frac{N_{cl} A_N}{N_{cm} - N_{cl}} \tag{28}$$

can be rewritten, to show the significance that upper and lower limit concentrations have in each parameter and ultimately seasonal plant response to nutrient application. Having upper

Table 6. Parametric factors invariant to site attributes, including soil type.

Plant Fraction	Parameter	Estimate
Both	Δb, kg ha^{-1}	0.660
Grain	$N_{cm\ g}$, g kg^{-1}	14.0
Total Plant	$N_{cm\ t}$, g kg^{-1}	9.18
Grain	$N_{cl\ g}$, g kg^{-1}	7.23
Total Plant	$N_{cl\ t}$, g kg^{-1}	4.75
Grain	$N_{cm\ t} - N_{cl\ t} = A_{Nt}/Y_{mt}$, g kg^{-1}	6.75
Total Plant	$N_{cm\ g} - N_{cl\ g} = A_{Ng}/Y_{mg}$, g kg^{-1}	4.43

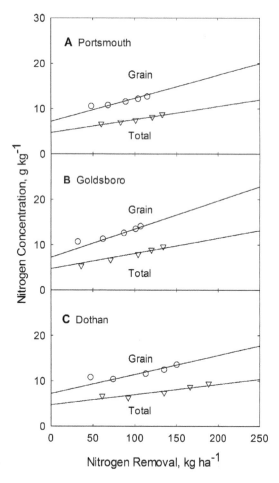

Figure 5. Dependence of grain and total plant N concentration on N uptake for corn grown at the Plymouth (A), Kinston (B), and Clayton (C) experiment stations in North Carolina. Curves are constructed from Eq. 11 and from parameters listed in Table 3.

and season length, the model reduces to three parameters (A_N, b_N, and c) when a crop and season length are chosen. From Overman & Scholtz [26] the logistic response originates within the soil's buffering capacity for P and for K, and the c parameter remains the same from the plant extractable logistic response to nutrient uptake logistic response, and to biomass production. It is here assumed that the c parameter for applied N also originates as the rate response parameter for the soil's buffering capacity of N. The c parameter can be modified by plant population [11]. A Future step should be to analyze various field studies to catalogue soil physical and chemical characteristics and the resulting impact on the c parameter. Mathematically b_N represents shifting parameter which in conjunction with the c parameter as

$$N_{0.5} = \frac{b_N}{c} \qquad (29)$$

$N_{0.5}$ represents the effective level of N necessary to achieve peak N uptake efficiency [10,16,17]. Ultimately, for environmental considerations, setting applied rate of N to the peak uptake level will result in the most N removed per unit N applied. Plus, provided the difference between b and b_N is greater than 0, then the yield will be on the upper portion of the logistic biomass curve to yield

$$Y = \frac{A_N}{N_{cm}[1 + \exp(-\Delta b)]}. \qquad (30)$$

The b_N parameter is affected by changes in plant population [11] and is also influenced by crop type [8,25]. The remaining parameter, A_N, is a linear parameter that is affected by the various environmental conditions, the crop type, the soil type, and various management practices [5–20]. Given Eq. (30), knowing the invariant Δb value for a given crop, and having a reasonable estimate for the background level of N already present in the soil, represented by Eq. (29), exists the beginning of a framework for a more reasonable and more sustainable nutrient management guide. Further analyses are being conducted to verify these findings with other annual propagated by seeding and with perennial crops.

Acknowledgments

The authors would like to acknowledge the hard work of field researchers, like Dr. Kamprath, whose tireless efforts have helped bring about greater understanding of the complexities of agricultural production.

Author Contributions

Conceived and designed the experiments: RVS ARO. Performed the experiments: RVS. Analyzed the data: RVS. Contributed reagents/materials/analysis tools: RVS. Wrote the paper: RVS. Mentorship and review: ARO.

and lower limits to plant nutrient concentration corresponds with plant physiology. Without a minimum level of a given required nutrient, there can be no yield, seasonal or otherwise. There should also be a maximum concentration that can be approached, as there should be diminishing yield increases as higher agronomic rates are applied, or there would be unbounded growth.

If this invariance with respect to soil type and water availability holds for all crops propagated by seeding, the model could be written in terms of parameters that have measurable physiological significance and could give further insight into relationships that govern plant development and nutrient removal. Initial evidence appears promising that perennial crops, such as ryegrass (*Lolium perenne* L.), when held to comparably the same seasonal management practices also exhibit very nearly the same conclusions with regard to both constant values of the N_{cm} and N_{cl} parameters [25].

Given that N uptake and biomass production can be described by five parameters, and if two are invariant to all but crop type

References

1. Overman AR, Martin FG, Wilkinson SR (1990) A logistic equation for yield response of forage grass to nitrogen. Commun Soil Sci Plan 21: 595–609.
2. Ferris T (1989) Coming of age in the milky way. New York: Bantam Doubleday Dell Publishing Group. 495 p.
3. France J, Thornley JHM (1984) Mathematical models in agriculture. London: Butterworth and Company. 352 p.
4. Overman AR, Wilkinson SR, Wilson DM (1994) An extended model of forage grass response to applied nitrogen. Agron J 86: 617–620.
5. Overman AR, Wilson DM, Kamprath EJ (1994) Estimation of yield and nitrogen removal by corn. Agron J 86: 1012–1016.
6. Reck WR, Overman AR (1996) Estimation of corn response to water and applied nitrogen. J Plant Nutr 19: 201-214.

7. Overman AR (1999) Model for accumulation of dry matter and plant nutrient elements by tobacco. J Plant Nutr 22: 1, 81–92.

8. Overman AR, Scholtz RV (2002) Mathematical models of crop growth and yield. New York: Marcel Decker. 342 p.

9. Overman AR, Scholtz RV (2002) Corn response to irrigation and applied nitrogen. Commun Soil Sci Plan 33: 3609–3619.

10. Overman AR, Brock KH (2003) Model analysis of corn response to applied nitrogen and tillage. Commun Soil Sci Plan 34: 2177–2191.

11. Overman AR, Scholtz RV, Brock KH (2006) Model analysis of corn response to applied nitrogen and plant population density. Commun Soil Sci Plan 37: 1157–1172.

12. Overman AR, Howard JC (2012) Model analysis of response to applied nitrogen by cotton. J Plant Nutr 35: 2118–2123.

13. Overman AR, Wilkinson SR (1995) Extended logistic model of forage grass response to applied nitrogen, phosphorus, and potassium. T ASAE 38: 103–108.

14. Overman AR, Stanley RL (1998) Bahiagrass responses to applied nitrogen and harvest interval. Commun Soil Sci Plan 29: 237–244.

15. Overman AR, Wilkinson SR (2003) Extended logistic model of forage grass response to applied nitrogen as affected by soil erosion. T ASAE 46: 1375–1380.

16. Overman AR, Brock KH (2003) Model comparison of coastal bermudagrass and pensacola bahiagrass response to applied nitrogen. Commun Soil Sci Plan 34: 2163–2176.

17. Overman AR, Scholtz RV, Taliaferro CM (2003) Model analysis of response of bermudagrass to applied nitrogen. Commun Soil Sci Plan 34: 1303–1310.

18. Overman AR, Rhoads FM, Brock KH (2005) Model analysis of response of pensacola bahiagrass to applied nitrogen, phosphorus, and potassium. I. Seasonal dry matter. J Plant Nutr 27: 1747–1756.

19. Overman AR, Rhoads FM, Brock KH (2005) Model analysis of response of pensacola bahiagrass to applied nitrogen, phosphorus, and potassium. II. Seasonal P and K uptake. J Plant Nutr 27: 1757–1777.

20. Overman AR, Scholtz RV (2005) Model analysis for response of dwarf elephantgrass to applied nitrogen and rainfall. Commun Soil Sci Plan 35: 2485–2494.

21. Monteith JL (1996) The quest for balance in crop modeling. Agron J 88: 695-697.

22. Kamprath EJ (1986) Nitrogen studies with corn on coastal plain soils, Technical Bulletin 282. Raleigh: North Carolina Agricultural Research Service, North Carolina State University. 15 p.

23. Draper NR, Smith H (1998) Applied regression analysis.3rd ed. New York: John Wiley and Sons. 736 p.

24. Nash JE, Sutcliffe JV (1970) River flow forecasting through conceptual models part I - A discussion of principles. J Hydrol 10: 282–290.

25. Scholtz RV (2002) Mathematical modeling of agronomic crops: analysis of nutrient removal and dry matter accumulation. Doctor of Philosophy Dissertation. Gainesville: University of Florida. 139 p.

26. Overman AR, Scholtz RV (2012) A memoir on: A model of yield and phosphorus uptake in response to applied phosphorus by potato. Gainesville: University of Florida. 16p. Available: http://ufdc.ufl.edu/IR00001234/00001 Accessed 11 December 2.

7

Reducing Insecticide Use in Broad-Acre Grains Production: An Australian Study

Sarina Macfadyen[1]*, Darryl C. Hardie[2,3], Laura Fagan[3], Katia Stefanova[4], Kym D. Perry[5], Helen E. DeGraaf[5], Joanne Holloway[6], Helen Spafford[3¤], Paul A. Umina[7,8]

1 CSIRO Ecosystem Sciences and Sustainable Agriculture Flagship, Canberra, Australia, 2 Department of Agriculture and Food, Irrigated Agriculture and Diversification, Perth, Australia, 3 The University of Western Australia, School of Animal Biology, Perth, Australia, 4 The University of Western Australia, The UWA Institute of Agriculture, Perth, Australia, 5 South Australian Research and Development Institute, Entomology Unit, Urrbrae, Australia, 6 New South Wales Department of Primary Industries, Wagga Wagga Agricultural Institute, Wagga Wagga, Australia, 7 The University of Melbourne, Department of Zoology, Melbourne, Australia, 8 cesar, Melbourne, VIC, Australia

Abstract

Prophylactic use of broad-spectrum insecticides is a common feature of broad-acre grains production systems around the world. Efforts to reduce pesticide use in these systems have the potential to deliver environmental benefits to large areas of agricultural land. However, research and extension initiatives aimed at decoupling pest management decisions from the simple act of applying a cheap insecticide have languished. This places farmers in a vulnerable position of high reliance on a few products that may lose their efficacy due to pests developing resistance, or be lost from use due to regulatory changes. The first step towards developing Integrated Pest Management (IPM) strategies involves an increased efficiency of pesticide inputs. Especially challenging is an understanding of when and where an insecticide application can be withheld without risking yield loss. Here, we quantify the effect of different pest management strategies on the abundance of pest and beneficial arthropods, crop damage and yield, across five sites that span the diversity of contexts in which grains crops are grown in southern Australia. Our results show that while greater insecticide use did reduce the abundance of many pests, this was not coupled with higher yields. Feeding damage by arthropod pests was seen in plots with lower insecticide use but this did not translate into yield losses. For canola, we found that plots that used insecticide seed treatments were most likely to deliver a yield benefit; however other insecticides appear to be unnecessary and economically costly. When considering wheat, none of the insecticide inputs provided an economically justifiable yield gain. These results indicate that there are opportunities for Australian grain growers to reduce insecticide inputs without risking yield loss in some seasons. We see this as the critical first step towards developing IPM practices that will be widely adopted across intensive production systems.

Editor: Nicolas Desneux, French National Institute for Agricultural Research (INRA), France

Funding: This project was funded by the Grains Research and Development Corporation (project UWA00134). www.grdc.com.au. The funders had no role in study design, data collection and analysis, decision to publish, or preparation of the manuscript.

Competing Interests: The authors declare that one of the co-authors (Dr. Paul Umina) is employed by both The University of Melbourne, and a private consultancy company, cesar. This does not compromise the objectivity or validity of the research, analyses, or interpretations in the paper.

* E-mail: sarina.macfadyen@csiro.au

¤ Current address: Department of Plant and Environmental Protection Sciences, College of Tropical Agriculture and Human Resources, University of Hawaii, Manoa, Honolulu, Hawaii, United States of America

Introduction

There are a range of management practices associated with the production of broad-acre grain crops, including the use of modern crop varieties, irrigation, fertiliser, and crop protectants to control losses from arthropod pests, disease and weeds. The availability and widespread use of agricultural pesticides since the 1950's is one factor that has enabled farmers to produce increasing yields of high quality food. However, these practices, either individually or cumulatively, have contributed to a substantial loss of biodiversity in agricultural landscapes around the world. Geiger *et al.* [1] assessed 13 components of intensification in European farmland and found that the use of insecticides and fungicides had consistent negative effects on biodiversity. This realisation has fuelled a policy debate around the use of pesticides, resulting in the loss of

pesticides in some countries due to regulatory reviews, and the introduction of legislation that mandates low pesticide-input farming in the European Union [2]. It is highly likely that there will be fewer pesticides available to farmers in the future and those that are available will be selective, more expensive, and will need to be used more strategically.

Australia is one of the larger grain producing countries in the world. Grain crops are grown primarily in dryland conditions in a large arc around the continent under a wide range of climates (Fig. 1). Wheat and barley account for 74% of total arable crop sowings and other crops such as canola, lupins, oats, sorghum and cotton are grown in smaller areas [3]. Grain crops are attacked by a diversity of arthropod pest species whose populations can be highly sporadic across space and time. Furthermore, the importance of particular pests has changed over recent years, with some becoming more problematic and others less so [4]. Farming practices that may have driven this change include the transition to minimum or no-tillage systems, changes to weed management, the introduction of GM cotton that expresses an insecticide, a significant increase in the total area sown to canola, and continued reliance on pesticides leading to resistance in some pest species [4,5]. Insecticide resistance has been recorded for several important arthropod pests, including the green peach aphid (*Myzus persicae*) [6], redlegged earth mite (*Halotydeus destructor*) [7], diamond back moth (*Plutella xylostella*) [8], *Helicoverpa* spp. [9], and the Western flower thrips (*Frankliniella occidentalis*). Several others species, such as *Balaustium medicagoense*, *Bryobia* sp. and *Sminthurus viridis* have a high natural tolerance to some insecticides [10,11]. How these pests can be effectively controlled in the future, under new cropping systems, without unacceptably high environmental costs, needs to be determined.

IPM has been the archetype model for controlling pests in a sustainable manner for over 50 years [12]. We have evidence that IPM can work in many farming systems and can reduce over-reliance on broad-spectrum insecticides. Pretty [13] examined 62 IPM initiatives in 26 countries and found a reduction in pesticide-use over time in the majority of these cases (around 50% reduction on average). A study using 539 wheat fields in Germany found that fields that used some IPM strategies had lower pesticide use [14]. In certain crops, IPM can also lead to economic savings. For

example, Reddy [15] calculated that a lower-input IPM strategy, that relied on biocontrol agents in cabbage, was almost US$100 per ha cheaper than a conventional pest control approach. In theory, IPM involves the use of cultural, biological and chemical control techniques with a full understanding of the relationship between pest ecology and abundance, plant damage and yield loss [16]. IPM requires a strong understanding of how beneficial and pests interact, move around the landscape, and use resources outside the field and between cropping seasons [17,18]. Insecticides may be used as part of an IPM strategy, however, in principle should only be applied as a last resort, after pest populations have reached a threshold, beyond which economic yield losses will be incurred. This threshold approach is fundamental to IPM practice.

Despite the longevity of IPM, the majority of grain growers in Australia continue to rely heavily on the use of cheap broad-spectrum insecticides to control pests [19]. IPM is more knowledge-intensive than a conventional approach that relies primarily on prophylactic applications of insecticides [20]. Monitoring to determine pest abundance takes time. Some checks for pest species in crops are conducted prior to spraying, but there are typically few regular scouting activities throughout the season. Furthermore, even if economic thresholds have been clearly defined they are not always adhered to for a variety of reasons [21]. Very few selective insecticides are available to growers, and

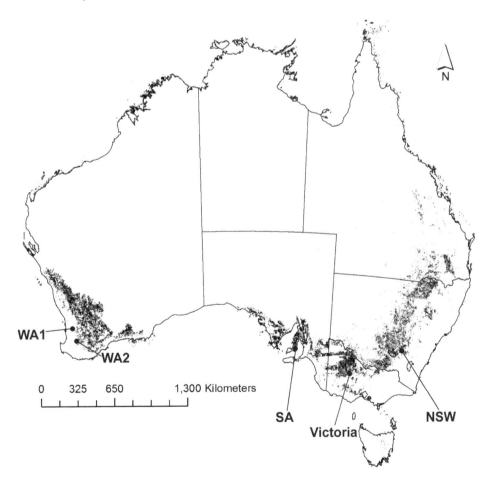

Figure 1. Map showing the location of trial sites throughout grain growing regions of southern Australia. Black shaded areas show where broad-acre cereals and oilseeds are grown. At each site large plots (50 m×50 m minimum) with three pest management approaches were assessed. Trials were conducted on canola in 2010 and wheat in 2011. Land use data comes from ABARES Land Use of Australia, Version 4, 2005/2006 (September 2010 release).

those that are available are relatively expensive [19,22]. Often IPM principles are not consistently implemented and true 'integration' of multiple control techniques is uncommon [13,22].

The aim of this study was to quantify the impact of different pest management approaches on arthropod pests and beneficial arthropods in grain crops grown across southern Australia. We test the hypothesis that a high levels of insecticide use (as seen in conventional practices) results in fewer pests in grain crops, less crop damage from pests, and no yield loss. We contrast the conventional practice with a low-input approach that uses monitoring of pest abundance to decide if an insecticide application is necessary. We test this hypothesis using replicated trials set in commercial fields, across five sites that span the diversity of contexts in which grain crops are grown, and across two cropping seasons. This implies our results can be used to highlight the situations when growers could reduce pesticide inputs without impacting on profitability. We see this as a critical first step towards developing sustainable pest management practices, like IPM, that can be widely adopted by Australian growers in broad-acre production systems.

Materials and Methods

Ethics Statement

All research was carried out on private properties. Permission to be on the land and conduct the research was given by the landholders. Sampling did not involve endangered or protected species. Data sets relating to the analysis presented here can be found at [23].

Description of Trial Sites

Five on-farm sites (labelled as Victoria, NSW, SA, WA1, and WA2 throughout) were established using a standardised experimental design and sampling protocols across the dryland grain growing regions of Australia (Fig. 1). Trials were undertaken in collaboration with a local farming-systems group and the landholder at each site. At three sites (Victoria, WA1 and WA2) these were long-term perennial pasture paddocks, while the previous years' crops were barley and lentils at the NSW and SA sites, respectively. In 2010, trials were established at each site using canola. In 2011, the trials were repeated using the same field plots, but with wheat. Each trial consisted of three comparative pest management strategies: a control (ideally no insecticides applied), a conventional "high-input" pest management approach (based on use of preventative and remedial insecticides currently used by many growers within the local region), and an alternative "low-input" pest management approach (where scouting information was used to decide when to apply an insecticide and, if possible, choice of a softer chemical option). Full insecticide details are listed in Table 1. All other chemical inputs (i.e. fertiliser, herbicide) and farming practices were reflective of the practices used by the landholder at each site. A randomised complete block design with 12 plots arranged in a three by four matrix was used across all sites. There were four replicate plots per treatment. Plot size varied across sites ranging from a minimum of 50 m by 50 m to 75 m by 75 m (up to 5625 m^2 per plot) depending on local seeding equipment. The plots were positioned within a larger field of the same crop and cultivar. At each site, plots were established 10–50 m from the nearest field edge. In some instances, the use of seed treated with insecticide could only be used across all plots due to the seeding practices being employed.

Arthropod Sampling

We used three sampling techniques to collect arthropods throughout the season; vacuum samples (to collect foliage and litter dwelling arthropods early in the season), sweep nets (to collect foliage dwelling arthropods late in the season) and pitfall traps (to collect ground dwelling, and night active arthropods). These sampling techniques are able to capture a range of arthropod pest and beneficial species in broad-acre crops, and are easily replicable across sites, however each will be more efficient at capturing some species than others [24,25]. All sampling was conducted at least 10 m from the edge of plots to account for edge effects caused by the movement of arthropods between plots. Vacuum sampling was conducted prior to crop sowing and at approximately 7, 14, 28 and 42 days after crop emergence (DAE) depending on the prevailing weather conditions. A modified, petrol-driven, leaf blower with a plastic tube inserted over the fan was used to collect the arthropods off the crop, other vegetation and soil surface. A bag or 100-micron fine cup sieve was fitted on to the end of the vacuum spout to capture arthropods. A rectangular frame (150 mm×600 mm) was placed on the soil surface over a row of plants and the nozzle of the suction sampler moved over the soil surface and plant material in this area for 5–10 seconds. A minimum of five samples and a maximum of 10 replicate samples were taken within each plot. Samples were taken at random locations within the plot at each time point (but usually no closer than 10 m from each other).

Sweep samples were used when the crop became too tall for vacuum sampling, generally from flowering to grain ripening. The total number of sampling points at each site varied depending on the crop development stage and weather conditions at the time of sampling. The sweep net consisted of a 380 mm diameter rigid aluminium hoop with a fine mesh net attached. Each sample consisted of 6 or 10 sweeps in canola (a single sweep was a 180° arc covering approximately 2 m with one stride per sweep) in five locations in each plot. In the wheat in 2011 each sample consisted of six sweeps. The sample contents were transferred to a plastic bag or a vial containing 70% (v/v) ethanol.

Pitfall sampling was conducted at several times throughout the growing season. The first sample was taken prior to crop emergence, with additional samples during crop establishment and spring. Each trap consisted of a polyvinyl chloride (PVC) sleeve that was placed in the ground (flush with the ground surface) at the start of the season using a solid steel pin or by excavating holes using a trowel depending on soil type. Vials (45 mm diameter, 120 mL volume) containing 30–60 mL of a propylene glycol (50%): water (50%) mixture, were placed inside the sleeves and left open for seven days. After this time the traps were collected and a lid placed over the sleeve until the next sampling interval. In NSW, SA, WA1 and WA2 nine pitfall traps, arranged in a 3×3 grid, were placed in each plot. However the numbers sorted ranged from three to nine traps per plot and were randomly chosen from those that had been sampled. In SA five traps were sorted per plot (the four outside corners of the grid and a central trap). In Victoria five pitfall traps were placed in a regular arrangement in the central 10 m x 10 m area of each plot. Four traps were placed in a square configuration, 5 m apart from each other, and the fifth placed centrally. All five traps were sorted from the Victorian site.

Direct visual observations of arthropod abundance were also made regularly throughout the season. This information was used to determine if an economic threshold had been reached in the low input treatment plots and therefore an insecticide application was necessary. This involved walking through the plots (from 3 starting positions) and examining sections that had missing plants, uneven

Table 1. Summary of insecticide inputs applied to each trial site across Australia.

Treatment	Crop (cultivar)	Insecticide seed treatment	Insecticide foliar treatment[$]
2010– Victoria (350 mm)			
Control	Canola (Clearfield 44C79)	–	–
Conventional	Canola (Clearfield 44C79)	–	alpha-cypermethrin (PSPE); omethoate (PE)
Low input	Canola (Clearfield 44C79)	imidacloprid	–
2011– Victoria (184 mm)			
Control	Wheat (Correll)	–	–
Conventional	Wheat (Correll)	–	alpha-cypermethrin (PE)
Low input	Wheat (Correll)	–	–
2010– NSW (337 mm)			
Control	Canola (Hybrid 46Y78)	imidacloprid	–
Conventional	Canola (Hybrid 46Y78)	imidacloprid	bifenthrin (PSPE); omethoate (PE)
Low input	Canola (Hybrid 46Y78)	imidacloprid	–
2011– NSW (203 mm)			
Control	Wheat (Sunvale)	–	–
Conventional	Wheat (Sunvale)	–	bifenthrin (PSPE)
Low input	Wheat (Sunvale)	–	–
2010– SA (406 mm)			
Control	Canola (Hybrid 46Y78)	imidacloprid	–
Conventional	Canola (Hybrid 46Y78)	imidacloprid	dimethoate+bifenthrin (PE)
Low input	Canola (Hybrid 46Y78)	imidacloprid	–
2011– SA (238 mm)			
Control[#]	Wheat (Mace)	–	–
Conventional[#]	Wheat (Mace)	–	omethoate+alpha-cypermethrin (PE)
Low input [#]	Wheat (Mace)	imidacloprid	–
2010 - WA1 (144 mm)			
Control	Canola (Argyle)	–	–
Conventional	Canola (Argyle)	–	bifenthrin+chlorpyrifos (PSPE); chlorpyrifos+dimethoate (PE)
Low input	Canola (Argyle)	–	dimethoate (PE); pirimicarb +Bt (LS)
2011 - WA1 (376 mm)			
Control	Wheat (Magenta)	imidacloprid	–
Conventional	Wheat (Magenta)	–	alpha-cypermethrin+chlorpyrifos (PS); alpha-cypermethrin (LS)
Low input	Wheat (Magenta)	–	–
2010– WA2 (139 mm)			
Control	Canola (Cobbler)	–	–
Conventional	Canola (Cobbler)	–	chlorpyrifos (PS); bifenthrin (PE)
Low input	Canola (Cobbler)	–	–
2011– WA2 (341 mm)			
Control	Wheat (Bullaring)	–	–
Conventional	Wheat (Bullaring)	–	cypermethrin (LS)
Low input	Wheat (Bullaring)	–	–

Growing season rainfall is shown in brackets. In 2010 the crop was canola and in the same location, wheat in 2011.
[$]PS = pre-sow; PSPE = post-sowing, pre-emergence; PE = post-emergence; LS = late season foliar treatments.
[#]An aerial application of metaldehyde was used across all plots to control snails late season.

plant growth, and 'hotspots' of chewing and/or sucking damage caused by pest feeding. If such an area occurred, the abundance of arthropod pests was determined using quadrat counts on the soil and plants in autumn and winter, or searching plants and taking sweep net samples in spring.

Arthropod Sorting and Identification

Samples were returned to the laboratory for sorting under a stereomicroscope. We did not identify all arthropods collected, but focussed on identifying key pest and beneficial species in each system. These species are known to cause damage to grain crops in southern Australia or are known to attack arthropod pests of grain crops. Other species were classified down to Family level where possible. Taxa were categorised into three groups: pest arthropods, beneficial arthropods, and other arthropods. Other arthropods were excluded from the analyses presented here. Examples of which taxa were included in each of these groups can be found in the supplementary material (Table S1). In WA pitfall traps all Collembola and Acari (mites) were excluded from the sorting owing to the extremely large numbers of these organic recyclers present in many of the samples.

Yield and Harvest Index Estimates

At all sites crop yield was estimated in each plot at the end of the cropping season using harvesting machines suitable for small-plots. The approach to estimate yield differed across sites. In Victoria there were three harvester passes (wheat approx. 80 m^2 and canola approx. 150 m^2 per plot) within the 30 m by 30 m centre area of each plot. In NSW canola and wheat, the harvester cut a single swath of 1.85 m wide by 75 m long in each plot (138.75 m^2 area per plot). In SA wheat, the harvester cut three swaths per plot of 1.8 m wide by 10 m long (54 m^2 area per plot). In SA canola a yield map of the entire plot was constructed using a GPS Trimble RTK system with 2 m accuracy linked to the farmer's harvester taking a reading every two seconds. A single "sample" per plot was estimated from this data by averaging all recorded yield points within the plot area. In WA1 canola and wheat the harvester cut five swaths of 10 m wide by 70 m long in each plot (3500 m^2 area per plot). In WA2 canola the harvester cut five swaths of 1.25 m wide by 70 m long in each plot (437.5 m^2 area per plot). In 2011, WA2 wheat, no yield data was collected.

At the end of the season, Harvest Index (HI) was estimated by hand-cutting and drying plants. At 6–10 locations in the 30 m by 30 m centre area of each plot a stick was placed along the ground (usually 1 m in length) and all plants cut at ground level. The plants were put into paper bags and allowed to air dry for at least seven days except in Western Australia where they were oven dried. For canola, once dried to <8% seed moisture content the pods were threshed to separate out the seed and all the seeds weighed. The remaining plant material was also weighed after drying. HI was calculated as a proportion of total seed weight by total plant biomass for each sampling location. The same was performed in 2011 for wheat. In WA2 canola, HI was not assessed.

Plant Assessments

All plant assessments were conducted at least 10 m from the plot edge. An assessment of feeding damage from arthropod pests on plants was conducted at 7, 14, 28, and 42 DAE. A maximum of 10 samples were taken at random locations within each plot (and usually no closer than 10 m from each other). At each sample location, a stick (usually 1 m in length) was placed on the ground along a row of plants, and the total number of plants counted. Row spacing was recorded so that plant density (per m^2) could be calculated. In SA the length sampled was adjusted based on row spacing to give a total sample area of 2.5 m^2 per plot (after 10 samples). The number of plants along a stick with chewing damage (Chew damage) and sucking damage (Suck damage) were recorded at each sample location. An overall feeding severity score was measured for the damaged plants along the stick, based on a 0–10 scale. Zero indicates no visible damage, five indicates approximately 50% leaf area damaged, averaged across all plants, and 10 indicates all plants dead or dying. This score has been validated by numerous authors working on grain pest arthropods [26,27–29]. The overall damage at each sample location was expressed as a proportion out of 10 using the formula:

$$Plant\ damage =$$
$$\frac{(Average\ severity\ score\ for\ damaged\ plants\ in\ stick\ length \times Number\ of\ plants\ with\ damage\ in\ stick\ length)}{Number\ of\ plants\ in\ stick\ length}$$

We also recorded the amount of crop cover and amount of weed cover at each location within the plot as an overall percentage.

Statistical Models and Analyses

Generalized linear mixed models [30] were used for the analyses of plant damage and arthropod count data. Due to the presence of DAE explanatory variable and consequently possible inclusion of polynomials of DAE in models, it is more precise to describe the fitted models as generalised additive mixed models (GAMMs). For all response variables presented as proportions, a binomial distribution was assumed and the link function used was the canonical link – logit. Similarly, all responses presented as counts were assumed Poisson distributed and the logarithmic (log) link function was used. The ratio of deviance (χ^2 approximation of residual deviance) to the degrees of freedom, called variance inflation factor c, was used to assess over-dispersion. The model selection aimed at getting an adequate fit without over-fitting and was, in general, based on Akaike Information Criterion (AIC or AIC$_c$ for small samples). In cases of over-dispersion the QAIC and Q AIC$_c$ [31] were used, respectively. For all trials the following response variables were analysed using the above explained statistical models: plant density, sucking damage, chewing damage, pest abundance, and beneficial abundance. In all models, Treatment factor (representing the three pest management approaches) and DAE were fitted as fixed, along with Treatment interactions with DAE and polynomials of DAE up to third degree. The blocking/plot structure was accounted for in the random part of the model. For the majority of fitted models the over-dispersion was due to outliers or the fit of covariates and was corrected. In the cases where the over-dispersion was due to the nature of the data (e.g. too many zeros or clustering of the data), the ASReml-R option for over-dispersion was used.

Yield and HI were analysed using linear mixed models (LMM) formulated using a randomization-model based approach. Typically, the model for each trait and trial included blocking terms to account for the randomization process, and additional terms to model the treatment effects, the covariates and the extra sources of variation, such as spatial trends and extraneous variation. Our methodology was based on the approach of Gilmour et al. [32], followed by additional diagnostics to assess the adequacy of the spatial model [33]. The initial mixed model for each trait by trial comprised random replicate, fixed treatment effect and a separable (column by row) autoregressive process of first order to account for the local spatial trend. After fitting this model, the residuals were checked (residual plots, variogram and faces of the variogram with 95% coverage intervals) to model additional spatial variation (global trend) and/or extraneous variation. We adopted this approach only for the SA canola yield, where the full spatial configuration of the trial was present. The analysis of the Victoria canola yield included a covariate to account for the percentage of lodging affected area and the angle of lodging for each sample. The significance of the fixed terms in the model, in this case the Treatment term, was assessed using Wald test statistic, which has an asymptotic chi-squared distribution (χ^2) with degrees of freedom equal to those of the treatment term.

Additionally, yield estimates from each trial were combined and statistical techniques for the analysis of a multi-environment trial (MET) were used to compare the different types of insecticide inputs used in each pest management approach. In this case the treatments were defined as a combination of foliar treatments during the early and late season, snail baits and insecticidal seed treatments. Treatment levels used in the MET analysis for canola were: ES for early season spray, ES_FS for combined early and late season sprays, ES_S for combined early season spray with seed treatment, S for seed treatment only, and N for no insecticide inputs. The combination of treatments applied to wheat trials were: ES for early season spray, ES_FS for combined early and late season sprays, S for seed treatment only, SB for snail baits, SB_ES for snail baits and early season spray, SB_S for snail baits and seed treatment, and N for no insecticide inputs. The data were analysed using ASReml-R [34], which facilitates joint modelling of blocking, treatment structure, spatial and extraneous terms and accommodates MET analyses.

Economic costs were calculated to estimate the price of insecticide inputs across each treatment at the different trial sites. This was performed using input prices (in Australian dollars) derived from chemical re-sellers in Victoria, Australia (current as of February 2013). Application costs, which assume all insecticides were applied via ground-rig, were included in the total economic price for each treatment.

Results

Using the three sampling techniques we collected large numbers of arthropods from a range of species (298,869 individuals, Table S2). Across the two years of the study pest pressure was generally low, and only one site reached established threshold levels for crop pests (WA1 canola late in the season suffered significant aphid attack). However, we still collected large numbers of pests (118,393 in canola, 116,614 in wheat, Table S2) and found variability in the numbers of pests at each trial site and within each plot. The low pest pressure led to very few insecticide applications on the low input plots, and, in seven out of the 10 trials, the insecticides applied to the low input plots were the same as for the control plots (Table 1). The sites in WA experienced drought conditions in 2010, with rainfall well below the growing season average. We summarise our results below by firstly highlighting the hypothesised pattern and then comparing this to the data collected in each trial. For those traits in which we found a significant treatment effect (overall P-value for treatment <0.05, and/or interaction between treatment and DAE <0.05) we have ranked the multiple comparison results for each trait (using the standard error of the difference). For example, "control>LI>conventional" indicates that for this trait, on average, the control plots had the greatest values, next the low input plots, and lastly the conventional plots. A bracket around two treatments, e.g. (control/LI)>conventional, indicates that there was no significant difference between these two treatments. Multiple brackets around all treatments, e.g. (control/(conventional)/LI), indicate a significant difference only between the highest and lowest treatments.

Impact of the Pest Management Approach on Pest and Beneficial Arthropods

We hypothesised that there should be lower abundance of pest and beneficial arthropods in plots that received greater insecticide inputs (control>LI>conventional). Overall we did find many significant effects of pest management approach on pest abundance across all trials and sampling techniques (Table 2). Often this was not consistent across the time period as evidenced

by a significant interaction between treatment and time (Table 2). For pest abundance in canola, out of the 15 possible combinations of site by sampling technique, seven showed significant effects of treatment and 12 showed significant interactions between treatment and time. Given the greater frequency of early-season insecticide applications across the trials we would expect the vacuum samples to show the clearest response to treatment (Figure C in File S1, Figure C in File S2, Figure C in File S3, Figure C in File S4, and Figure C in File S5). Four out of five canola trials showed a significant effect of treatment on pest abundance in vacuum samples that matched our expectations of greater pest abundance in the control or low input plots. The NSW canola trial (Figure C in File S2) is a good example of this pattern, with decreased pest abundance across time in the conventional plots. In the wheat trials, eight models showed significant effects of treatment, and 10 showed significant interactions between treatment and time (Table 2). Only one trial (WA1), showed a non-significant effect of treatment on pest abundance in wheat vacuum samples. Only two of the five trials matched our expectation of greater pest abundance in the control or low input plots.

For beneficial arthropod abundance, there were fewer significant effects relating to pest management approach (Table 3). In the canola trials, three models showed significant effects of treatment, and seven showed significant interactions between treatment and time. These significant results were seen for the ground-dwelling species collected using pitfall traps and species collected from the plant foliage using the vacuum sampler. In wheat trials, three models showed significant effects of treatment, and five showed significant interactions between treatment and time (Table 3). The pest and beneficial abundance at one site (Victoria, canola) in vacuum samples (Figure C in File S1) most clearly supported our hypothesis with conventional plots having the lowest numbers of both pest and beneficial arthropods at multiple sample dates. Additional graphs showing the pest and beneficial abundance in each trial across time using pitfalls and sweep net sampling can be found in Figures A & B in File S1, Figures A & B in File S2, Figures A & B in File S3, Figures A & B in File S4, and Figures A & B in File S5.

Impact of the Pest Management Approach on Crop Plant Damage

We hypothesised that there should be lower levels of crop plant damage from arthropod pests (control>LI>conventional) and higher plant density (conventional>LI>control) in plots that received greater insecticide inputs. We found that whilst there were some significant effects of pest management approach on plant damage estimates (Table 4), overall the amount of plant damage was relatively low. Chewing damage was more prevalent than sucking damage, but only at one site did we see very high levels of chewing damage (over 35% at WA1 canola). The pattern of plant damage generally supported our hypothesis in the canola trials, with control plots having the greatest relative amount of chewing damage (Table 4). However, in wheat our hypothesis was not supported with many trials showing similar levels of plant damage across the three pest management approaches. We measured plant density to account for plants that were completely removed at the early growth stage by arthropod pests. The pest management approach used had a significant impact on plant density in the Victoria, WA1 and WA2 sites with canola and all wheat trials (Table 4). However, the patterns between treatments did not always match the hypothesis. For example, in only two canola trials did the conventional plots have higher plant density,

Table 2. The effect of different pest management approaches on pest arthropod abundance.

Site	Sampling technique	Treatment P-value[§]	treatment × DAE interaction P-value[§]	Ranking[#]
Canola				
Victoria	Pitfall	0.0044**	<0.001***	Control>LI>conven [B]
NSW	Pitfall	<0.001***	<0.001***	Conven>control>LI [A]
SA	Pitfall	0.78	<0.001***	LI>control>conven [A]
WA1	Pitfall	0.41	0.027*	(Control/LI/conven) [C]
WA2	Pitfall	0.24	<0.001***	(Control/LI/conven) [C]
Victoria	Vacuum	<0.001***	<0.001***	(Control/LI)>conven [B]
NSW	Vacuum	<0.001***	0.0026**	(LI/control)>conven [B]
SA	Vacuum	0.0058**	0.010*	(LI/control)>conven [B]
WA1	Vacuum	0.020*	<0.001***	Control>LI>conven [B]
WA2	Vacuum	0.48	0.065	NP [C]
Victoria	Sweep	0.012*	<0.001***	Control>(conven/LI) [A]
NSW	Sweep	0.70	0.69	NP [C]
SA	Sweep	0.83	0.56	NP [C]
WA1	Sweep	0.092	<0.001***	Conven>(LI/control) [A]
WA2	Sweep	0.44	<0.001***	Control>(LI/conven) [A]
Wheat				
Victoria	Pitfall	0.035*	<0.001***	(Control/LI)>conven [B]
NSW	Pitfall	0.15	<0.001***	Conven>(control/LI) [A]
SA	Pitfall	0.15	<0.001***	(Control/conven)>LI [A]
WA1	Pitfall	0.31	0.37	NP [C]
WA2	Pitfall	0.0017**	0.26	Conven>LI>control [A]
Victoria	Vacuum	0.0039**	<0.001***	(LI/control)>conven [B]
NSW	Vacuum	<0.001***	0.0074**	Control>LI>conven [B]
SA	Vacuum	0.047*	<0.001***	Control>(conven/LI) [A]
WA1	Vacuum	0.56	0.0015**	Conven>(LI/control) [A]
WA2	Vacuum	0.0094**	0.049*	(LI/(control)/conven) [A]
Victoria	Sweep	0.79	0.80	NP [C]
NSW	Sweep	<0.001***	–	(LI/control)>conven [B]
SA	Sweep	<0.001***	0.10	Control>LI>conven [B]
WA1	Sweep	0.090	<0.001***	(Control/LI)>conven [B]
WA2	Sweep	0.38	<0.001***	(LI/(control)/conven) [A]

A GAMM analysis was used to assess the effect of three pest management approaches (treatment: conventional, low input (LI), or control) and time (DAE, days after crop emergence) on the abundance of all arthropod pests collected using three different sampling techniques.
[#]A, a significant difference between treatments but the pattern does not follow what we expect; B, a significant difference between treatments and abundance was highest in control (or low input) and lowest in the conventional (control> LI>conven); C, no difference in pest abundance between the treatments. In this case no ranking was provided (NP). [§] P-value of *<0.05, ** <0.01, *** <0.001.

Table 3. The effect of different pest management approaches on beneficial arthropod abundance.

Site	Sampling technique	Treatment P-value[§]	treatment × DAE interaction P-value[§]	Ranking[#]
Canola				
Victoria	Pitfall	<0.001***	<0.001***	(Control/LI)>conven [B]
NSW	Pitfall	0.041*	<0.001***	(LI/control)>conven [B]
SA	Pitfall	0.43	<0.001***	(LI/(conven)/control) [A]
WA1	Pitfall	0.50	<0.001***	(Control/(LI)/conven) [A]
WA2	Pitfall	0.24	0.41	NP [C]
Victoria	Vacuum	0.0011**	0.23	(LI/control)>conven [B]
NSW	Vacuum	0.14	0.022*	(LI/control)>conven [B]
SA	Vacuum	0.82	0.0025**	(Conven/(control)/LI) [A]
WA1	Vacuum	0.38	0.55	NP [C]
WA2	Vacuum	0.81	0.68	NP [C]
Victoria	Sweep	0.19	0.31	NP [C]
NSW	Sweep	0.70	0.11	NP [C]
SA	Sweep	0.90	0.019*	(Conven/(LI)/control) [A]
WA1	Sweep	0.35	0.098	NP [C]
WA2	Sweep	0.32	0.89	NP [C]
Wheat				
Victoria	Pitfall	0.94	0.053	NP [C]
NSW	Pitfall	0.68	<0.001***	(LI/(conven)/control) [A]
SA	Pitfall	0.21	<0.001***	(Control/LI)>conven [B]
WA1	Pitfall	0.90	0.85	NP [C]
WA2	Pitfall	0.40	0.067	NP [C]
Victoria	Vacuum	0.96	0.0036**	(Control/(conven)/LI) [A]
NSW	Vacuum	0.0031**	0.71	(Control/(LI)/conven) [A]
SA	Vacuum	0.012*	<0.001***	(Control/(conven)/LI) [A]
WA1	Vacuum	0.78	0.53	NP [C]
WA2	Vacuum	0.52	0.78	NP [C]
Victoria	Sweep	0.48	0.43	NP [C]
NSW	Sweep	0.72	0.64	NP [C]
SA	Sweep	0.42	0.014*	(Conven/(control)/LI) [A]
WA1	Sweep	0.082	0.61	NP [C]
WA2	Sweep	<0.001***	0.52	(LI/control)>conven [B]

A GAMM analysis was used to assess the effect of three pest management approaches (treatment: conventional, low input (LI), or control) and time (DAE, days after crop emergence) on the abundance of all beneficial arthropods (predators and parasitoids) collected using three different sampling techniques.
[#]A, a significant difference between treatments but the pattern does not follow what we expect; B, a significant difference between treatments and abundance was highest in control (or low input) and lowest in the conventional (control> LI>conven); C, no difference in pest abundance between the treatments. In this case no ranking was provided (NP). [§] P-value of *<0.05, ** <0.01, *** <0.001.

suggesting that more plants had been damaged by the activities of arthropod pests in the control and low input plots.

Impact of the Pest Management Approach on Crop Yield

We hypothesised there would be higher crop yield in plots that received greater insecticide inputs due to less damage from arthropod pests (conventional>LI>control). The results from the analyses typically showed no significant treatment effects (Fig. 2) in

relation to yield (8 out of 10 trials). In WA1 canola there was a marginally significant effect on yield ($P = 0.049$, (conventional/ LI)>control). There was a significant difference in yield in the SA wheat trial, however this result was sensitive to the addition or removal of one sample point (with outlier $P = 0.116$; outlier deleted $P = 0.003$, conventional>(control/LI)). An estimate of HI was made at the end of the season in each plot to examine the ratio of grain yield to plant biomass. We found less consistent effects of pest management approach on HI. Two out of four canola trials

Table 4. The effect of pest management approach on estimates of crop plant damage.

Site	Treatment P-value[§]	treatment×DAE interaction P-value[§]	Ranking[#]
Sucking damage			
Canola			
Victoria	<0.001***	<0.001***	Control>LI>conven[B]
NSW	0.071	0.75	NP [C]
SA	–	–	–
WA1	<0.001***	0.69	(Control/(LI)/conven) [A]
WA2	0.78	0.96	NP [C]
Wheat			
Victoria	0.36	0.020*	Control>(conven/LI) [A]
NSW	0.013*	0.088	LI>(control/conven) [A]
SA	–	–	–
WA1	0.74	0.78	NP [C]
WA2	0.92	0.71	NP [C]
Chewing damage			
Canola			
Victoria	<0.001***	<0.001***	Control>(LI/conven) [B]
NSW	0.0017**	0.54	(Control/LI)>conven[B]
SA	0.023*	0.93	(Control/(LI)/conven) [B]
WA1	0.17	0.88	NP [C]
WA2	0.0017**	0.69	(Control/LI)>conven[B]
Wheat			
Victoria	NA	NA	NA
NSW	0.47	<0.001***	(Conven/(LI)/control) [A]
SA	<0.001***	0.77	Control>conven>LI[A]
WA1	0.21	0.93	NP [C]
WA2	0.20	0.78	NP [C]
Plant density			
Canola			
Victoria	<0.001***	<0.001***	LI>conven>control[A]
NSW	0.095	0.0051**	LI>(control/conven) [A]
SA	0.27	<0.001***	Control>(LI/conven) [A]
WA1	<0.001***	<0.001***	Conven>(LI/control) [B]
WA2	<0.001***	0.045*	Conven>control>LI[B]
Wheat			
Victoria	<0.001***	<0.001***	Control>conven>LI[A]
NSW	0.0029**	0.37	Control>(LI/conven) [A]
SA	<0.001***	<0.001***	LI>conven>control[A]
WA1	<0.001***	<0.001***	(Conven/control)>LI[A]
WA2	<0.001***	0.44	Control>(conven/LI) [A]

A GAMM analysis was used to assess the effect of three pest management approaches (treatment: conventional, low input (LI), or control) and time (DAE) on plant damage from feeding by pest herbivores. A dash indicates that data was not collected during that trial and NA indicates that a model couldn't be fitted due to zeros in data set.
[#]A, a significant difference between treatments but the pattern does not follow what we expect; B, a significant difference between treatments and damage was highest in control (or low input) and lowest in the conventional (control> LI>conven), or for plant density we expect greatest density in the conventional and lowest in the control (or low input) (conven>LI>control); C, no difference in plant damage or density between the treatments. In this case no ranking was provided (NP). [§] P-value of *<0.05, ** <0.01, *** <0.001.

showed a significant effect (Victoria $P = 0.001$ control>(conventional/LI), WA1 $P = 0.034$ (LI/conventional)>control)). Of the five wheat trials, only WA1 showed significant differences in relation to pest management approach ($P = 0.012$, (LI/conventional)>control).

Due to the method used to harvest the canola in SA, yield was analysed using a spatial linear mixed model. The aim of the spatial analysis is to adjust for the natural variation (by fitting autocorrelations for the local trend and regressions on row/column number for the global row/column trends, respectively). Yield did not exhibit significant linear trends across the rows or columns and the column autocorrelation was moderate (0.37). In the Victoria canola trial a covariate describing crop lodging was fitted in the models for yield and HI without identifying a significant effect for either trait.

To summarise the impact of different types of insecticide-inputs used across all trials (regardless of pest management approach used) we conducted a MET analysis for yield (Table 5). We grouped the insecticide applications according to whether they were applied early season during crop establishment or later in the season, and if they were foliar sprays or applied as seed treatments. The results revealed a significant treatment effect for canola ($P = 0.009$) and no significant treatment effect for wheat ($P = 0.104$) (Table 5). The results for canola varied across treatments. The seed treatment alone and in combination with early season foliar spray showed the highest predicted yield, 2.4 and 2.8 t/ha respectively. For the other treatments (early spray alone, early spray in combination with late spray, and no treatment), the yield ranged from 0.25–1.0 t/ha. Still, one should take into account that early spray in combination with late spray treatment was only used at the WA sites, which were both very low yielding; therefore there is a confounding effect of treatment with climatic conditions. The results for wheat suggest that regardless of the pest management approach used, additional insecticide inputs did not increase crop yield. Still, there is an interesting pattern in the predicted yield means. The yield was 4 t/ha or higher for trials where seed treatment, snail baits or a combination of both were used (Table 5). The highest yield (4.8 t/ha) was observed for a treatment combination of snail baits and early season foliar spray, however snail baits were only used in SA so we cannot say how much influence snail baits alone would have. For trials where early season spray alone or in combination with late season spray was used or no treatment was applied, the predicted yield ranged from 3.5–3.9 t/ha.

We estimated the economic cost of insecticide inputs across the different treatments at each trial site (Table 6). Control treatments always had the lowest, or equal lowest, economic cost across all trials. The conventional treatments were more expensive than the low input treatments in nine trials. Only in one trial, WA1 in 2010, was the low input treatment ($61.61/ha) more expensive than the conventional treatment ($6.85/ha). The main expense for this low input treatment was the addition of a *Bt* spray (Table 1). When the economic costs were divided by the mean crop yield in each treatment, the conventional wheat plots had the highest average cost ($3.21/t/ha), followed by the low input ($2.38/t/ha) and control ($1.17/t/ha) treatments. In canola the low input treatment had the highest average costs ($61.97/t/ha) compared to the conventional ($24.35/t/ha) and control ($0.05/t/ha) treatments. However, the WA1 site heavily biased these estimates. When this trial was removed from the analysis, the low input treatment ($7.01/t/ha) was $17/t/ha less than the conventional treatment.

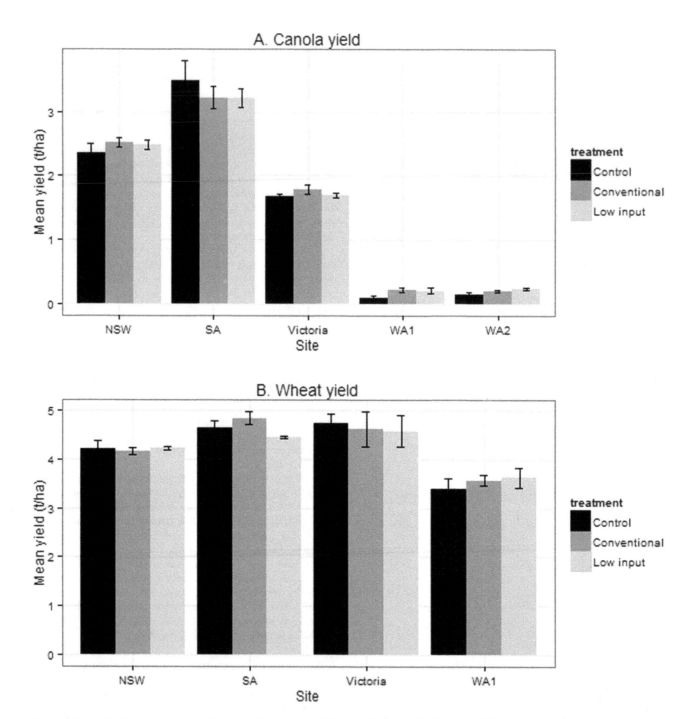

Figure 2. Impact of pest management approach on crop yield in small-plot trials of canola (A) in 2010 and wheat (B) in 2011. Trials were conducted at five sites across the grain growing regions of Australia. There were three pest management approaches assessed (conventional, low input, or control). Overall we found no significant effect of pest management approach on crop yield. In SA wheat (B) we found a significant effect but this was sensitive to the presence or absence of one sample point. In WA1 canola there was a marginally significant effect on yield ($P = 0.049$, (conventional/LI)>control). Bars indicate the mean of 4 replicate plots and $1 \times SE$.

Discussion

A change of practice toward agroecosystem-based IPM requires three progressive steps [35,36]. Firstly, an increased efficiency of pesticide inputs; secondly, input substitution with more benign chemicals or tactics; and thirdly, a system re-design that ensures the cropping landscape is less susceptible to pest-outbreaks. Our study addresses the first step in this process, by testing whether crops with greater inputs of insecticides experience higher crop yield. In theory, an application of an insecticide should lead to fewer pests, therefore less feeding damage to crop plants, and ultimately higher grain yields which would cancel out the economic cost of the insecticides (and the costs of applying chemicals). In reality this is a more complex process with some pest species able to withstand or avoid insecticide exposure [37], natural pest control services being lost as beneficial populations are

Table 5. The effect of different insecticide inputs on crop yield analysed using a multi-environment approach.

Crop	Treatment[#]	Predicted mean yield (t/ha)	Standard error[$]
Canola	Early season foliar sprays	1.000	0.594
	Combined early and late season foliar sprays	0.253	0.839
	Combined early season foliar sprays and seed treatment	2.831	0.596
	Seed treatment only	2.392	0.486
	No insecticide applications	0.674	0.485
Wheat	Early season foliar sprays	3.533	0.450
	Combined early and late season foliar sprays	3.928	0.476
	Seed treatment only	3.994	0.476
	Snail baits	4.646	0.893
	Combined snail baits and early season foliar sprays	4.847	0.893
	Combined snail baits and seed treatment	4.453	0.893
	No insecticide applications	3.765	0.444

[#]See Table 1 for details about insecticide chemicals used.
[$]Averaged SED for canola is 0.858 and for wheat 0.772.

reduced by the insecticide [38], secondary pest outbreaks occurring later in the season [39], plants being able to compensate for damage [40], and defend themselves against future damage [41], and differing costs of insecticide products and commodity prices. Given these complexities, the only way to adequately assess the effect of insecticides is to test them in as near to a commercial situation as possible, as we have done here using large replicated plots embedded in commercial cropping landscapes. What our empirical results show is that while insecticide use did impact the numbers of pests collected, we did not generally see higher yields in the conventionally managed plots (i.e. our hypothesis was not universally supported). Although there was clear evidence of feeding damage by arthropod pests in plots that were unsprayed, this did not translate into a lower yield compared to the conventional plots at the end of the season. Despite overall low abundance of beneficial arthropods we saw evidence of insecticide application reducing numbers of ground-dwelling and foliage-

dwelling beneficials early in the season in Victoria and NSW canola. We suspect the mechanism underlying these patterns is a combination of greater tolerance to insecticides in particular pest species, lack of or loss of beneficial arthropods in some plots, and crop plants compensating for damage throughout the season.

Throughout our study there are few examples where yield loss in crops can be directly attributed to the activities of arthropod pests. If the results of these trials can be extended to commercial situations then there are opportunities for growers to reduce insecticide-inputs without risking yield loss. The yield loss observed within control plots at a single trial (WA1 canola) was due to the activities of mite pest species (*Penthaleus major* and *Halotydeus destructor*) early in the season (Figure C in File S4). This led to a significantly greater proportion of sucking damage and lower plant density in the control plots (Table 4) and ultimately contributed to the loss in yield (Fig. 2). However, it is important to note that yield potential in this trial was already greatly suppressed due to low rainfall (0.21 t/ha in the conventional and low input plots, compared with 0.09 t/ha in the control plots). Furthermore, there was no significant difference in yield between the conventional and low input plots suggesting the additional sprays within the conventional plots provided no yield benefit. The MET analysis results further support our conclusion that increased insecticide inputs does not necessarily lead to a yield gain (Table 5). Our results suggest that growers planting canola using insecticidal treated seed are likely to see some yield gain; however other insecticide inputs appear to be unnecessary in years with low-pest pressure. For growers planting wheat none of the insecticide inputs provided an economically justifiable yield gain in our trials.

Previous semi-field trials have shown that an IPM approach can lead to reduced pest populations, crop damage and higher yield in other broad-acre crops such as cotton and horticultural crops [15,42]. Yet the application of IPM to grain crops in Australia has been limited. A similar statement was made by Wratten *et al.* [43] in 1995 regarding wheat in the UK, Netherlands, the USA and New Zealand. The low-input approach implemented in our study typically consisted of insecticide seed treatments, withholding insecticide applications if pest density was low, and replacing the conventional insecticides with a more target-specific or "low-risk" insecticide when pests reached a critical threshold. This is only one aspect of an IPM approach, ignoring cultural strategies imple-

Table 6. Economic cost of insecticide inputs across the different treatments at each trial site, including the costs for the application of chemicals.

Crop	Site	Treatment		
		Control	Conventional	Low input
Canola	Victoria	0	13.67	0.70
	NSW	0.36	13.85	0.36
	SA	0.30	9.44	0.30
	WA1	0	6.85	61.61
	WA2	0	14.09	0
Wheat	Victoria	0	6.39	0
	NSW	0	5.98	0
	SA	21.80	30.75	30.20
	WA1	0	13.19	9.92
	WA2	0	6.64	0

All values expressed in AU$/ha.

mented outside the season/field to encourage beneficial populations and reduce carry-over of pests [20] that would be difficult to include in a field-plot trial. The large size of our field plots (minimum of 2500 m^2) allowed us to assess both direct and indirect insecticide effect patterns across the whole season for a variety of pest and beneficial species. These large plots are particularly useful when mobile arthropods are involved and for examining season-long effects of insecticide application [44]. Replicating the trials across five sites over two years allowed us to use these results to generalise across a wide area. However, repeating these trials in years with high pest pressure is required to assess thresholds at which the number and type of insecticides applied switches from providing crop protection to offering little advantage in terms of yield benefit. Alternatively, simulating high pest pressure by artificially infesting plots with pest species is possible (e.g., [45]), however undesirable if plots are located on commercial properties.

The incentive for grain growers to move towards a reduction in insecticide inputs is often based on short-term economic factors. In our study, we show that growers could potentially save money by altering their current insect pest management approaches in canola and wheat crops (Table 6). In four canola trials the cost of insecticide inputs was greatest in the conventional treatments, with an average of $12.67/ha. In comparison, the low input treatments cost an average of $0.34/ha. In the five wheat trials the cost of insecticides was greatest in the conventional treatments, with an average of $12.59/ha, compared to an average of $8.02/ha in the low input treatments. Despite the additional costs of the conventional farmer approach, there were no significant yield benefits over the low input approach in any trials. At one trial (WA1 canola), the economic cost of using insecticides in the low input plots was substantially greater (almost 10 times) compared with the conventional approach. This was principally due to the late season application of a selective insecticide, highlighting one of the largest barriers to widespread adoption of IPM among Australian grain growers – the high economic price of many selective chemicals. Although a reduction in insecticide inputs will lead to some direct cost savings, we anticipate that the additional costs associated with the implementation of IPM (e.g. the cost of selective insecticides, pest monitoring costs) will cancel out these savings. In the Australian context pesticide reduction strategies will be driven more by risk minimisation (also see [46]), sustainability and regulatory changes rather than direct economic benefits to the grower.

On only one occasion during our study were arthropod pest pressures above established economic spray thresholds (late season aphids in WA1 canola); so the low input plots were sprayed with insecticides considered low-risk to beneficials. Unfortunately, there are currently few soft insecticides, which are less disruptive to beneficial species registered for use in grain crops in Australia [22] and relatively little R&D investment into newer chemistries. It is encouraging then, that our results have shown, that in certain seasons with low pest pressure, grain growers can avoid insecticide spray applications altogether. However, for growers to be confident about abstaining from insecticide applications we must develop methods to forecast low pest pressure seasons in advance, and cost-effective in-season monitoring strategies that can be implemented across wide geographic areas. The results we have presented show that grain growers in a variety of agricultural landscapes in Australia can potentially farm with fewer pesticides. These growers have the potential to improve sustainability and environmental performance without a reduction in productivity. However this change of practice will not occur until greater emphasis is placed on developing new risk management tools and research into how IPM can be integrated in farm businesses.

Supporting Information

Table S1   Taxa that were included in the study and the functional groups into which they were classified. Only Arthropoda were included (i.e. slugs, snails and earthworms ignored). 'Pest' included any common pest of grain crops across Australia, 'Beneficial' included natural enemies of these pests such as predators and parasitoids, 'Other' included other arthropods that could not be easily grouped into the previous categories. A summary value known as 'all.arthropod' included Pests, Beneficials and Others.

Table S2   Summary of the total number of pest and beneficial individuals captured using each sampling technique and included in the analyses.

File S1   This file contains Figure A, B, and C. Figure A, Pest and beneficial arthropods collected using pitfall traps (mean number per sample) at the Victoria trial site. At the trial site large plots (50 m×50 m minimum) were allocated to one of three pest management approaches; Conventional, Low Input, and Control with minimal insecticide inputs. Each dot represents the average of multiple samples collected within a plot. DAE = days after emergence, 0 and 2 indicates a pre-sow sample. Figure B, Pest and beneficial arthropods collected using sweep net sampling (mean number per sample) at the Victoria trial site. At the trial site large plots (50 m×50 m minimum) were allocated to one of three pest management approaches; Conventional, Low Input, and Control with minimal insecticide inputs. Each dot represents the average of multiple samples collected within a plot. DAE = days after emergence. Figure C, Pest and beneficial arthropods collected using vacuum sampler (number per sample) at the Victoria site. At the trial site large plots (50 m×50 m minimum) with three pest management approaches were assessed (treatment: conventional, low input, or control). Each dot represents the average of multiple samples collected within a plot. DAE = days after emergence, 0 indicates a pre-sow sample. One large outlier was removed from bottom RHS chart to improve clarity of the graphs.

File S2   This file contains Figure A, B, and C. Figure A, Pest and beneficial arthropods collected using pitfall traps (mean number per sample) at the NSW trial site. At the trial site large plots (50 m×50 m minimum) were allocated to one of three pest management approaches; Conventional, Low Input, and Control with minimal insecticide inputs. Each dot represents the average of multiple samples collected within a plot. DAE = days after emergence, 0 and 2 indicates a pre-sow sample. Figure B, Pest and beneficial arthropods collected using sweep net sampling (mean number per sample) at the NSW trial site. At the trial site large plots (50 m×50 m minimum) were allocated to one of three pest management approaches; Conventional, Low Input, and Control with minimal insecticide inputs. Each dot represents the average of multiple samples collected within a plot. DAE = days after emergence. Figure C, Pest and beneficial arthropods collected using a vacuum sampler (number per sample) at the NSW trial site. At the trial site large plots (50 m×50 m minimum) were allocated to one of three pest management approaches (treatment: conventional, low input, or control). Each dot represents the average of multiple samples collected within a plot. DAE = days after emergence, 0 indicates a pre-sow sample.

File S3   This file contains Figure A, B, and C. Figure A, Pest and beneficial arthropods collected using pitfall traps (mean number per sample) at the SA trial site. At the trial site large plots (50 m×50 m minimum) were allocated to one of three pest management approaches; Conventional, Low Input, and Control with minimal insecticide inputs. Each dot represents the average of multiple samples collected within a plot. DAE = days after emergence, 0 and 2 indicates a pre-sow sample. Figure B, Pest and beneficial arthropods collected using sweep net sampling (mean number per sample) at the SA trial site. At the trial site large plots (50 m×50 m minimum) were allocated to one of three pest management approaches; Conventional, Low Input, and Control with minimal insecticide inputs. Each dot represents the average of multiple samples collected within a plot. DAE = days after emergence. Figure C, Pest and beneficial arthropods collected using a vacuum sampler (number per sample) at the SA trial site. At the trial site large plots (50 m×50 m minimum) were allocated to one of three pest management approaches (treatment: conventional, low input, or control). Each dot represents the average of multiple samples collected within a plot. DAE = days after emergence, 0 indicates a pre-sow sample.

File S4   This file contains Figure A, B, and C. Figure A, Pest and beneficial arthropods collected using pitfall traps (mean number per sample) at the WA1 trial site. At the trial site large plots (50 m×50 m minimum) were allocated to one of three pest management approaches; Conventional, Low Input, and Control with minimal insecticide inputs. Each dot represents the average of multiple samples collected within a plot. DAE = days after emergence, 0 and 2 indicates a pre-sow sample. In these pitfall traps all Collembola and Acari (mites) were excluded from the sorting. Figure B, Pest and beneficial arthropods collected using sweep net sampling (mean number per sample) at the WA1 trial site. At the trial site large plots (50 m×50 m minimum) were allocated to one of three pest management approaches; Conventional, Low Input, and Control with minimal insecticide inputs. Each dot represents the average of multiple samples collected within a plot. DAE = days after emergence. Figure C, Pest and beneficial arthropods collected using vacuum sampler (number per sample) at the WA1 site. At the trial site large plots (50 m×50 m minimum) with three pest management approaches were assessed (treatment: conventional, low input, or control). Each dot represents the average of multiple samples collected within a plot. DAE = days after emergence, 0 indicates a pre-sow sample. One

large outlier was removed from bottom RHS chart to improve clarity of the graphs.

File S5   This file contains Figure A, B, and C. Figure A, Pest and beneficial arthropods collected using pitfall traps (mean number per sample) at the WA2 trial site. At the trial site large plots (50 m×50 m minimum) were allocated to one of three pest management approaches; Conventional, Low Input, and Control with minimal insecticide inputs. Each dot represents the average of multiple samples collected within a plot. DAE = days after emergence, 0 and 2 indicates a pre-sow sample. In these pitfall traps all Collembola and Acari (mites) were excluded from the sorting. Figure B, Pest and beneficial arthropods collected using sweep net sampling (mean number per sample) at the WA2 trial site. At the trial site large plots (50 m×50 m minimum) were allocated to one of three pest management approaches; Conventional, Low Input, and Control with minimal insecticide inputs. Each dot represents the average of multiple samples collected within a plot. DAE = days after emergence. Figure C, Pest and beneficial arthropods collected using vacuum sampler (number per sample) at the WA2 site. At the trial site large plots (50 m×50 m minimum) with three pest management approaches were assessed (treatment: conventional, low input, or control). Each dot represents the average of multiple samples collected within a plot. DAE = days after emergence, 0 indicates a pre-sow sample. One large outlier was removed from each of the canola and wheat pest graphs to improve clarity.

Acknowledgments

We would like to thank the landholders for allowing us access to their properties and people who helped us from the farming-systems groups: BCG (Victoria), FarmLink Research (NSW), Yorke Peninsula Alkaline Soils Group (SA), Living Farm and Facey group (WA). Valuable assistance was provided by additional staff at each site. Our thanks goes to Stuart McColl, Samantha Strano, Valerie Caron (Victoria), Adam Shephard, John Lester, Mick Neave (NSW), Sarah Mantel, Latif Salehi, Mark Barrett (SA), Danica Collins, Mary van Wees, Alonso Calvo Araya, Peter Mangano, Alan Lord, Dusty Severtson, John Botha, Cameron Brumley and Paul Yeoh (WA).

Author Contributions

Conceived and designed the experiments: SM DH LF KP HD JH HS PU. Performed the experiments: SM LF KP HD JH PU. Analyzed the data: SM KS. Contributed reagents/materials/analysis tools: SM DH LF KP HD JH PU. Wrote the paper: SM KS PU KP JH. Edited drafts of the manuscript: SM DH LF KS KP HD JH HS PU.

References

1. Geiger F, Bengtsson J, Berendse F, Weisser WW, Emmerson M, et al. (2010) Persistent negative effects of pesticides on biodiversity and biological control potential on European farmland. Basic and Applied Ecology 11: 97–105.

2. Hillocks RJ (2012) Farming with fewer pesticides: EU pesticide review and resulting challenges for UK agriculture. Crop Protection 31: 85–93.

3. Unkovich M, Baldock J, Marvanek S (2009) Which crops should be included in a carbon accounting system for Australian agriculture? Crop & Pasture Science 60: 617–626.

4. Hoffmann AA, Weeks AR, Nash MA, Mangano GP, Umina PA (2008) The changing status of invertebrate pests and the future of pest management in the Australian grains industry. Australian Journal of Experimental Agriculture 48: 1481–1493.

5. Gu H, Edwards OR, Hardy AT, Fitt GP (2008) Host plant resistance in grain crops and prospects for invertebrate pest management in Australia: an overview. Australian Journal of Experimental Agriculture 48: 1543–1548.

6. Edwards OR, Franzmann B, Thackray DJ, Micic S (2008) Insecticide resistance and implications for future aphid management in Australian grains and pastures: a review Australian Journal of Experimental Agriculture 448: 1523–1530.

7. Umina PA, Weeks AR, Roberts J, Jenkins S, Mangano GP, et al. (2012) The current status of pesticide resistance in Australian populations of the redlegged earth mite (Halotydeus destructor). Pest Management Science 68: 889–896.

8. Endersby NM, Ridland PM, Hoffmann AA (2008) The effects of local selection versus dispersal on insecticide resistance patterns: longitudinal evidence from diamondback moth (Plutella xylostella (Lepidoptera: Plutellidae)) in Australia evolving resistance to pyrethroids. Bull Entomol Res 98: 145–157.

9. Downes S, Mahon R (2012) Evolution, ecology and management of resistance in Helicoverpa spp. to Bt cotton in Australia. Journal of Invertebrate Pathology 110: 281–286.

10. Arthur AL, Hoffmann AA, Umina PA, Weeks AR (2008) Emerging pest mites of grains (Balaustium medicagoense and Bryobia sp.) show high levels of tolerance to currently registered pesticides. Australian Journal of Experimental Agriculture 48: 1126–1132.

11. Roberts JMK, Umina PA, Hoffmann AA, Weeks AR (2009) The tolerance of the lucerne flea, Sminthurus viridis, to currently registered pesticides in Australia. Australian Journal of Entomology 48: 241–246.

12. Kogan M (1998) Integrated Pest Management: Historical perspectives and contemporary developments. Annual Review of Entomology 43: 243–270.

13. Pretty J (2005) Sustainability in agriculture: Recent progress and emergent challenges. In: Hester RE, Harrison RM, editors. Sustainability in Agriculture. 1–15.

14. Bürger J, de Mol F, Gerowitt B (2012) Influence of cropping system factors on pesticide use intensity – A multivariate analysis of on-farm data in North East Germany. European Journal of Agronomy 40: 54–63.

15. Reddy GVP (2011) Comparative effect of integrated pest management and farmers' standard pest control practice for managing insect pests on cabbage (*Brassica* spp.). Pest Management Science 67: 980–985.

16. Dent D (1995) Integrated Pest Management. London: chapman & Hall. 356 p.

17. Brewer MJ, Goodell PB (2012) Approaches and Incentives to Implement Integrated Pest Management that Addresses Regional and Environmental Issues. In: Berenbaum MR, editor. Annual Review of Entomology, Vol 57. 41–59.

18. Schellhorn NA, Macfadyen S, Bianchi F, Williams DG, Zalucki MP (2008) Managing ecosystem services in broadacre landscapes: what are the appropriate spatial scales? Australian Journal of Experimental Agriculture 48: 1549–1559.

19. Nash MA, Hoffmann AA (2012) Effective invertebrate pest management in dryland cropping in southern Australia: The challenge of marginality. Crop Protection 42: 289–304.

20. Thomas MB (1999) Ecological approaches and the development of "truly integrated" pest management. Proceedings of the National Academy of Sciences of the United States of America 96: 5944–5951.

21. Whitehouse MEA (2011) IPM of mirids in Australian cotton: Why and when pest managers spray for mirids. Agricultural Systems 104: 30–41.

22. Jenkins S, Hoffmann AA, McColl S, Tsitsilas A, Umina PA (2013) Synthetic pesticides in agro-ecosystems: are they as detrimental to nontarget invertebrate fauna as we suspect? Journal of Economic Entomology 106: 756–775.

23. Macfadyen S, Hardie DC, Fagan L, Stefanova K, Perry KD, et al. (2013) Reducing insecticide use in broad-acre grains production: an Australian study. v1. CSIRO. Data Collection. DOI: 10.4225/08/52B24099AC7D7, permalink: http://dx.doi.org/10.4225/08/52B24099AC7D7.

24. Wade MR, Scholz BCG, Lloyd RJ, Cleary AJ, Franzmann BA, et al. (2006) Temporal variation in arthropod sampling effectiveness: the case for using the beat sheet method in cotton. Entomologia Experimentalis Et Applicata 120: 139–153.

25. Kharboutli MS, Mack TP (1993) Comparison of three methods for sampling arthropod pests and their natural enemies in peanut fields. Journal of Economic Entomology 86: 1802–1810.

26. Chapman R, Ridsdill-Smith TJ, Turner NC (2000) Water stress and redlegged earth mites affect the early growth of seedlings in a subterranean clover/capeweed pasture community. Australian Journal of Agricultural Research 51: 361–370.

27. Gillespie DJ (1991) Identification of resistance to redlegged earth mite (*Halotydeus destructor*) in pasture legumes. Plant Protection Quarterly 6: 170–171.

28. Liu A, Ridsdill-Smith TJ (2000) Feeding by redlegged earth mite (*Halotydeus destructor*) on seedlings influences subsequent plant performance of different pulse crops. Australian Journal of Agricultural Research 40: 715–723.

29. Umina PA, Hoffmann AA (2004) Plant host associations of *Penthaleus* species and *Halotydeus destructor* (Acari: Penthaleidae) and implications for integrated pest management. Experimental and Applied Acarology 33: 1–20.

30. McCullagh P, Nelder JA (1989) Generalized Linear Models. London: Chapman and Hall.

31. Burnham KP, Anderson DR (2002) Model Selection and Multi-model Inference. A Practical Information-Theoretic Approach. New York: Springer.

32. Gilmour AR, Cullis BR, Verbyla AP (1997) Accounting for natural variation in the analysis of field experiments. Journal of Agricultural, Biological, and Environmental Statistics 2: 269–273.

33. Stefanova KT, Smith AB, Cullis BR (2009) Enhanced diagnostics for the spatial analysis of field trials. Journal of Agricultural, Biological, and Environmental Statistics 14: 392–410.

34. Butler DG, Cullis BR, Gilmour AR, Gogel BJ (2009) ASReml-R Reference Manual, Release 3. Queensland DPI.

35. Nicholls CI, Altieri MA (2007) Agroecology: contributions towards a renewed ecological foundation for pest management. In: Kogan M, Jepson P, editors. Perspectives in ecological theory and Integrated Pest Management. Cambridge: Cambridge University Press. 431–468.

36. Letourneau DK (2012) Integrated Pest Management - outbreaks prevented, delayed, or facilitated? In: Barbosa P, Letourneau DK, Agrawal AA, editors. Insect Outbreaks Revisited. Chicester: Blackwell Publishing Ltd. 371–394.

37. Martini X, Kincy N, Nansen C (2012) Quantitative impact assessment of spray coverage and pest behavior on contact pesticide performance. Pest Management Science 68: 1471–1477.

38. Jepson PC, Thacker JRM (1990) Analysis of the spatial component of pesticide side-effects on nontarget invertebrate populations and its relevance to hazard analysis. Functional Ecology 4: 349–355.

39. Wilson LJ, Bauer LR, Lally DA (1999) Insecticide-induced increases in aphid abundance in cotton. Australian Journal of Entomology 38: 242–243.

40. Wilson LJ, Lei TT, Sadras VO, Wilson LT, Heimoana SC (2009) Undamaged cotton plants yield more if their neighbour is damaged: implications for pest management. Bulletin of Entomological Research 99: 467–478.

41. War AR, Paulraj MG, Ahmad T, Buhroo AA, Hussain B, et al. (2012) Mechanisms of plant defense against insect herbivores. Plant signaling & behavior 7: 1306–1320.

42. Furlong MJ, Shi ZH, Liu YQ, Guo SJ, Lu YB, et al. (2004) Experimental analysis of the influence of pest management practice on the efficacy of an endemic arthropod natural enemy complex of the diamondback moth. Journal of Economic Entomology 97: 1814–1827.

43. Wratten SD, Elliott NC, Farrell JA (1995) Integrated pest management in wheat. In: Dent D, editor. Integrated Pest Management. London: chapman & Hall. 241–279.

44. Macfadyen S, Banks JE, Stark JD, Davies AP (2013) The use of semi-field studies for examining the effect of pesticides on mobile terrestrial invertebrates. Annual Review of Entomology In Press.

45. Wilson LJ, Bauer LR, Lally DA (1998) Effect of early season insecticide use on predators and outbreaks of spider mites (Acari : Tetranychidae) in cotton. Bulletin of Entomological Research 88: 477–488.

46. Reisig DD, Bacheler JS, Herbert DA, Kuhar T, Malone S, et al. (2012) Efficacy and Value of Prophylactic vs. Integrated Pest Management Approaches for management of cereal leaf beetle (Coleoptera: Chrysomelidae) in wheat and ramifications for adoption by growers. Journal of Economic Entomology 105: 1612–1619.

8

Injury Profile SIMulator, a Qualitative Aggregative Modelling Framework to Predict Crop Injury Profile as a Function of Cropping Practices, and the Abiotic and Biotic Environment. I. Conceptual Bases

Jean-Noël Aubertot[1,2]*, Marie-Hélène Robin[1,3]

1 Institut National de la Recherche Agronomique, Unité Mixte de Recherche 1248 Agrosystèmes et Agricultures, Gestion des Ressources, Innovations et Ruralités, Castanet-Tolosan, France, **2** Université Toulouse, Institut National Polytechnique de Toulouse, Unité Mixte de Recherche 1248 Agrosystèmes et Agricultures, Gestion des Ressources, Innovations et Ruralités, Castanet-Tolosan, France, **3** Université de Toulouse, Institut National Polytechnique de Toulouse, Ecole d'Ingénieurs de Purpan, Toulouse, France

Abstract

The limitation of damage caused by pests (plant pathogens, weeds, and animal pests) in any agricultural crop requires integrated management strategies. Although significant efforts have been made to i) develop, and to a lesser extent ii) combine genetic, biological, cultural, physical and chemical control methods in Integrated Pest Management (IPM) strategies (vertical integration), there is a need for tools to help manage Injury Profiles (horizontal integration). Farmers design cropping systems according to their goals, knowledge, cognition and perception of socio-economic and technological drivers as well as their physical, biological, and chemical environment. In return, a given cropping system, in a given production situation will exhibit a unique injury profile, defined as a dynamic vector of the main injuries affecting the crop. This simple description of agroecosystems has been used to develop IPSIM (Injury Profile SIMulator), a modelling framework to predict injury profiles as a function of cropping practices, abiotic and biotic environment. Due to the tremendous complexity of agroecosystems, a simple holistic aggregative approach was chosen instead of attempting to couple detailed models. This paper describes the conceptual bases of IPSIM, an aggregative hierarchical framework and a method to help specify IPSIM for a given crop. A companion paper presents a proof of concept of the proposed approach for a single disease of a major crop (eyespot on wheat). In the future, IPSIM could be used as a tool to help design *ex-ante* IPM strategies at the field scale if coupled with a damage sub-model, and a multicriteria sub-model that assesses the social, environmental, and economic performances of simulated agroecosystems. In addition, IPSIM could also be used to help make diagnoses on commercial fields. It is important to point out that the presented concepts are not crop- or pest-specific and that IPSIM can be used on any crop.

Editor: Matteo Convertino, University of Florida, United States of America

Funding: This study was carried out within a PhD project co-funded by INRA and INPT EI Purpan, by the project MICMAC design (ANR-09-STRA-06) supported by the French National Agency for Research (ANR), and by the Programme "Assessing and reducing environmental risks from plant protection products (pesticides)", funded by the French Ministry in charge of Ecology and Sustainable Development (project "ASPIB"). The funders had no role in study design, data collection and analysis, decision to publish, or preparation of the manuscript.

Competing Interests: The authors have declared that no competing interests exist.

* E-mail: Jean-Noel.Aubertot@toulouse.inra.fr

Introduction

Third millennium agriculture must reconcile environmental protection and productivity. The world population is projected to reach 8.7–10 billion by 2050 and annual production will need to increase by 200 million tons by then to meet the projected 470 million ton demand [1]. Several authors attribute the spectacular increase of agricultural production in the second half of the twentieth century to the massive use of products resulting from chemical synthesis [2]; but this intensive production model is nowadays questioned because of public health, agronomic, environmental, and sometimes socio-economic issues. Concepts in crop protection in intensive agricultural production systems changed from destruction of pests (by which we mean, plant pathogens and animal pests in this paper) by the use of pesticides

to pest management with techniques based on the improved knowledge of pest dynamics and their natural enemies and the interaction between pests and crops under the influence of Cropping Practices [3]. It is therefore necessary to combine cultural, genetic, biological, physical and chemical control methods to manage pests through Integrated Pest Management (IPM) strategies in order to maintain the pest population levels below those causing economic losses [4].

True IPM is quite different from the practices recommended up to now [5] and is still faced with agronomic and technical difficulties which can curb its development. Its impact on pests is difficult to estimate because of their multiplicity and of their many interactions within agroecosystems. Studies on the effects of alternative control methods mostly concern a major pest (monospecific approach) while farmers have to manage an injury profile in a given field, i.e. a combination of injury levels caused by multiple pests (multi-specific approach) [6]. Similarly, the research has focused on the effect of one (or a few) control method(s), but farmers usually combine several operations (which may have only partial effects) to limit pest development. Each technical operation is likely to modify the sanitary status of a crop [7]. In addition, not only do cultural practices interact with each other, but also, one technique can be detrimental to some pests and favourable to others. Pest populations are characterised by a very high level of diversity and complexity because of multiple interactions within and between populations and with biological, physical, and chemical environments. This complexity is one of the constraints to the implementation of IPM, in addition to others [8]. In order to reduce the reliance of cropping systems on pesticides, it is therefore necessary to develop tools to help the "vertical integration" (combination of several control methods) and the "horizontal integration" (simultaneous management of several pests) of IPM strategies. Dynamics of pest populations can lead to combinations of injuries on a crop which can in turn lead to quantitative or qualitative damage, which usually results in economic losses for farmers and more generally for society as a whole. However, these relationships are not linear and depend on the production situation as shown by several authors [6,9,10]. In this paper, we will assume that the production situation is defined by the physical, chemical and biological components, except for the crop, of a given field (or agroecosystem) and its environment, as well as socio-economic drivers that affect farmer's decisions (adapted from [11]). In this definition, the term "environment" refers to the climate and the territory (i.e. landscape and the associated actors) that can directly or indirectly influence the considered field. In a given production situation, a farmer can design several cropping systems according to his goals, his perception of the socio-economic context and his environment, farm organisation, knowledge and his cognition. However, a given cropping system in a given production situation will be assumed to lead to a unique injury profile.

In order to help design cropping systems, modelling is a key tool [12]. However, because of the complexity of agroecosystems, models usually only address a limited part of agroecosystems. Crop models have been developed for decades but do not take into account interactions with pests (e.g. [13,14]). Epidemiological models *sensu lato* have been developed to represent pest dynamics, often to help decision making for pesticide treatments. However, these models usually take into account rather poorly the critical effects of cropping practices [15] due to their multiple consequences on the crop-pest-environment dynamics [16]. In addition, the majority of these models address single pests (except for models such as EPIPRE [17,18]). So far, the only models that consider injury profiles are damage models [19,20]. However these models do not predict injury profiles but the quantitative damage that they cause. There is thus a strong need to develop an innovative approach to predict injury profiles as a function of production situations and cropping practices. Because of the complexity of the considered systems [21], and the lack of representation of the effects of cropping practices and their interactions, the linkage of available crop models to epidemiological models seems unlikely to happen when considering multiple pests [3]. Even if a crop model was available, together with epidemiological models for diseases, weeds and animal pests, taking into account the crop status and the effects of cropping practices, attempting to link them would certainly lead to a dead end because of the propagation error phenomenon as well as the large number of parameters and input variables needed. Alternatively, one could consider statistical approaches to cope with the impossibility of addressing these issues when using mechanistic models. However, datasets with observed injury profiles, cropping systems and production situation are scarce and statistical approaches are thus even more unlikely to succeed than mechanistic modelling approaches. As an alternative, a generic modelling framework, called IPSIM for Injury Profile SIMulator is proposed. It is deliberately simple in the way mechanisms are represented because the system being described, i.e. the agroecosystem, is far too complex for a truly mechanistic representation. It is based on a simple qualitative hierarchical aggregative approach to represent the effects of various factors affecting injury profiles. This paper presents the basic principles of IPSIM, describing its implementation in a software program and providing an example of its specification for a given crop. A companion paper [22] provides a proof of concept of this innovative modelling approach in the field of crop protection for an important disease of wheat.

Materials and Methods

Basic Principles of IPSIM

Figure 1 is a schematic representation of an agroecosystem. This figure is the conceptual basis of IPSIM, although its scope is broader than the system directly addressed by IPSIM. According to the farmer's goals, his farm features, his perception of the environment and of the socio-economic context, as well as his knowledge and cognition, he designs cropping systems that will achieve social, economic and environmental performances, as a function of the production situation. These performances will be highly dependent on the injury profile encountered. The term "cropping system" refers here to "a set of management procedures applied to a given, uniformly treated area, which may be a field, part of a field or a group of fields" [23]. This covers many technical operations, for instance, the choice of the crop sequence, cover cropping, cultivar, tillage practices, date and density of sowing, rate of fertilisation and chemical pest control. The term "system" is used here because these technical choices are interdependent [24].

IPSIM is embedded in Figure 1, where its output variable is the injury profile. Input variables of IPSIM are embedded within the three following components: cropping practices, field environment, and physical, chemical and biological components of the field (crop, pests, beneficial and harmless living organisms). An injury profile can thus be seen as the result of hierarchical interactions among the cropping practices and the production situation. Qualitative aggregative hierarchical approaches have been used in several fields to help assess the performances of various options when managing a system: industry (e.g. [25,26]), soil science (e.g. [27]), tourism (e.g. [28,29]). In the field of agronomy, qualitative aggregative hierarchical models have been used for the assessment of the sustainability of cropping systems *ex-ante* or *ex-post* [30–32], the assessment of organic systems [33], the management of Genetically Modified crops (e.g. [34]), the assessment of less-favoured areas for agricultural production (e.g. [35]), the evaluation of energy crops for biogas production (e.g. [36]), the assessment of varieties or cultivars (e.g. [37,38]) and the assessment of the effects of market-gardening cropping systems on soil borne pathogens and animal pests using expert knowledge of advisors [39]. We used this approach to summarise available knowledge in the literature for a given crop and to develop a generic modelling framework for IPM.

Implementation of IPSIM with a Software Program

IPSIM was developed using the DEX method, and is implemented with the DEXi software ([40], http://www-ai.ijs.si/

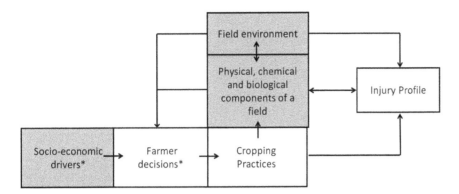

Figure 1. Schematic representation of an agroecosystem and its drivers. In green: components defining the Production Situation (except for the crop). The injury profile is the output variable of IPSIM, whereas its input variables are included within the three following components: cropping practices, field environment, and physical, chemical and biological (crop, pests, beneficials and harmless living organisms) components of the field. *Not taken into account in IPSIM.

MarkoBohanec/dexi.html). DEX is a method for qualitative hierarchical multi-attribute decision modelling and support based on a breakdown of a complex decision problem into smaller and less complex sub-problems. This tool is generally used to evaluate and analyse decision problems [27,29,41]. In this study, it is used for the first time to develop a simulation model that represents the behaviour of an agroecosystem and which quality of prediction can be assessed. The modelling framework has the following features [40]. The sub-problems are hierarchically structured into a tree of attributes that represents the "skeleton" of the model. Terminal nodes of the tree, i.e. leaves or basic attributes, represent input variables of the model (and must be specified by the user). The root node represents the main output: an overall assessment of the evaluated scenarios (an injury profile which is defined by cropping practices and elements of the production situation in this case). The internal nodes of the model are called aggregated attributes. All the attributes in the model are qualitative (ordinal and nominal) rather than quantitative (interval) variables. They take only discrete symbolic values usually represented by words. In the DEX method, the aggregation of values up the tree is defined by "utility functions" based on a set of "if-then" aggregation rules. In our approach, we renamed these functions "aggregating tables" since they are not related to the concept of "utility" in decision theory.

IPSIM Structure

The process of building a DEXi model usually involves the following four steps [40]: (1) identifying the attributes, (2) structuring the attributes, (3) defining attribute scales, and (4) defining the aggregating tables. These steps should be followed for the development of IPSIM using the diagram presented Figure 1. However, only the first three steps can be carried out in a generic way. Only the generic aggregating tables will be described here since most of them are crop-specific.

Structure of the attributes used to predict injury profiles. The structure of attributes that predict injury profiles is presented in Figure 2. Each injury can take a limited number of severity levels. For instance, 5 classes (very low, low, medium, high, very high) or 7 classes (nil, very low, low, medium, high, very high and maximum) can be considered in IPSIM. Even if only 10 pests and 5 severity levels are considered for a given crop, a theoretical number of $5^{10} = 9.765625 \times 10^6$ possible injury profiles could thus be simulated with IPSIM. This number is only theoretical since some of these injury profiles are impossible due to

interactions among pests. In order to take into account these interactions, IPSIM first calculates the severity for single pests independently, as if one pest only was present (Figure 2). Then, interactions between pests are taken into account according the level of each pest and a simple typology of interaction between two pests: high facilitation, low facilitation, no interaction, low reduction, high reduction (Table 1). Table 1 is used to calculate the overall effect of all other pests on the considered pest. Then, the number of pests with high facilitation, low facilitation, no effect, low reduction, high reduction is calculated (Figure 2) and the overall interactions are calculated according to the aggregating tables presented Table 2. Ultimately, the severity of each pest is calculated using the generic aggregating table presented in Table 3 as a function of the severity that would occur without any other pest, and the overall interactions calculated with the aggregating table presented in Table 2.

Structure of the attributes used to predict the severity of a single pest. The input attributes of IPSIM describe cropping

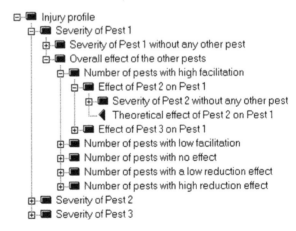

Figure 2. Overall output attributes of IPSIM: description of an injury profile (screenshot of the DEXi software). For the sake of simplicity, only 3 pests are represented in this figure. The severity of a given pest is first calculated independently by IPSIM as if no other pest was present. The aggregated severity of a given pest is then calculated by taking into account the combined effects of all other pests. This is done by considering the theoretical effect of one pest on another according to five levels: high facilitation, low facilitation, no effect, low reduction, high reduction.

Table 1. Generic aggregating table used to represent the effect of one pest on another in IPSIM.

Severity of Pest 2 without any other pest	Theoretical effect of Pest 2 on Pest 1	Actual effect of Pest 2 on Pest 1
Maximum, very high or high	High facilitation	High facilitation
Maximum, very high or high	Low facilitation	Low facilitation
Maximum, very high or high	No effect	No effect
Maximum, very high or high	Low reduction	Low reduction
Maximum, very high or high	High reduction	High reduction
Medium	High and low facilitation	Low facilitation
Medium	No effect	No effect
Medium	High and low reduction	Low reduction
Low or very low	High facilitation	Low facilitation
Low or very low	Low facilitation, no effect, low reduction	No effect
Low or very low	High reduction	Low reduction
Nil	Any	No effect

practices, soil and climate (physical and chemical components of the field which partly define the considered production situation), and biological interactions at the territory level (Figure 1). Figure 3 represents the sub-tree used in IPSIM to calculate the severity of a single pest without any interaction with other pests for a given crop. In this sub-tree, cropping practices are composed of cultural, genetic, biological, physical and chemical control actions. Inocula *sensu lato* are supposed to be non limiting in order to keep basic attributes as simple as possible. The most detailed level is cultural control. It is composed of actions for the management of primary inoculum (through the interaction between crop sequence and tillage for arable crops and prophylactic measures for perennial crops); escape strategies through the choice of the sowing date (some crops are less susceptible to some pests after or before some phenological stages) and mitigation through crop status (as a function of the sowing rate, fertilisation, irrigation, pruning for perennial crops, and application of crop growth regulators). The genetic control represents the level of resistance of the cultivar (or the cultivar mixture) to the considered pest. For some pests, biological control can be applied using living organisms released at the field or greenhouse scale. Physical control consists of using any mechanical, thermal, or electromagnetic actions to limit the pest population. Finally, the attribute "Chemical control" describes the efficacy of pesticide treatments and/or use of non-lethal chemicals such as pheromones or repellents. The effect of soil and climate are described independently and later aggregated in a "Soil and climate" attribute. Finally, the effects of elements (e.g. other fields, hedges, forests) at the territory level are taken into account by describing sources of primary inoculum and beneficials at the territory level, as well as the presence of physical barriers that might limit these interactions between the considered field and its surrounding environment. Harmless living organisms (i.e. neither pests nor beneficials) are not specifically represented in the model.

The scales and the aggregating tables used for the attributes presented in Figure 3 cannot be determined in a generic way. They have to be defined according to experimental results, literature, models, or expert knowledge and are specific to the considered crop and pests.

Typology of simulated injury profiles. So far, IPSIM was presented as a simulator of the severity levels for single pests interacting in an injury profile (Figure 2). This detailed information is valuable to researchers, advisers and even farmers to

characterise the agronomic performance of cropping practices in a given production situation with regard to potential losses that various pests may cause. However, IPSIM can provide other information, less precise for the injury profile description, but more pertinent for the diagnosis of the overall effects of cropping practices and the biological environment of the considered field on injury dynamics. We chose to categorise pests according to a simple characteristic that describes their level of dependency to the cropping system: their level of endocyclism (high and low). The term "endocyclic" refers to an organism whose development is mostly restricted to a field and highly depends on the field endo-inoculum. The level of endocyclism of a given pest is therefore directly defined by the level of persistence of primary endo-inoculum *sensu lato* in a given field and its dispersal ability. Pests with a high level of persistence and low dispersal ability are highly endocyclic. Pests with a low level of persistence are slightly endocyclic, regardless their dispersal ability. Pests with a high level of persistence and a high dispersal ability are moderately endocyclic. The inoculum produced by an endocyclic pest in one season can be carried over to the next, thus building up a cumulative inoculum reservoir over the years. Endocyclic organisms are thus highly dependent on field history. The categorisation of pests into two groups (high/medium and low levels of endocyclism) can help identify the main level to address to control them: the field or territory level.

For example, root-knot nematodes (*Meloidogyne* spp.) on horticultural crops, wireworms on potato (*Agriotes* spp.), wheat common bunt (*Tilletia* spp.), take-all on wheat (*Gaeumannomyces graminis* var. *tritici*), dicotyledonous weeds such as *Chenopodium album* and *Fallopia convolvulus* are highly endocyclic pests. However, highly endocyclic pests can sometimes be spread to other fields by anthropic activities (e.g. via agricultural machinery, pruning tools, clothes and boots of greenhouse technicians). This dispersal mechanism will not be taken into account in the model. Aphids on several crops (e.g. *Brevicoryne brassicae*), powdery mildew on grapevine (*Erysiphe necator*), rusts on cereals (e.g. *Puccinia recondita*), codling moth on apple tree (*Cydia pomonella*), and weeds such as some Asteraceae (e.g. *Taraxacum dens leonis*) or grassy weeds (e.g. *Bromus sterilis*) are slightly endocyclic pests.

Two aggregating tables were designed to summarise the distribution of final injury levels of single pests using two aggregated variables: the overall final severity of i) highly/

Table 2. Generic aggregating table used to calculate the overall effect on a given pest caused by all the other pests in an injury profile.

Number of pests with high facilitation	Number of pests with low facilitation	Number of pests with no effect	Number of pests with low reduction	Number of pests with high reduction	Overall effects of all other pests
0	0	0	0	0	No effect
0	0	0	0	≥1	High reduction
0	0	0	≥1	0	Low reduction
0	0	0	≥1	≥1	High reduction
0	0	≥1	0	0	No effect
0	0	≥1	0	≥1	High reduction
0	0	≥1	≥1	0	Low reduction
0	0	≥1	≥1	≥1	High reduction
0	≥1	0	0	0	Low facilitation
0	≥1	0	0	≥1	High reduction
0	≥1	0	≥1	0	Low reduction
0	≥1	0	≥1	≥1	Low reduction
0	≥1	≥1	0	0	Low facilitation
0	≥1	≥1	0	≥1	High reduction
0	≥1	≥1	≥1	0	Low reduction
0	≥1	≥1	≥1	≥1	High reduction
≥1	0	0	0	0	High facilitation
≥1	0	0	0	≥1	Low reduction
≥1	0	0	≥1	0	No effect
≥1	0	0	≥1	≥1	Low reduction
≥1	0	≥1	0	0	High facilitation
≥1	0	≥1	0	≥1	Low reduction
≥1	0	≥1	≥1	0	Low reduction
≥1	0	≥1	≥1	≥1	Low reduction
≥1	≥1	0	0	0	High facilitation
≥1	≥1	0	0	≥1	Low reduction
≥1	≥1	0	≥1	0	Low reduction
≥1	≥1	0	≥1	≥1	High reduction
≥1	≥1	≥1	0	0	High facilitation
≥1	≥1	≥1	0	≥1	Low reduction
≥1	≥1	≥1	≥1	0	No effect
≥1	≥1	≥1	≥1	≥1	Low reduction

moderately and ii) slightly endocyclic pests (Table 4). Considering three levels of final injury (low, medium, high) for each of the two endocyclism groups, a range of nine possible generic injury profiles was proposed (Figure 4) for any agricultural productions world-wide (i.e. major crops; vegetables; vineyard; orchards; horticulture; industrial crops, aromatic and medicinal plants; grassland; in field or in Controlled Environment Agriculture). For production situations where injury profile has high final injury levels of highly endocyclic pests (IP7, IP8; IP9; Table 5), a better management of primary inoculum production at the field level should be undertaken (e.g. interaction between by crop sequence and tillage; stubble management, volunteer management, stale seedbeds and sanitation measures for perennial crops). For production situations with injury profiles with high levels of slightly endocyclic pests (IP3, IP6; IP9; Table 5), special attention should be paid to i) the management of inoculum production at the territory level (e.g. spatial distribution of cropping systems,

management of primary inoculum production in the neighbouring fields or waste piles, management of interstitial spaces to promote beneficials); ii) escape strategies (sowing date adaptation); iii) mitigation through the crop status (e.g. cultivar choice, sowing rate, nitrogen fertilisation, irrigation).

Results

Implementation of IPSIM Generic Framework into a Simulation Model, an Example

This article aims to present the whole modelling process: i) development of a conceptual framework; ii) implementation of this conceptual scheme into a simulation model for a simple case; iii) simulation to exemplify potential uses of IPSIM models. The specification of IPSIM will be performed for a simple injury profile on wheat: two highly endocyclic diseases (eyespot and sharp eyespot) and a slightly endocyclic disease (brown rust). Eyespot,

Table 3. Generic aggregating table used to calculate the severity of one pest in interaction with the other pests of an injury profile.

Severity of the considered pest without any other pests	Overall effect of the other pests	Severity of the considered pest under the influence of other pests
Maximum	High facilitation	Maximum
Maximum	Low facilitation	Maximum
Maximum	No effect	Maximum
Maximum	Low reduction	Very high
Maximum	High reduction	High
Very high	High facilitation	Maximum
Very high	Low facilitation	Maximum
Very high	No effect	Very high
Very high	Low reduction	High
Very high	High reduction	Medium
High	High facilitation	Maximum
High	Low facilitation	Very high
High	No effect	High
High	Low reduction	Medium
High	High reduction	Low
Medium	High facilitation	Very high
Medium	Low facilitation	High
Medium	No effect	Medium
Medium	Low reduction	Low
Medium	High reduction	Very low
Low	High facilitation	High
Low	Low facilitation	Medium
Low	No effect	Low
Low	Low reduction	Very low
Low	High reduction	Very low
Very low	High facilitation	Medium
Very low	Low facilitation	Low
Very low	No effect	Very low
Very low	Low reduction	Very low
Very low	High reduction	Very low
Nil	High facilitation	Nil
Nil	Low facilitation	Nil
Nil	No effect	Nil
Nil	Low reduction	Nil
Nil	High reduction	Nil

caused by the necrotrophic and soil-borne fungi *Oculimacula yallundae* and *O. acuformis*, anamorph *Pseudocercosporella herpotrichoides* is considered to be the most important stem-base disease of cereals in temperate countries. In France, sharp eyespot, another soil-borne fungus caused *by Rhizoctonia cerealis*, is one of the minor diseases of the foot disease complex of winter wheat, but is thought to interact strongly with eyespot. The two pathogens show distinct antagonistic behaviour within the infected stem base, which translates into a negative correlation between sharp eyespot and eyespot incidence [42–44]. Finally, brown rust, caused by *Puccinia triticina*, is the most common rust disease of wheat and is now recognised as an important pathogen in wheat production worldwide, causing significant yield losses over large geographical

areas [45]. As opposed to the first two soil-borne diseases which are disseminated over short distances, brown rust is an obligate, airborne disease with conidia which are wind-dispersed over hundreds of kilometres, resulting in rust epidemics on a continental scale [46].

The design of IPSIM-Wheat-Eyespot and the evaluation of its predictive quality is described in a companion paper [22]. For the sake of simplicity and readability, the other two models will not be presented in detail, but their development was similar to the one presented in [22]. The two models for eyespot and sharp eyespot are similar in terms of structure (tree) and aggregating tables because the impact of cropping practices on sharp eyespot is similar to that on eyespot [42]. However, since brown rust is an

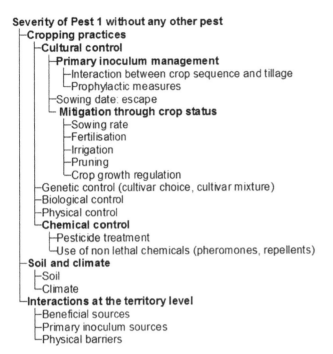

Severity of Pest 1 without any other pest
├─**Cropping practices**
│ ├─**Cultural control**
│ │ ├─**Primary inoculum management**
│ │ │ ├─Interaction between crop sequence and tillage
│ │ │ └─Prophylactic measures
│ │ ├─Sowing date: escape
│ │ └─ **Mitigation through crop status**
│ │   ├─Sowing rate
│ │   ├─Fertilisation
│ │   ├─Irrigation
│ │   ├─Pruning
│ │   └─Crop growth regulation
│ ├─Genetic control (cultivar choice, cultivar mixture)
│ ├─Biological control
│ ├─Physical control
│ └─**Chemical control**
│   ├─Pesticide treatment
│   └─Use of non lethal chemicals (pheromones, repellents)
├─**Soil and climate**
│ ├─Soil
│ └─Climate
└─**Interactions at the territory level**
   ├─Beneficial sources
   ├─Primary inoculum sources
   └─Physical barriers

Figure 3. Hierarchical sub-tree to predict the severity of a single pest without any interaction with other pests (screenshot of the DEXi software).

airborne disease, the effects of primary inoculum management at the field scale are less important than for the two soil-borne diseases. For this airborne disease, the main control methods are: i) mitigation through crop status (using a resistant cultivar for instance) and; ii) the management of primary inoculum sources at the territory level. For this schematic injury profile, we will assume that no direct interactions occur between the two soil-borne, stembase diseases and this airborne, foliar disease.

Simulation Scenarios

The use of the model presented in the first sub-section of the "results" section is exemplified for three contrasting cropping practices in a given production situation (Figure 5). The three cropping practices considered were: intensive, integrated and organic systems. The intensive system is a wheat monoculture with a high level of inputs and a high-yielding cultivar susceptible to diseases, aiming at a high yield level. The integrated system is characterised by a limited use of inputs, with a lower-yielding cultivar than the former system, but less susceptible to diseases, a short wheat rotation, and a satisfactory yield level. The organic system is characterised by low inputs, with a disease-resistant cultivar with a limited yield, associated with a long wheat rotation and appropriate crop management. The three systems were tested in the same production situation, with a weather scenario favourable to the development of the considered diseases.

Simulation Results

The DEXi software computed the aggregated attribute values of the model presented in the first sub-section of the "results". In the same production situation, the three cropping practices led to contrasting injury profiles. In the absence of estimates of potential yield losses caused by these injury profiles, it is difficult to provide direct recommendations for cropping practices adaptations. However, these simulations enable a diagnosis in terms of pest

development for the three simulated systems. The intensive system led to IP4, i.e. a medium final injury level for highly endocyclic pests associated with a low final injury level for slightly endocyclic pests (Figure 5). For this system, the model suggests that a better management of primary inoculum of the pathogen responsible for eyespot injury should be considered. The integrated system led to IP2, i.e. a low final injury level for highly endocyclic pests associated with a medium final injury level for slightly endocyclic pests (Figure 5). For this system, the model suggests that a better control of brown rust through the use of a more resistant cultivar or the use of a low-dose fungicide, provided that it would be economically sound. The organic system led to IP1, i.e. a low final injury level for highly endocyclic pests associated with a low final injury level for slightly endocyclic pests (Figure 5). This is consistent with the associated cropping practices which aims at minimising pest development by combining prophylactic measures with partial effects. It is important to underline that this diagnosis did not address yield losses, but focused only on injury.

Discussion

Potential Uses of IPSIM Models

These simulations illustrate how IPSIM can be used to assess *ex-ante* the performance of various cropping systems with regard to the control of pest injury on a given crop. This information is useful when designing innovative cropping systems, either by prototyping, e.g. [47], simulation, e.g. [48], or expert knowledge, e.g. [12]. Since climate significantly affects injury profiles, weather frequency analyses are needed, using a set of input variables describing a wide range of climatic scenarios so that the information provided by IPSIM is robust in the face of weather variability. However, IPSIM cannot be seen as a model to design innovative cropping systems *in silico* for two major reasons. First, crop damage is not simulated by IPSIM, which makes it difficult to rank pests with respect to the crop losses they cause. Second, the social, economic and environmental performance of the simulated cropping systems are not calculated. To tackle this problem, IPSIM could be coupled to a damage model (such as RICEPEST [6,19,20] or WHEATPEST [20]) that would predict yield losses as a function of the injury profiles encountered and other relevant variables. Alternatively, a crop model (e.g. STICS [14]) could be used, with a set of single damage functions (such as the ones used in WHEATPEST [20]), and coupled with IPSIM. Then, once the damage caused by a given injury profile in a given production situation has been predicted, a more general framework, such as MASC, [30] or DEXiPM, [49], could be used to predict the social, economic and environmental performance of the tested systems in a given production situation. This approach will help design innovative cropping systems less vulnerable to pests. Using that modelling framework, IPSIM would be the missing link to fill the gap between crop models that can help predict performance of pest-free cropping systems and epidemiological models that generally do not represent the effects of crop status under the influence of cropping practices. In addition, models developed with IPSIM could be used to create typologies of injury profiles at a regional, national, continental or even worldwide scale, using a schematic description of soil and climate, together with a description of the diversity of cropping practices. This should reveal the main injury profiles encountered and help design strategies to control them with better vertical and horizontal integration of IPM. If the corresponding damage models were available, the typology produced could help prioritise objectively research efforts on the main harmful pests.

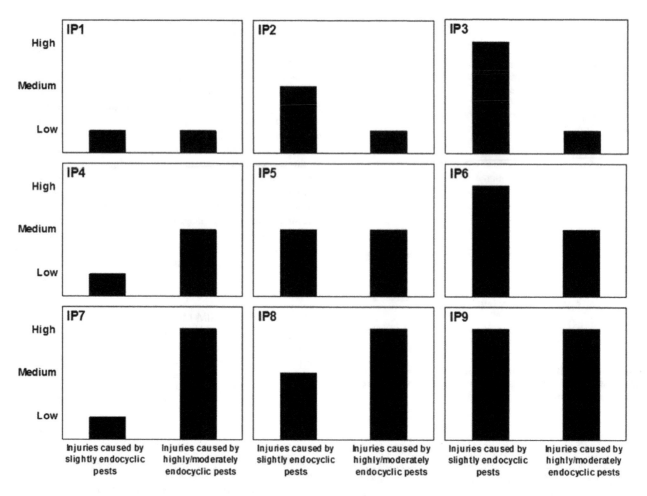

Figure 4. Typology of injuries caused by multiple pests on a crop for given Cropping Practices in a given Production Situation using nine generic Injury Profiles (IP1–IP9). These Injury Profiles are determined by the final levels of the injuries caused by slightly and highly/moderately endocyclic pests (plant pathogens, weeds and animal pests). They can be used to perform cross-cutting analyses for a wide range of agricultural productions.

IPSIM could also be used in an *ex-post* analysis to understand the behaviour of commercial field or experimental plots. Finally, it can be viewed as a communication tool for groups, as well as to teach practitioners and students. Knowledge of several scientific fields involved in crop protection, as well as several types of expertise (of scientists, extension engineers, or farmers) can be built into IPSIM, offering a framework for these various communities to interact and combine their knowledge.

Table 4. Generic aggregating table used to define the level of severity of slightly endocyclic pests in an Injury Profile as a function of the final injury level of single pests.

Number of slightly endocyclic pests with a very high or maximum final injury level	Number of slightly endocyclic pests with a low, medium or high final injury level	Number of slightly endocyclic pests with a null or very low final injury level	Overall severity of slightly endocyclic pests in the Injury Profile
>1	>1	>1	High
>1	>1	0	High
>1	0	>1	High
>1	0	0	High
0	>1	>1	Medium
0	>1	0	Medium
0	0	>1	Low
0	0	0	Low

The same aggregating table is used to define the level of severity of highly/moderately endocyclic pests.

Table 5. Equivalence between features of qualitative models developed within the IPSIM framework and quantitative simulation models.

Feature	Qualitative simulation models such as the ones developed with the IPSIM framework	Quantitative simulation models
Type of input variables	Nominal, ordinal, or interval	Interval
Type of state variables	Ordinal	Interval
Type of output variables	Ordinal (can be transformed into static interval or even dynamic interval)	Interval
Model structure	Aggregation tree	Equations
Specification of the model structure	Aggregating tables	Parameters
Analysis of model's behaviour	Table of local and global weights for each input and aggregated attributes	Sensitivity analyses to input variables and parameters
Measures of agreement (non exhaustive)	Proportion of correctly predicted ordinal classes; non parametric Wilcoxon signed rank test to analyse if the distribution of errors is significantly biased or not; matched marginal distribution analysis or joint distribution analysis in a square contingency table	Bias; Mean Absolute Error; Root Mean Squared Error; Efficiency

Limitations of the Approach

Like any other model, the predictive quality of IPSIM should be assessed prior to its use. This highlights the urgent need to collect data in commercial fields describing the input and output variables of IPSIM (*i.e.* cropping practices, soil and climate, field environment, injury profiles), along with the social, economic and environmental performances of the monitored agroecosystems. It is important to also add measurements of state variables characterising the crop status (e.g. biomass per area unit, Leaf Area Index) in order to better describe important state variables of the agroecosystem for other possible future analyses of the created datasets. However, due to the lack of datasets containing a description of injury profiles, the confidence that users may have in IPSIM models could also be enhanced by comparing simulation outputs with their own expertise to identify any mismatches. All the information contained within IPSIM models is held in the hierarchical trees and the associated aggregating tables. One of the consequences of this specificity of models developed with the IPSIM framework is that, once developed, the predictive quality of the models can be enhanced easily using experimental datasets by modifying aggregating tables, and, if need be, the structure of the model.

The possible injury profiles that IPSIM models can simulate are numerous. However, observations tend to show that the diversity of injury profiles encountered in commercial fields is much less than the structure of IPSIM models can generate. This results from two mechanisms. First, pests can interact directly (through facilitation, predation, competition for the same ecological niche) or indirectly (though modification of the biotope). This implies that not all potential theoretical injury levels could occur simultaneously. This is a limitation of IPSIM which does not account for the impact of injuries on crop growth. Secondly, the soil, climate, cropping practices and landscape occurring in a given territory might not be diverse enough to lead to all theoretical injury levels (for instance, the theoretical injury profile with all the forms of injury at their maximum level does not exist in reality). Another limitation of IPSIM is the way that interactions between pests are represented. If n pests are considered, n (n-1) interactions are to be described. This is similar to the three-body (or n-body) problem in physics, which has a global analytical solution in the form of convergent power series [50], but that has to be approximated in

practice because they converge too slowly. IPSIM models approximate interactions among injuries by arbitrarily calculating the global interaction that would occur between a given injury and the rest of the injury profile defined as the sum of single injuries simulated without taking into account interactions among pests. However, this approximation certainly appears negligible as compared to other necessary simplification hypotheses.

From the conceptual viewpoint, it could be asked why the crop which is entered in the field biological component (Figure 1) does not appear at the first level of the IPSIM tree. After all, pests only experience physical, chemical and biological interactions within agroecosystems and a description of i) the crop status, ii) soil and climate, and iii) the neighbouring environment of the field are indeed the true drivers of pest dynamics. This option was tried when developing IPSIM structure, but led to too complicated a structure, the effect of single cultural operations being overlooked among the numerous levels of the tree. In addition, datasets with a description of cropping practices and injury profiles are extremely scarce. The requirement of additional variables describing the crop status (e.g. in terms of phenology, architecture, biomass, Leaf Area Index) would also lead to greater difficulties in developing IPSIM models and in evaluating its predictive quality.

We recommend to develop models with no more than 7 final injury levels for a single pest. The lack of precision of IPSIM models could be seen as a drawback as compared to quantitative epidemiological models. Firstly this is because these latter models address a much simpler system: a single pest, rather than an injury profile. Secondly, when developing models of complex systems, accuracy should be sought rather than precision. Searching for better precision would certainly lead to an increase in the model's complexity and possibly to a dead end. We believe that the proposed precision of the models that will be developed with the IPSIM framework is more than enough for the main ultimate purpose of the model: helping the design of innovative cropping systems less vulnerable to pests.

Points for Reflection

The presented structure of IPSIM is not exhaustive in terms of control methods that can be undertaken. However, developers of models within the IPSIM framework can always easily modify its structure in order to take into account the effects of control

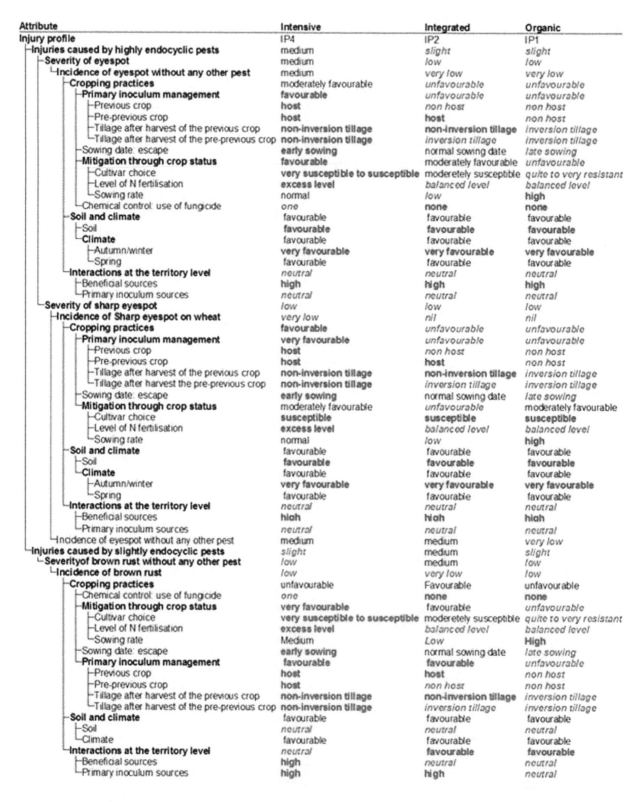

Attribute	Intensive	Integrated	Organic
Injury profile	IP4	IP2	IP1
─Injuries caused by highly endocyclic pests	medium	*slight*	*slight*
└─Severity of eyespot	medium	*low*	*low*
└─Incidence of eyespot without any other pest	medium	*very low*	*very low*
┌─Cropping practices	moderately favourable	*unfavourable*	*unfavourable*
├─Primary inoculum management	favourable	*unfavourable*	*unfavourable*
├─Previous crop	host	*non host*	*non host*
├─Pre-previous crop	host	host	*non host*
├─Tillage after harvest of the previous crop	non-inversion tillage	non-inversion tillage	*inversion tillage*
└─Tillage after harvest of the pre-previous crop	non-inversion tillage	*inversion tillage*	*inversion tillage*
─Sowing date: escape	early sowing	normal sowing date	*late sowing*
├─Mitigation through crop status	favourable	moderately favourable	*unfavourable*
├─Cultivar choice	very susceptible to susceptible	moderately susceptible	*quite to very resistant*
├─Level of N fertilisation	excess level	*balanced level*	*balanced level*
└─Sowing rate	normal	*low*	high
└─Chemical control: use of fungicide	one	none	none
─Soil and climate	favourable	favourable	favourable
├─Soil	favourable	favourable	favourable
├─Climate	favourable	favourable	favourable
├─Autumn/winter	very favourable	very favourable	very favourable
└─Spring	favourable	favourable	favourable
└─Interactions at the territory level	*neutral*	*neutral*	*neutral*
├─Beneficial sources	high	high	high
└─Primary inoculum sources	*neutral*	*neutral*	*neutral*
└─Severity of sharp eyespot	*low*	*low*	*low*
└─Incidence of Sharp eyespot on wheat	*very low*	*nil*	*nil*
┌─Cropping practices	favourable	*unfavourable*	*unfavourable*
├─Primary inoculum management	very favourable	*unfavourable*	*unfavourable*
├─Previous crop	host	*non host*	*non host*
├─Pre-previous crop	host	host	*non host*
├─Tillage after harvest of the previous crop	non-inversion tillage	non-inversion tillage	*inversion tillage*
└─Tillage after harvest the pre-previous crop	non-inversion tillage	*inversion tillage*	*inversion tillage*
─Sowing date: escape	early sowing	normal sowing date	*late sowing*
├─Mitigation through crop status	moderately favourable	*unfavourable*	moderately favourable
├─Cultivar choice	susceptible	susceptible	susceptible
├─Level of N fertilisation	excess level	*balanced level*	*balanced level*
└─Sowing rate	normal	*low*	high
─Soil and climate	favourable	favourable	favourable
├─Soil	favourable	favourable	favourable
├─Climate	favourable	favourable	favourable
├─Autumn/winter	very favourable	very favourable	very favourable
└─Spring	favourable	favourable	favourable
└─Interactions at the territory level	*neutral*	*neutral*	*neutral*
├─Beneficial sources	high	high	high
└─Primary inoculum sources	*neutral*	*neutral*	*neutral*
└─Incidence of eyespot without any other pest	medium	medium	*very low*
└─Injuries caused by slightly endocyclic pests	*slight*	medium	*slight*
└─Severity of brown rust without any other pest	*low*	medium	*low*
└─Incidence of brown rust	*low*	*very low*	*low*
┌─Cropping practices	unfavourable	Favourable	unfavourable
├─Chemical control: use of fungicide	one	none	none
├─Mitigation through crop status	very favourable	favourable	*unfavourable*
├─Cultivar choice	very susceptible to susceptible	moderetely susceptible	*quite to very resistant*
├─Level of N fertilisation	excess level	*balanced level*	*balanced level*
└─Sowing rate	Medium	Low	High
─Sowing date: escape	early sowing	normal sowing date	*late sowing*
├─Primary inoculum management	favourable	favourable	*unfavourable*
├─Previous crop	host	host	*non host*
├─Pre-previous crop	host	*non host*	*non host*
├─Tillage after harvest of the previous crop	non-inversion tillage	non-inversion tillage	*inversion tillage*
└─Tillage after harvest of the pre-previous crop	non-inversion tillage	*inversion tillage*	*inversion tillage*
─Soil and climate	favourable	favourable	favourable
├─Soil	*neutral*	*neutral*	*neutral*
├─Climate	favourable	favourable	favourable
└─Interactions at the territory level	*neutral*	favourable	favourable
├─Beneficial sources	high	*neutral*	*neutral*
└─Primary inoculum sources	high	high	*neutral*

Figure 5. Example of simulation outputs for wheat obtained for three cropping systems (intensive, integrated and organic) in a given production situation (screenshot of the DEXi software). Three pests in interaction were taken into account in these simulations: eyespot, sharp eyespot and brown rust.

measures not present in Figure 3. For instance, the effect of cultivar mixtures or intercrops could be implemented, provided that the required knowledge is available.

The main breakthrough of IPSIM is to be able to handle complexity in a simple way. Input variables of the IPSIM models should be simple to provide. Most of these input variables will be static variables, except for weather variables that will be dynamic. The price to pay to handle the level of ecological complexity (such as defined by Li [51]) addressed by IPSIM is that IPSIM models are static. This is certainly not a problem to predict the consequences of technical options in a given production situation, but could hamper the linkage with dynamic models as suggested earlier. This limitation could easily be overcome by associating the level of final injury predicted by IPSIM models with generic dynamics. In order to do so, exponential, monomolecular, logistic, Gompertz, or Richards models [52] could be used with generic parameters chosen to represent the qualitative ordinal different injury levels predicted by IPSIM models.

The choice of qualitative variables to describe agroecosystems is relevant for several reasons. Firstly, farmers generally rely on a qualitative perception of their environment to make decisions. This suits the formalism of IPSIM. Secondly, because of the complexity of the system, few datasets are available to describe its components, i.e. the production situation, cropping practices and the injury profile. Using qualitative variables enables one to gather and use various existing datasets that were not acquired for the development of IPSIM models. For instance, datasets from diagnoses of commercial fields or even from experiments may not have used the same severity scale for a given disease. The use of qualitative classes allows data from different origins and/or with different precision to be combined. It is possible to associate interval classes with qualitative attributes. For instance, the 7 levels "nil"; "very low"; "low"; "medium"; "high"; "very high"; and "maximum" can be transformed into [0]; [0–20]; [20–40]; [40–60]; [60–80]; [80–100]; [100] intervals of percentage of diseased foliage respectively, if one wants to compare these outputs with observed severities of a disease for instance. Thus, data acquired on various scales can still be combined to strengthen the dataset used to estimate the predictive quality of the model or to improve the aggregating tables.

Table 5 presents the equivalence between features of models developed within the IPSIM framework and more common quantitative simulation models. Input attributes of IPSIM models can be nominal, ordinal or interval variables, unlike quantitative simulation models, which require only interval input variables. The state variables (aggregated attributes) of IPSIM models, including output variables, are ordinal. However, if need be, output variables of IPSIM models can be transformed into interval variables. This transformation can be performed by associating each possible ordinal value with a quantitative value (e.g. static final value of an injury level, or quantitative intervals) or with an injury dynamic. The relationship between variables is described by a tree in aggregative qualitative models, whereas quantitative models use equations. The DEXi software [40] provides a table with the respective weights of input and aggregated attributes on the value of the root node (main output). This table can be seen as an equivalent to a simple sensitivity analysis to input variables for quantitative models, prior to more detailed ones [53]. It is notable that IPSIM models have no parameters. The equivalents of parameters that specify relationships among variables in quanti-

tative models are the aggregating tables. The proportion of situations correctly simulated is a criterion that can be used to characterise the agreement between values simulated with an IPSIM model and observations. In addition, a non-parametric Wilcoxon signed rank test can be used to analyse whether the distribution of errors is significantly biased or not. These criteria can be seen as equivalent to common statistical criteria for quantitative models (Bias, Mean Absolute Error; Root Mean Square Error; Efficiency [54]). At last, methods specific to matched-pairs data with ordered categories can be used. In order to do so, various models comparing matched marginal distributions or analysing the joint distribution in a square contingency table can be applied [55].

The qualitative attributes of IPSIM models can lead to threshold effects. In order to cope with this limitation, a tool, named proDEX was developed to model uncertain expert knowledge [56]. This software offers the definition of probabilistic aggregating tables, where each combination of descendants' values maps to a probability distribution of the aggregated attributes, rather than a single value. In this approach, input values must be categorised prior to their use in the model. Since this process is time-consuming, proDEX allows categorisations to be part of the model definition and the inputs to be entered as interval variables. In combination with probabilistic aggregating tables, categorisations can be made to transform numerical values into probabilistic distributions, eliminating the problem of crisp interval boundaries. Eventually, the proDEX method could permit a useful extension of the modelling approach presented in this paper.

Finally, a website giving online access to all the functionalities of IPSIM is planned. This website will enable researchers, advisors, farmers and students to develop their own models for a wide range of crops.

Conclusion

We believe that IPSIM is a useful innovative modelling framework to help vertical and horizontal integrations for IPM. Its output attributes include nine generic injury profiles that are based on a two-level categorisation of the degree of endocyclism of harmful organisms. These nine injury profiles can be seen as a tool to perform cross-cutting typologies of agroecosytems for various types of crop (arable crops, vegetables, orchards, vineyards, Controlled Environment Agriculture), with regard to the main pests that have to be managed. IPSIM will generate new knowledge by combining various sources of information from experiments, diagnoses of commercial field, models, and expert panels in a simple way, despite the high ecological complexity of the system addressed. The associated companion paper provides a proof of concept of the proposed method for a single pest.

Author Contributions

Analyzed the data: JNA MHR. Contributed reagents/materials/analysis tools: JNA MHR. Wrote the paper: JNA MHR.

References

1. Fess TL, Kotcon JB, Benedito VA (2011) Crop breeding for low input agriculture: a sustainable response to feed a growing world population. Sustainability 3: 1742–1772.

2. Kropff MJ, Bouma J, Jones JW (2001) Systems approaches for the design of sustainable agro-ecosystems. Agricultural Systems 70: 369–393.

3. Kropff MJ, Teng PS, Rabbinge R (1995) The challenge of linking pest and crop models. Agricultural Systems 49: 413–434.

4. Birch ANE, Begg GS, Squire GR (2011) How agro-ecological research helps to address food security issues under new IPM and pesticide reduction policies for global crop production systems. Journal of Experimental Botany 62: 3251–3261.

5. Ferron P, Deguine JP (2005) Crop protection, biological control, habitat management and integrated farming. A review. Agronomy for Sustainable Development 25: 17–24.

6. Savary S, Willocquet L, Elazegui FA, Teng PS, Pham Van D, et al. (2000) Rice pest constraints in tropical Asia: characterization of injury profiles in relation to production situations. Plant Disease 84: 341–337.

7. Zadoks JC (1993) Modern crop protection: developments and perspectives; Zadoks JC, editor. Wageningen Netherlands: Wageningen Pers. ix +309 p.

8. Jeger MJ (2000) Bottlenecks in IPM. Crop Protection 19: 787–792.

9. Zadoks JC (1984) A quarter century of disease warning, 1958–1983. Plant Disease 68: 352–355.

10. Daamen RA, Wijnands FG, Vliet Gvd (1989) Epidemics of diseases and pests of winter wheat at different levels of agrochemical input. A study on the possibilities for designing an integrated cropping system. Journal of Phytopathology 125: 305–319.

11. Breman H, de Wit CT (1983) Rangeland Productivity and Exploitation in the Sahel. Science 221: 1341–1347.

12. Debaeke P, Munier-Jolain N, Bertrand M, Guichard L, Nolot JM, et al. (2009) Iterative design and evaluation of rule-based cropping systems: methodology and case studies. A review. Agronomy for Sustainable Development 29: 73–86.

13. Stockle CO, Martin SA, Campbell GS (1994) CropSyst, a cropping systems simulation model: water/nitrogen budgets and crop yield. Agricultural Systems 46: 335–359.

14. Brisson N, Mary B, Ripoche D, Jeuffroy MH, Ruget F, et al. (1998) STICS: a generic model for the simulation of crops and their water and nitrogen balances. I. Theory and parameterization applied to wheat and corn. Agronomie 18: 311–346.

15. Aubertot JN, Salam MU, Diggle AJ, Dakowska S, Jedryczka M (2006) SimMat, a new dynamic module of Blackleg Sporacle for the prediction of pseudothecial maturation of *L. maculans/L. biglobosa* species complex. Parameterisation and evaluation under Polish conditions. In: Koopmann B, Cook S, Evans N, Ulber B, editors. Bulletin OILB/SROP. 277–285.

16. Bergez JE, Colbach N, Crespo O, Garcia F, Jeuffroy MH, et al. (2010) Designing crop management systems by simulation. European Journal of Agronomy 32: 3–9.

17. Rabbinge R, Rijsdijk FH (1983) EPIPRE: a disease and pest management system for winter wheat, taking account of micrometeorological factors. Bulletin, OEPP 13: 297–305.

18. Zadoks JC (1988) EPIPRE: research, development and application of an integrated pest and disease management system for wheat. Bulletin SROP. 82–90.

19. Willocquet L, Savary S, Fernandez L, Elazegui FA, Castilla N, et al. (2002) Structure and validation of RICEPEST, a production situation-driven, crop growth model simulating rice yield response to multiple pest injuries for tropical Asia. Ecological Modelling 153: 247–268.

20. Willocquet L, Aubertot JN, Lebard S, Robert C, Lannou C, et al. (2008) Simulating multiple pest damage in varying winter wheat production situations. Field Crops Research 107: 12–28.

21. Savary S, Mille B, Rolland B, Lucas P (2006) Patterns and management of crop multiple pathosystems. European Journal of Plant Pathology 115: 123–138.

22. Robin MH, Colbach N, Lucas P, Monfort F, Cholez C, et al. (2013) Injury Profile SIMulator, a hierarchical aggregative modelling framework to predict an injury profile as a function of cropping practices, and the abiotic and biotic environment. II. Proof of concept: design of IPSIM-Wheat-Eyespot. PLoS ONE. In press.

23. Sebillotte M (1990) Systèmes de culture, un concept opératoire pour les agronomes, L. Combe, D. Picard, Editors, *Les Systèmes de Culture*, INRA, Paris (1990), 165–196.

24. Meynard JM, Dore T, Lucas P (2003) Agronomic approach: cropping systems and plant diseases. Comptes Rendus Biologies 326: 37–46.

25. Oblak L, Novak B, Lipuscek I, Kropivsek J (2007) Launching work orders into production of wood enterprise with the multi-criteria decision-making method. Zbornik Gozdarstva in Lesarstva: 33–39.

26. Rozman C, Pazek K, Bavec F, Bavec M, Turk J, et al. (2006) A multi-criteria analysis of spelt food processing alternatives on small organic farms. Journal of Sustainable Agriculture 28: 159–179.

27. Griffiths BS, Ball BC, Daniell TJ, Hallett PD, Neilson R, et al. (2010) Integrating soil quality changes to arable agricultural systems following organic matter addition, or adoption of a ley-arable rotation. Applied Soil Ecology 46: 43–53.

28. Rozman C, Potocnik M, Pazek K, Borec A, Majkovic D, et al. (2009) A multi-criteria assessment of tourist farm service quality. Tourism Management 30: 629–637.

29. Ars MS, Bohanec M (2010) Towards the ecotourism: a decision support model for the assessment of sustainability of mountain huts in the Alps. Journal of Environmental Management 91: 2554–2564.

30. Sadok W, Angevin F, Bergez JE, Bockstaller C, Colomb B, et al. (2009) MASC, a qualitative multi-attribute decision model for ex ante assessment of the sustainability of cropping systems. Agronomy for Sustainable Development 29: 447–461.

31. Pelzer E, Fortino G, Bockstaller C, Angevin F, Lamine C, et al. (2012) Assessing innovative cropping systems with DEXiPM, a qualitative multi-criteria assessment tool derived from DEXi. Ecological Indicators 18: 171–182.

32. Pazek K, Rozman C, Bavec F, Borec A, Bavec M (2010) A multi-criteria decision analysis framework tool for the selection of farm business models on organic mountain farms. Journal of Sustainable Agriculture 34: 778–799.

33. Pazek K, Rozman C (2007) The decision support system for supplementary activities on organic farms. Agricultura (Slovenia) 5: 15–20.

34. Bohanec M, Messean A, Scatasta S, Angevin F, Griffiths B, et al. (2008) A qualitative multi-attribute model for economic and ecological assessment of genetically modified crops. Ecological Modelling 215: 247–261.

35. Pazek K, Rozman C, Irgolic A, Turk J (2012) Multicriteria decision model for evaluating less-favoured areas for agricultural production. 47th Croatian and 7th International Symposium on Agriculture, Opatija, Croatia, 13–17 February 2012 Proceedings: 222–226.

36. Vindis P, Stajnko D, Berk P, Lakota M (2012) Evaluation of energy crops for biogas production with a combination of simulation modeling and DEX-i multicriteria method. Polish Journal of Environmental Studies 21: 763–770.

37. Pazek K, Rozman C, Pavlovic V, Cerenak A, Pavlovic M (2009) The multi-criteria decision model aid for assessment of the hop cultivars (*Humulus lupulus* L.). In: Florijancic T, Luzaic R, editors. Zbornik Radova 44 Hrvatski i 4 Medunarodni Simpozij Agronoma, Opatija, Hrvatska, 16–20 Veljace 2009. Osijeku, Hrvatska: Poljoprivredni Fakultet Sveucilista Josipa Jurja Strossmayera u Osijeku. 360–364.

38. Pazek K, Rozman C, Cejvanovic F, Par V, Borec A, et al. (2005) Multi attribute decision model for orchard renewal - case study in Bosnia and Herzegovina. Agricultura (Slovenia) 3: 13–20.

39. Tchamitchian M, Collange B, Navarrete M, Peyre G (2011) Multicriteria evaluation of the pathological resilience of soil-based protected cropping systems. In: Dorais M, editor. Acta Horticulturae. Leuven, Belgium: International Society for Horticultural Science (ISHS). 1239–1246.

40. Bohanec B (2003) Decision support. In: Mladeniæ D, Lavraè N, Bohanec M, Moyle S (Eds.). Data mining and decision support: Integration and collaboration, Kluwer Academic Publishers 23–35.

41. Delmotte S, Gary C, Ripoche A, Barbier JM, Wery J (2009) Contextualization of on farm ex-ante evaluation of the sustainability of innovative cropping systems in viticulture, using a multiple criteria assessment tool (DEXi). Wageningen University and Research Centre. 180–181.

42. Colbach N, Lucas P, Cavelier N, Cavelier A (1997) Influence of cropping system on sharp eyespot in winter wheat. Crop Protection 16: 415–422.

43. Reinecke P, Fehrmann H (1979) Rhizoctonia cerealis van der Hoeven on cereals in the Federal Republic of Germany. Zeitschrift fur Pflanzenkrankheiten und Pflanzenschutz 86: 190–204.

44. Cavelier N, Lucas P, Boulch G (1985) Evolution of the parasite complex *Rhizoctonia cerealis* Van der Hoeven and *Pseudocercosporella herpotrichoides* (Fron) Deighton, fungi attacking the stem base of cereals. Agronomie 5: 693–700.

45. Kolmer JA (2005) Tracking wheat rust on a continental scale. Current Opinion in Plant Biology 8: 441–449.

46. Bolton MD, Kolmer JA, Garvin DF (2008) Wheat leaf rust caused by *Puccinia triticina*. Molecular Plant Pathology 9: 563–575.

47. Lançon J, Wery J, Rapidel B, Angokaye M, Gerardeaux E, et al. (2007) An improved methodology for integrated crop management systems. Agronomy for Sustainable Development 27: 101–110.

48. Ould-Sidi MM, Lescourret F (2011) Model-based design of integrated production systems: a review. Agronomy for Sustainable Development 31: 571–588.

49. Pelzer E, Fortino G, Bockstaller C, Angevin F, Lamine C, et al. (2012) Assessing innovative cropping systems with DEXiPM, a qualitative multi-criteria assessment tool derived from DEXi. Ecological Indicators 18: 171–182.

50. Sundman KE (1912) Mémoire sur le problème des trois corps, Acta Mathematica 36: 105–179.

51. Li BL (2010) Editorial. Ecological Complexity 7: 421–422.

52. Madden LV (2006) Botanical epidemiology: some key advances and its continuing role in disease management. European Journal of Plant Pathology 115: 3–23.

53. Carpani M, Bergez JE, Monod H (2012) Sensitivity analysis of a hierarchical qualitative model for sustainability assessment of cropping systems. Journal Environmental Modelling & Software 27–28: 15–22.

54. Wallach D (2011) Crop model calibration: a statistical perspective. Agronomy Journal 103: 1144–1151.

55. Agresti A (2010) Matched-pairs data with ordered categories. In: Analysis of ordinal categorical data. Second edition. Wiley& Sons, Inc. Hoboken, New-Jersey, USA. 225–261.

56. Žnidaršič M, Bohanec M, Zupan B (2006) proDEX – A DSS tool for environmental decision-making. Environmental Modelling & Software 21: 1514–1516.

9

Spatial Variation in Carbon and Nitrogen in Cultivated Soils in Henan Province, China: Potential Effect on Crop Yield

Xuelin Zhang[1]*, Qun Wang[1], Frank S. Gilliam[2], Yilun Wang[1], Feina Cha[3], Chaohai Li[1]

1 The Incubation Base of the National Key Laboratory for Physiological Ecology and Genetic Improvement of Food Crops in Henan Province, Zhengzhou, China; Agronomy College of Henan Agricultural University, Zhengzhou, China, 2 Department of Biological Sciences, Marshall University, Huntington, West Virginia, United States of America, 3 Meteorological Bureau of Zhengzhou, Zhengzhou, China

Abstract

Improved management of soil carbon (C) and nitrogen (N) storage in agro-ecosystems represents an important strategy for ensuring food security and sustainable agricultural development in China. Accurate estimates of the distribution of soil C and N stores and their relationship to crop yield are crucial to developing appropriate cropland management policies. The current study examined the spatial variation of soil organic C (SOC), total soil N (TSN), and associated variables in the surface layer (0–40 cm) of soils from intensive agricultural systems in 19 counties within Henan Province, China, and compared these patterns with crop yield. Mean soil C and N concentrations were 14.9 g kg^{-1} and 1.37 g kg^{-1}, respectively, whereas soil C and N stores were 4.1 kg m^{-2} and 0.4 kg m^{-2}, respectively. Total crop production of each county was significantly, positively related to SOC, TSN, soil C and N store, and soil C and N stock. Soil C and N were positively correlated with soil bulk density but negatively correlated with soil porosity. These results indicate that variations in soil C could regulate crop yield in intensive agricultural systems, and that spatial patterns of C and N levels in soils may be regulated by both climatic factors and agro-ecosystem management. When developing suitable management programs, the importance of soil C and N stores and their effects on crop yield should be considered.

Editor: Dafeng Hui, Tennessee State University, United States of America

Funding: This study was supported by grants from Henan Science and Technology Department of China under the Key Research Project (30200051). The funder had no role in study design, data collection and analysis, decision to publish, or preparation of the manuscript.

Competing Interests: The authors have declared that no competing interests exist.

* Email: xuelinzhang1998@163.com

Introduction

Safeguarding food security and ensuring sustainable development are two fundamental goals of intensive agriculture in China [1,2]. Increasing soil C and N sequestration while reducing C and N emissions from agricultural fields are important aspects of sustainable farming and these goals can be achieved through improvement in soil quality [1,3]. This requires a better understanding of the functional relationship between crop yield and soil organic C and N stores.

Indeed, variations in soil C and N stores may closely regulate crop yield, although published data on the relationship between these parameters are inconsistent. Some studies have reported a positive correlation between soil C and N and crop yield [4,5], whereas other studies have found no significant relationship between these parameters [6,7]. Lal (2006) reported that the relationship between soil organic C and crop yield may vary between patterns that are sigmoidal, linear, or exponential [8]. Clearly, the existence of such variability warrants further investigation.

Soil C and N stores in crop lands, especially in the topsoil layer, are potentially greatly affected by human activity; thus, understanding the spatial pattern of soil C and N stores on a regional scale is crucial to developing a management strategy for improving soil fertility [1,2]. Spatial variation in soil C and N stores in agro-ecosystems has been widely reported [9,10,11], including from the northern [12,13], eastern [14], and southern [15,16] regions of China. Since these reports from China were based on two national surveys from 1960 and 1983, such data may have limited use in helping to develop management strategies based on current practices [17]. Therefore, in order to better understand the spatial patterns and their relationship to crop yield, it is necessary to update regional soil organic C and N information with contemporary measurements, especially for intensively-used crop land.

Henan Province is the second largest area of crop production in China (China National Bureau of Statistics). To produce an adequate supply of food for the domestic population, unsustainable production methods have often been used in this province. Historically, intensive production based on an annual wheat-maize system has been used to achieve high crop yield. This practice, however, has resulted in badly degraded agricultural soils, causing erosion and a loss of good soil structure. More than 600 kg N ha^{-1} annually has been applied in this production area, resulting in an increase in soil acidity [18]. Based on the determination that crop yields in China will need to increase from 50 billion in 2010 to 65 billion kg in 2020, the provincial crop lands in Henan Province will continue to play an important role in food production. Such goals create the challenge of improving soil quality, enhancing soil fertility, and mitigating C and N loss, while achieving food security and practicing sustainable agriculture. A better understanding of the spatial variability of soil organic C and N, and their relationship to crop yield, should help to develop management practices that are designed to meet this challenge [1,19].

The objective of the present study was to characterize the spatial distribution of C and N stores in intensively cultivated counties within the Henan Province of China and to determine the relationship between crop yield and soil organic C and N.

Materials and Methods

Statement: We have field permits for sampling soil in each of the field sites within each county of Henan Province, China. All of the sampling sites are privately owned, and there was no potential impact on any endangered or protected species among these sampling sites.

Study site

The study was carried out in 19 counties within Henan Province, located in central China (Figure 1). Map data were obtained from the National Geomatics Center of China (http://ngcc.sbsm.gov.cn/) using ArcGIS software. As of 2009, the human population of Henan was about 9.9×10^7 persons. The Province is

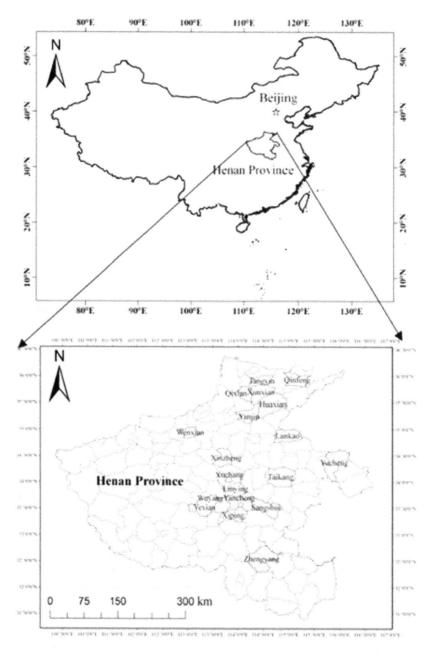

Figure 1. Map of China (top) showing location of Henan Province and counties (bottom) within Henan Province used in this study.

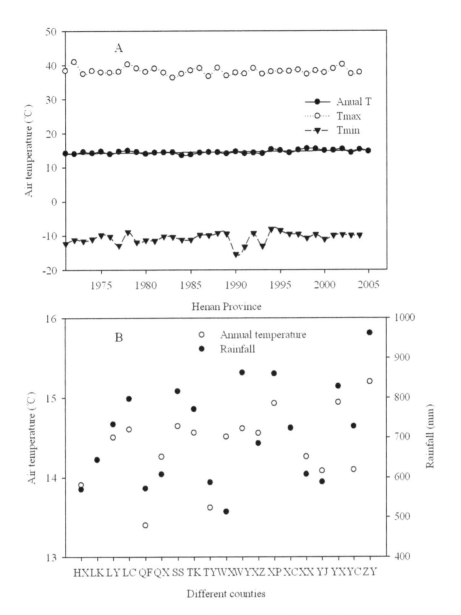

Figure 2. Average annual, maximum, and minimum temperature from 1971 to 2004 in Henan Province, China (A), and (B) average temperature and rainfall from 1975 to 2006 in different counties within Henan Province, China (B). See Table 1 for key to county name abbreviations. All these counties were arranged in English alphabetical order.

approximately 167,000 km^2 in land area, lying within the monsoonal temperate zone. It has a cultivated land area of 79, 260 km^2 for the production of wheat and maize. There are three dominant soil types in Henan Province: Yellow-cinnamon soil (Eutric Cambisols in FAO taxonomy), Sajiang black soil (Eutric Vertisols/Gleyic Cambisol), and Fluvo-aquic soil (Fluvisols in FAO taxonomy) [20]. Mean annual precipitation ranges from 400 to 1000 mm among the counties of the study, with ~70% of it occurring from July to September; mean annual temperature ranges from 13.6 to 15°C (Figure 2). Cultivated agricultural fields are the predominant land use, representing 60% of the total land area in Henan Province. A double cropping system of winter wheat (early October-early June) and maize (mid-June–later September) is the most common planting system used in this region.

Collection of crop yield and soil sampling and analysis

Data on total crop production (including wheat, maize and millet) and wheat yield from 1978–2009 (Figure. 3A) were obtained from the Henan Statistical Yearbook 2010 (13–17) (http://www.ha.stats.gov.cn/hntj/index.htm). Annual yield data for winter wheat and total crop production in 2009 were also obtained from Henan Statistical Yearbook 2010 (29-7) and the Agricultural Bureau of each of the 19 counties in which soil sampling took place (Figure. 3B). These counties, along with basic climatic information, are listed in Table 1. Climatic data of each county were obtained from Meteorological Bureau of Zhengzhou. All counties will be referred to by the two-letter codes presented in Table 1.

The 19 counties were selected as representative of the main agro-ecosystems of Henan Province. Soil samples were collected during June 1–15, 2009 following the wheat harvest but prior to the sowing of maize. Six representative, replicate field plots,

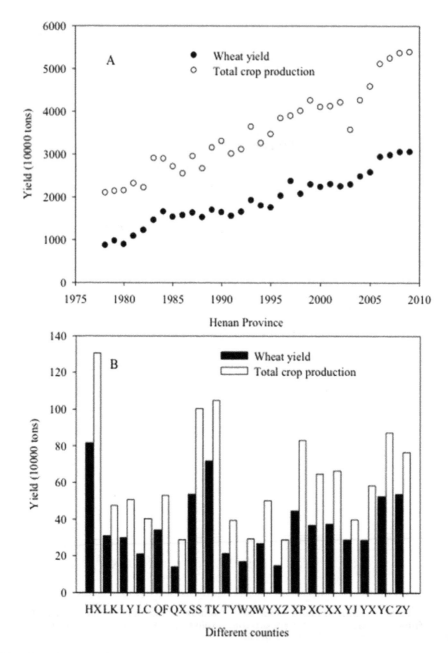

Figure 3. Wheat yield and total crop production (including wheat, maize, millet,) in Henan Province from 1978–2009 (A), and wheat yield and total crop production in different counties within Henan Province in 2009 (B). See Table 1 for key to county name abbreviations.

located at least 6 km apart, were selected within each county based on four criteria: (1) the field plots had been continuously cultivated for at least 30 yr with a native variety, (2) the cropland area was located within 5 km of native vegetation with a similar landscape, soil type and texture, and a relatively flat terrain, and (3) all of the sampling sites are privately owned, and (4) there was no potential impact on any endangered or protected species in the sampling site. Geographic coordinates of each sampling site was recorded by handed GPS of Magellan eXplorist 210(USA), and all of these data were attached in the supporting information.

Sample areas of ∼1300 m^2 were established in each plot, with sixteen sampling points taken at random in each of two layers (0–

20 cm and 20–40 cm) using a 70 mm - diameter auger. All of the soil samples taken at each layer within a sample plot were mixed together and treated as one sample to represent the value of the plot, yielding 114 soil samples at each layer.

Residual plant material was removed from the soil samples after the samples were air-dried at room temperature. The soil samples were then ground to pass a 2 mm sieve, and a portion of the ground sample was subsequently ground again in a porcelain mortar in order to pass through a 0.15–mm sieve. Organic C and total N measurements were obtained from the twice-ground soil samples. Soil organic C (SOC) was measured using a modified Mebius method. Briefly, 0.1 g soil samples were digested for 5 min

Table 1. Basic geographic coordinates for each county, along with climate data for 19 counties within Henan Province, China.

County	Latitude	Longitude	Sea level (m)	Average Temp (°C)	Rainfall (mm)	Sunshine (h)
Huaxian (HX)	35°44′	114°28′	68	13.9	570.0	2060.9
Lankao (LK)	34°55′	114°46′	70	14.2	644.5	2183.2
Linying (LY)	33°55′	113°55′	63	14.5	732.9	2141.3
Luoheyancheng (LC)	33°35′	114°02′	65	14.6	797.2	2273.0
Qingfeng (QF)	35°53′	115°06′	51	13.4	571.9	2209.1
Qixian (QX)	35°35′	114°12′	72	14.3	607.5	2133.8
Shangshui (SS)	33°39′	114°34′	52	14.6	815.8	1902.0
Taikang (TK)	34°05′	114°50′	53	14.6	770.9	1998.4
Tangyin (TY)	36°03′	114°19′	103	13.6	587.1	2159.3
Wenxian (WX)	35°01′	113°03′	109	14.5	513.2	2302.2
Wuyang (WY)	33°36′	113°32′	77	14.6	862.3	2060.4
Xinzheng (XZ)	34°30′	113°39′	159	14.6	684.6	2058.7
Xiping (XP)	33°29′	113°59′	65	14.9	859.8	2084.7
Xuchang (XC)	34°04′	113°52′	72	14.6	722.7	1959.8
Xunxian (XX)	35°40′	114°32′	59	14.3	607.5	2133.8
Yanjin (YJ)	35°13′	114°11′	69	14.1	588.0	2287.8
Yexiang (YX)	33°38′	113°21′	88	14.9	827.8	1972.4
Yucheng (YC)	34°25′	115°52′	46	14.1	727.3	2244.6
Zhengyang (ZY)	32°37′	114°24′	70	15.2	961.8	2004.4

Note: Counties are arranged in English alphabetical order.

with 5 mL of 1N $K_2Cr_2O_7$ and 10 mL of concentrated H_2SO_4 at 150°C, followed by titration of the digests with standardized $FeSO_4$. Total soil N (TSN) was measured using a modified Kjeldahl wet digestion procedure and a Tector Kjeltec System 1026 distilling unit. Soil available N was analyzed using a micro-diffusion technique after alkaline hydrolysis (1.8 mol L^{-1} NaOH). The Olsen method was used to determine available soil phosphorus (P), and available soil potassium (K) was measured in 1 mol L^{-1} NH_4OAc extracts by flame photometry (Table 2).

Three sampling points were used to determine soil bulk density in each plot. Samples were collected separately from four layers within a depth of 0–40 cm in each sampling point. Soil bulk density was measured using 100-cm^3 soil cores obtained from the four layers. Soil porosity was calculated from soil bulk density and specific gravity, with any stone material removed and not considered in bulk density calculations.

Calculation of soil organic C and N stores and SOC and TSN

Total soil organic C store (TSOCS) and total soil N stores (TSNS) at 0–40 cm depth were calculated as follows:

$$TSOC(g.m^{-2}) =$$
Soil organic $C(g.kg^{-1}) \times$ soil bulk density $(g.cm^{-3}) \times$ sampling depth(cm)

$$TSN(g.m^{-2}) =$$
Soil total $N(g.kg^{-1}) \times$ soil bulk density $(g.cm^{-3}) \times$ sampling depth(cm)

Given the cultivated area, the total cultivated topsoil (0–40 cm) C and N stocks of each county were estimated by the equation:

$$CS = \sum area_i \times TSOC$$
$$NS = \sum area_i \times TSN$$

where *area* is the given total cultivated area of each county, and CS and NS are C and N stocks, respectively. SOC and TSN were means of six sampling sites of each county.

Statistics

Analysis of variance was used to assess the significance of location (county) on soil C and N concentration and storage; means were compared using Duncan's multi-range test at $\alpha = 0.05$. Linear regression was used to determine the relationships between C and N stock versus wheat and total crop production. Principle components analysis was used to assess patterns of similarity/dissimilarity among counties with respect to several environmental variables [21]. All statistical analyses were performed using SPSS 10.0 (Chicago IL, USA).

Results

Wheat yields increased more than 250% from 1978 to 2009 while total annual crop production in Henan Province increased from 21 to 54 million tons over the same time period (Figure 3A). Wheat yield varied from 143 to 729 thousand tons among the different counties in 2009 (Figure 3B).

The absolute value of SOC concentration in the top 40 cm of soil varied from 8.13 to 27.89 g kg^{-1} among the 19 counties in 2009 (Table 2) while TSN concentration varied from 0.84 to 2.2 g kg^{-1}. Soil C/N varied from 6.4 to 20 (Table 2). Soil organic C stores (TSOCS) in the 0–40 cm soil layer varied from 2,322 g m^{-2} to 8,038 g m^{-2}, whereas total N stores (TSNS) varied from 221 to

Table 2. Spatial variation in soil (0–40 cm depth) properties, soil organic C (SOC), total soil N (TSN) concentration (g kg^{-1}), and C/N in the 0–40 cm soil layer in 19 counties within Henan province, China.

	Alkaline-extractable N (mg kg^{-1})	Olsen-extractable P (mg kg^{-1})	NH$_4$OAc-extractable K (mg kg^{-1})	Bulk density (g cm^{-3})	Soil porosity (%)	SOC (g kg^{-1})	TSN (g kg^{-1})	C/N
HX	48.9±3.2abc	1.8±0.7a	80.1±9.2abc	1.44±0.03de	38.3±1.4abcd	12.4±0.9abc	1.4±0.05abcd	8.8±0.7abc
LK	56.5±2.4abcd	7.6±1.9ab	71.9±11.9abc	1.42±0.02bcde	40.7±1.0bcdef	11.2±0.7ab	1.4±0.09abcd	7.9±0.5a
LY	49.9±3.1abc	4.2±0.7a	145.4±24.1ef	1.36±0.02abc	41.7±1.4cdefg	15.5±1.0bcd	1.1±0.07a	14.2±0.2efg
LC	49.0±1.6abc	11.5±2.1abc	103.6±8.9abcde	1.39±0.02bcd	38.7±1.6abcd	14.6±1.4abcd	1.4±0.14bcd	10.5±1.1abcd
QF	47.5±2.8abc	10.9±4.9abc	71.7±6.3abc	1.39±0.01bcd	41.8±0.4cdefg	11.8±0.6ab	1.4±0.08abcd	8.7±0.8abc
QX	51.9±3.7abc	6.3±2.2ab	82.1±10.8abc	1.44±0.01de	38.7±0.3abcd	21.1±1.8f	1.5±0.21cd	16.2±3.1g
SS	45.1±1.9abc	11.7±3.7abc	169.3±33.9f	1.35±0.02abc	37.9±1.3abcd	14.5±0.9abcd	1.3±0.08abc	11.4±0.8bcde
TK	59.3±4.5cd	17.7±8.2bcd	140.6±23.9def	1.35±0.02ab	41.5±1.5cdef	13.4±1.0abcd	1.1±0.05ab	11.9±0.7cde
TY	59.2±2.4cd	6.4±2.6ab	110.9±12.8bcde	1.45±0.03de	38.6±1.3abcd	15.0±0.4abcd	1.7±0.09de	8.9±0.5abc
WX	56.8±3.5abcd	11.0±1.9abc	82.1±7.7abc	1.3±0.03a	43.6±1.5fg	17.1±1.9de	1.5±0.11bcd	11.5±0.6bcde
WY	47.9±1.1abc	10.3±2.9ab	84.2±10.5abc	1.38±0.02bcd	36.8±1.2ab	14.9±1.6abcd	1.6±0.09cd	9.5±0.6abc
XZ	72.9±7.4e	7.8±2.1ab	95.8±18.9abcd	1.47±0.02e	38.3±0.7abcd	16.1±1.6cd	1.1±0.12ab	14.5±1.1efg
XP	72.2±4.2e	27.3±5.8d	117.5±17.7cde	1.43±0.01cde	37.8±1.2abc	19.9±1.9ef	1.3±0.07abc	15.4±1.3fg
XC	43.2±3.4a	17.2±5.8bcd	66.2±9.6ab	1.42±0.03bcde	42.9±0.9efg	16.6±2.3cde	1.3±0.18abc	12.7±0.7def
XX	53.1±4.2abcd	5.1±0.5ab	89.9±3.0abc	1.41±0.02bcde	35.3±1.0a	14.8±0.6abcd	1.9±0.11e	7.9±0.3a
YJ	49.5±2.9abc	12.8±4abc	77.0±9.6abc	1.39±0.02bcd	38.2±1.0abcd	10.8±0.6a	1.3±0.07abc	8.2±0.4ab
YX	44.5±2.1ab	22.9±5.9cd	89.5±11.4abc	1.3±0.04a	45.5±1.9g	15.4±0.9bcd	1.2±0.07ab	13.6±1.1defg
YC	66.4±11.3de	3.5±1.1a	59.9±8.3a	1.44±0.02de	39±1.3abcde	14.8±0.9abcd	1.1±0.08ab	13.1±0.2defg
ZY	58.7±2.4bcd	6.3±1.4ab	73.8±5.7abc	1.35±0.02ab	41.9±0.9defg	12.7±0.3abc	1.5±0.06bcd	8.8±0.5abc

Different letters indicate significant differences ($p = 0.05$) among the 19 counties. Counties are arranged in English alphabetical order.

Table 3. Total C (TSOCS) and N (TSNS) stores in the surface soil layer (0–40 cm) of soils in 19 counties in Henan Province, China.

	C store (g m^{-2})	N store (g m^{-2})
HX	3541±261.8 abcd	410.1±18.9 cde
LK	3118.9±189.3 ab	399.7±23.6 bcde
LY	4106.7±294.1 abcd	290.4±22.2 a
LC	4023.8±372 abcd	398.6±35.6 bcde
QF	3229.7±140 abc	381.8±22.3 abcd
QX	5977.9±524.3 e	429.1±63.1 de
SS	3881.8±219 abcd	348.3±24.0 abcd
TK	3605.1±328.1 abcd	303.4±19.2 ab
TY	4300.4±98.5 bcd	494.3±24.9 ef
WX	4396.2±451.9 cd	379.4±23.4 abcd
WY	4081.3±455.2 abcd	429.3±25.3 de
XZ	4528.2±516.5 d	320.3±40.1 abc
XP	5709.2±582.9 e	369.8±21.9 abcd
XC	4614.5±609.9 d	366.3±47.7 abcd
XX	4323.8±224.7 bcd	558.2±35.1 f
YJ	3072.5±178.3 a	378.7±19.2 abcd
YX	4190.7±241.3 abcd	315.4±17.4 abc
YC	3926.2±304.8 abcd	299.3±26.3 ab
ZY	3413.4±104.3 abcd	398.1±21.7 bcde

Counties are arranged in English alphabetical order.

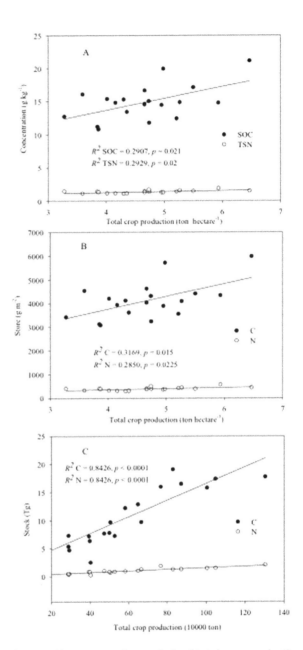

Figure 4. Linear regression analysis of total crop production in each county (ton ha^{-1}) with SOC and TSN (A) and with soil C and N store (0–40 cm) (B), and the total crop production of each county (10000 ton) with their soil C and N stock (C) (n = 19).

659 g m^{-2}. The highest value was in XX County and the lowest in LY County in N reserves (Table 3).

Linear regression analysis indicated that total crop production was significantly and positively correlated with SOC and TSN (Figure 4A), soil C and N store (Figure 4B), and soil C and N stocks (Figure 4C). Soil bulk density was significantly and positively correlated with soil N concentration ($r = 0.25$, $p = 0.008$, n = 114), soil C ($r = 0.21$, $p = 0.03$, n = 114) and N store ($r = 0.43$, $p = 0.001$, n = 114). While soil porosity was significantly and negatively correlated with soil N concentration ($r = -0.19$, $p = 0.05$, n = 114), soil C ($r = -0.25$, $p = 0.007$, n = 114) and N store ($r = -0.32$, $p = 0.001$, n = 114).

Principle components analysis revealed that Axis 1, which explained 98% of the variation in all data (eigenvalue = 0.98), was highly correlated with soil C, whereas Axis 2, explaining 1% of the variation (eigenvalue = 0.09), was highly correlated with soil N. Thus, counties such as QX and XP located highly positive on Axis 1 with high levels of soil C, but other counties, such as LK, YJ, and QF, occupied positions toward the negative end of Axis 1 with low soil C (Figure 5).

Discussion

Potential influences on crop yield

It is notable that 14 environmental (e.g., mean annual temperature and precipitation –Table 1) and soil variables (including extractable nutrients-Table 2) examined in our analysis of the data from the 19 counties in Henan Province were correlated with either wheat or total crop yield (data not shown), and total crop production were significantly, positively related to SOC and TSN, soil C and N store, and soil C and N stock

(Figure 4). Part of this is likely related to the highly integrated nature of the measures of C and N stocks, i.e., their calculations combine soil concentrations of C and N, soil bulk density, sampling depth, and area of cultivation. However, all of these have been shown to directly influence crop performance. For example, increases in soil C have been shown to increase crop yield in other studies. Lal (2004, 2006) reported increases in yield from 20 to 70 kg ha^{-1} and 10 to 300 kg ha^{-1} for wheat and maize, respectively, following increases of 1 MT of C in agricultural soils in Africa [1,8]. Similarly, loss of soil C has been shown to decrease yield in agricultural soils of Canada and the U.S. [4,5].

Soil C-mediated increases of crop yields also may arise from improvements in soil structure and available water-holding

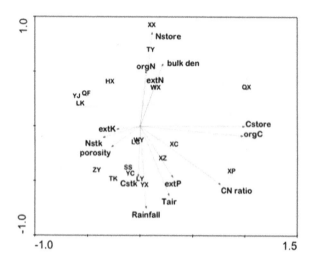

Figure 5. Principle components analysis of environmental and soil variables for agricultural soils in 19 counties within Henan Province. Length of arrows is directly proportional to their importance in explaining spatial patterns in the counties. Direction of the arrows indicates increasing values. Thus, the x-axis is primarily a gradient in soil C, whereas the y-axis is primarily a gradient in soil N and rainfall and secondarily a gradient in soil N. See Table 1 for key to county name abbreviations.

capacity. Enhanced soil structure, via increased soil C, generally arises from several processes, including increasing stability of soil aggregates [22,23,24]. As a result of the increased stability of the aggregates, soils become less prone to crusting, compaction, and erosion [25,26,28]. Emerson (1995) demonstrated that an increase of 1 g of soil organic matter (~50% of which is C) can increase available soil moisture by up to 10 g [27], which is enough to maintain crop growth between periods of rainfall of 5 to 10 days [8].

Spatial variation in cultivated soils

In this study, soil organic C concentration averaged 14.9 g kg^{-1} and total N averaged 1.4 g kg^{-1} in the 0–40 cm layer across all sites, while soil C and N stores averaged 4.1 kg C m^{-2} and 0.38 kg N m^{-2}, respectively. These values are comparable to published values from other regions of China, including 9–15 g C kg^{-1} and 1.2–1.8 g N kg^{-1} in northern China [12,29], and 16.1 g C kg^{-1} and 1.04 g N kg^{-1} in eastern and southern China [14,16,30]. Liu et al. (2011) reported soil C stores of 4.57 kg C m^{-2} in the Loess Plateau region in northwestern China [13].

Principal components analysis separated the 19 counties primarily along a gradient in soil C, with counties LK, YJ, QF, ZY, HX, and TK (mean soil C = 12.1 g C kg^{-1}) toward the lower end and XP and QX (mean soil C = 20.5 g C kg^{-1}) toward the

upper end of Axis 1, which accounted for nearly 80% of the variation in soil and environmental data (Figure 5). Spatial variation in soil organic C in agricultural systems can be influenced by several factors, including microclimate, soil type, topography, and especially human activity [31].

Spatial variation in soil N was essentially orthogonal to that of soil C. This was surprising since typically, the two are highly correlated in terrestrial ecosystems [32]. As a result, the secondary gradient (i.e., Axis 2) was one of soil N, with counties TK, YC, SS, LY, YX, and XP (mean soil N = 1.15 g N kg^{-1}) located toward the lower end of Axis 2 (accounting for <10% of variation) and XX and TY (mean soil N = 1.81 g N kg^{-1}) located toward the upper end of Axis 2 (Figure 5). Although C and N are often correlated through their organic forms in plant detritus, spatial variation of N in soils of agro-ecosystems can also be greatly influenced by the extensive use of N fertilizers.

Management methods used in crop production systems, including tillage practices and fertilizer use, can affect soil C and N on broad spatial scales, including that of an entire Province [33]. Over the course of repeated seasons of crop growth in Henan Province, agricultural fields are repeatedly subjected to soil tillage, planting, fertilization, irrigation, and harvest, all of which potentially influence soil C and N stores [30,34]. In contrast, Zhang et al. (2012) reported that raised-bed planting, a viable alternative to conventional tillage, can significantly enhance the yield of summer maize while simultaneously improving soil structure, as well as the structure and function of microbial communities essential to the quality of agricultural soils [22].

Results presented in the current study underscore the complexity of factors that can impact agricultural soils and their ability to produce crops to meet the ever-increasing demand in China resulting from population growth. Some of the spatial pattern exhibited in ordination space (Figure 5) is clearly related to regional factors, such as microclimate. For example, WY and LC are adjacent to each other in Henan Province (Figure 1) and are also closely clustered in ordination space, indicating that they are very similar with respect to environmental and soil characteristics. XP and SS, however, are also adjacent counties; yet occur distant from each other in ordination space, indicating great dissimilarity in environmental and soil factors. Agronomists should take into account the large spatial variability in important components of the soils in Henan Province, especially in the variation of soil C and N, when considering appropriate agronomic management practices.

Author Contributions

Conceived and designed the experiments: XLZ QW CHL. Performed the experiments: XLZ QW YLW. Analyzed the data: XLZ QW FSG. Contributed reagents/materials/analysis tools: XLZ QW YLW FNC. Contributed to the writing of the manuscript: XLZ FSG CHL.

References

1. Lal R (2004) Soil carbon sequestration impacts on global climate change and food security. Science 304: 1623–1627.
2. Liu DW, Wang ZM, Zhang B, Song KS, Li XY, et al. (2006) Spatial distribution of soil organic carbon and analysis of related factors in croplands of the black soil region, Northeast China. Agriculture, Ecosystems and Environment 113: 73–81.
3. Smith WN, Desjardins RL, Pattey E (2000) The net flux of carbon from agricultural soils in Canada 1970–2010. Global Change Biology 6: 557–568.
4. Bauer A, Black AL (1994) Quantification of the effect of soil organic matter content on soil productivity. Soil Science Society of America Journal 58: 185–193.
5. Larney FJ, Janzen HH, Olson BM, Lindwall CW (2000) Soil quality and productivity response to simulated erosion and restorative amendments. Canadian Journal of Soil Science 80: 515–522.

6. Hairiah K, Van Noordwijk M, Cadisch G (2000) Crop yield, C and N balance of the three types of cropping systems on an Ultisol in northern Lampung. Netherland Journal of Agricultural Science 48: 3–17.
7. Duxbury JM (2001) Long-term yield trends in the rice-wheat cropping system: results from experiments in Northwest India. Journal of Crop Production 3: 27–52.
8. Lal R (2006) Enhancing crop yields in the developing countries through restoration of the soil organic carbon pool in agricultural lands. Land Degradation and Development 17: 197–209.
9. Batjes NH (2002) Carbon and nitrogen stocks in the soils of Central and Eastern Europe. Soil Use and Management 18: 324–329.

10. Maia SMF, Ogle SM, Cerri CC, Cerri CEP (2010) Changes in soil organic carbon storage under different agricultural management systems in the Southwest Amazon Region of Brazil. Soil and Tillage Research 106: 177–184.

11. Piao SL, Fang JY, Ciais P, Peylin P, Huang Y, et al. (2009) The carbon balance of terrestrial ecosystems in China. Nature 458, doi:10.1038/nature 07944.

12. Wang ZM, Zhang B, Song KS, Liu DW, Ren CY (2010) Spatial variability of soil organic carbon under maize monoculture in the Song-Nen plain, Northeast China. Pedosphere 20: 80–89.

13. Liu ZP, Shao MA, Wang YQ (2011) Effect of environmental factors on regional soil organic carbon stocks across the Loess Plateau region, China. Agriculture, Ecosystems and Environment 142: 184–194.

14. Liao QL, Zhang XH, Li ZP, Pan GX, Smith P, et al. (2009) Increase in soil organic carbon stock over the last two decades in China's Jiangsu Province. Global Change Biology 15: 861–875.

15. Zhang HB, Luo YM, Wong MH, Zhao QG, Zhang GL (2007) Soil organic carbon storage and changes with reduction in agricultural activities in Hong Kong. Geoderma 139: 412–419.

16. Feng S, Tan S, Zhang A, Zhang Q, Pan G, et al. (2011) Effect of household land management on cropland topsoil organic carbon storage at plot scale in a red earth soil area of South China. Journal of Agricultural Science 149: 557–566.

17. Harper RJ, Gilkes RJ (1995) Some factors affecting the distribution of carbon in soils of a dry land agricultural system in southwestern Australia. In: Lal R, Kimble JM, Follett RF, Stewart BA (editors). Assessment Methods for Soil Carbon. CRC Press. Boca Raton, FL, USA. PP.577–591.

18. Guo JH, Liu XJ, Zhang Y, Shen JL, Han WX, et al. (2010) Significant Acidification in Major Chinese Croplands. Science 327: 1008–1010.

19. Pan GX, Li LQ, Wu LS, Zhang XH (2003) Storage and sequestration potential of topsoil organic carbon in China's paddy soils. Global Change Biology 10: 79–92.

20. Wu HB, Guo ZT, Gao Q, Peng CH (2009) Distribution of soil inorganic carbon storage and its changes due to agricultural land use activity in China. Agriculture, Ecosystems and Environment 129: 413–421.

21. Gilliam FS, Saunders NE (2003) Making more sense of the order: A review of Canoco for Windows 4.5, PC-ORD version 4 and SYN-TAX 2000. Journal of Vegetation Science 14: 297–304.

22. Zhang XL, Ma L, Gilliam FS, Wang Q, Liu T, et al. (2012) Effects of raised-bed planting for enhanced summer maize yield on soil microbial functional groups and enzyme activity in Henan Province, China. Field Crops Research 130: 28–37.

23. Feller C, Beare MH (1997) Physical control of soil organic matter dynamics in tropics. Geoderma 79: 69–116.

24. Haynes RJ, Naidu R (1998) Influence of lime, fertilizer and manure applications on soil organic matter content and soil physical conditions: a review. Nutrient Cycling in Agroecosystems 51: 123–137.

25. Diaz-Zorita M, Grosso GA (2000) Effect of soil texture, organic carbon and water retention on the compatibility of soils from the Argentinean Pampas. Soil and Tillage Research 54: 121–126.

26. Schertz DL, Moldenhauer WC, Livingston SJ, Weeisies GA, Hintz AE (1989) Effect of past soil erosion on crop productivity in Indiana. Journal of Soil and Water Conservation 44: 604–608.

27. Emerson WW (1995) Water-retention, organic-carbon and soil texture. Australian Journal of Soil Research 33: 241–251.

28. Powlson DS, Hirsch PR, Brookes PC (2001) The role of soil micro-organisms in soil organic matter conservation in the tropics. Nutritional Cycling in Agroecosystems 61: 41–51.

29. Du ZL, Ren TS, Hu CS (2010) Tillage and residue removal effects on soil carbon and nitrogen storage in the North China Plain. Soil Science Society of American Journal 74: 196–202.

30. Pan GX, Li LQ, Zhang Q, Wang XK, Sun XB, et al. (2005) Organic carbon stock in topsoil of Jiangsu Province, China, and the recent trend of carbon sequestration. Journal of Environmental Sciences 17: 1–7.

31. Post WM, Pastor J, Zinke PJ, Stangenberger AG (1985) Global patterns of soil nitrogen storage. Nature 317: 613–616.

32. Gilliam FS, Dick DA, Kerr ML, Adams MB (2004) Effects of silvicultural practices on soil carbon and nitrogen in a nitrogen saturated Central Appalachian (USA) hardwood forest ecosystem. Environmental Management 33: S108–S119.

33. Pan GX, Zhao QG (2005) Study on evolution of organic carbon stock in agricultural soils of China: facing the challenge of global change and food security. Advances in Earth Science 20: 384–393 (in Chinese).

34. Dersch G, Böhm K (2001) Effects of agronomic practices on the soil carbon storage potential in arable farming in Austria. Nutrient Cycling in Agroecosystems 60: 49–55.

10

Environmental Fate of Soil Applied Neonicotinoid Insecticides in an Irrigated Potato Agroecosystem

Anders S. Huseth[1], Russell L. Groves[2]*

1 Department of Entomology, Cornell University, New York State Agricultural Experiment Station, Geneva, New York, United States of America, **2** Department of Entomology, University of Wisconsin-Madison, Madison, Wisconsin, United States of America

Abstract

Since 1995, neonicotinoid insecticides have been a critical component of arthropod management in potato, *Solanum tuberosum* L. Recent detections of neonicotinoids in groundwater have generated questions about the sources of these contaminants and the relative contribution from commodities in U.S. agriculture. Delivery of neonicotinoids to crops typically occurs as a seed or in-furrow treatment to manage early season insect herbivores. Applied in this way, these insecticides become systemically mobile in the plant and provide control of key pest species. An outcome of this project links these soil insecticide application strategies in crop plants with neonicotinoid contamination of water leaching from the application zone. In 2011 and 2012, our objectives were to document the temporal patterns of neonicotinoid leachate below the planting furrow following common insecticide delivery methods in potato. Leaching loss of thiamethoxam from potato was measured using pan lysimeters from three at-plant treatments and one foliar application treatment. Insecticide concentration in leachate was assessed for six consecutive months using liquid chromatography-tandem mass spectrometry. Findings from this study suggest leaching of neonicotinoids from potato may be greater following crop harvest in comparison to other times during the growing season. Furthermore, this study documented recycling of neonicotinoid insecticides from contaminated groundwater back onto the crop via high capacity irrigation wells. These results document interactions between cultivated potato, different neonicotinoid delivery methods, and the potential for subsurface water contamination via leaching.

Editor: Christopher J. Salice, Texas Tech University, United States of America

Funding: This research was supported by the Wisconsin Potato Industry Board and the National Potato Council's State Cooperative Research Program FY11-13. The funders had no role in study design, data collection and analysis, decision to publish, or preparation of the manuscript.

Competing Interests: RLG has received research funding, not related to this project, from Bayer CropScience, DuPont, Syngenta, and Valent U.S.A.

* E-mail: groves@entomology.wisc.edu

Introduction

The neonicotinoid group of insecticides is among the most broadly adopted, conventional management tools for insect pests of annual and perennial cropping systems [1]. Benefits of the neonicotinoid group of compounds include flexibility of application, diversity of active ingredients, and broad spectrum activity [2]. Moreover, growers have readily adopted neonicotinoids for two specific reasons: first, these compounds are fully systemic in plants after soil application and second, several new generic formulations have recently become available which have incentivized their continued use in many crops [1–3]. Since 2001, the United States Environmental Protection Agency (EPA) has classified several neonicotinoids as either conventional, reduced-risk pesticides, or as organophosphate alternatives [4],[5]. EPA certification often requires replacement of older, broad-spectrum pesticides with newer, more specific products for management of key economic pests. Critical attributes of replacement insecticides include documented reductions in human and environmental risk when compared to older, broad-spectrum pesticides [5]. Despite acceptance of neonicotinoid insecticides as reduced-risk by growers and regulatory agencies, nearly two decades of widespread, repetitive use has resulted in several insecticide resistance

issues, impacts on native and domestic pollinators, and unanticipated environmental impacts [6–9].

The environmental fate of several neonicotinoid active ingredients have been previously assessed. Previous studies focused on degradation and movement processes in soil, leachate, and runoff [10–15]. The leaching potential of the neonicotinoids into groundwater, as well as persistence in the plant canopy, is related to properties of the chemicals and delivery method of the compound to the crop (Fig. S1)[12],[15],[16]. Soil application (e.g., seed treatment or in-furrow) has been adopted as the principal form of insecticide delivery in potato production as it provides the longest interval of pest control, while also reducing non-target impacts, and limits exposure to workers when compared to foliar application methods. Since 1995, soil-applied neonicotinoids (i.e., clothianidin, imidacloprid, thiamethoxam) have been the most common pest management strategy used to control infestation of Colorado potato beetle, *Leptinotarsa decemlineata* Say; potato leafhopper, *Empoasca fabae* Harris; green peach aphid, *Myzus persicae* Sulzer; and potato aphid, *Macrosiphium euphorbiae* Thomas. The now widespread and extensive use of these systemic neonicotinoid insecticides, coupled with the recent detection of thiamethoxam in groundwater [17],[18], supports the hypothesis that potato pest management may contribute a portion of the documented neonicotinoid contaminants reported in

Wisconsin, USA. Furthermore, we hypothesized that neonicotinoid insecticides applied to potato are most vulnerable to leaching in the spring season when the root system of the plant has yet to fully exploit all of the active ingredient applied directly in the seed furrow. Large rain events at this time could drive insecticide leaching from potato and subsequent groundwater contamination at large scales. In this study, we examined how neonicotinoid concentrations in leachate were altered in response to different insecticide delivery methods using potatoes grown under commercial production practices. We also report the patterns of historic neonicotinoid insecticide detections in groundwater using water quality surveys collected by the Wisconsin Department of Agriculture, Trade and Consumer Protection-Environmental Quality Section (WI DATCP-EQ). Second, using potato as a model system, we analyzed leachate captured below different seed treatments, soil-applications, and foliar delivery treatments for thiamethoxam using liquid chromatography-tandem mass spectrometry (LC/MS/MS) over two consecutive field seasons. In this experiment, thiamethoxam was chosen as one representative insecticide in a broader group of water-soluble neonicotinoids. Moreover, this active ingredient represented the majority of positive neonicotinoid detections in groundwater monitoring surveys conducted by the WI DATCP-EQ [17], [18]. Third, using identical quantitative methods, we measured thiamethoxam concentration in irrigation water collected from operating, high-capacity irrigation wells at two time points in each sampling year. And finally, we characterize irrigation use and production trends of crops that may contribute to neonicotinoid detection in groundwater. Results of this study increase our understanding about the influence of insecticide delivery method on the neonicotinoid insecticides leaching from potato into the surrounding environment.

Materials and Methods

Ethics Statement

No specific permits were required for the field study described here. Access to field sites was granted by the private landholder to conduct leaching experiments. No specific permissions were needed to present publically available records provided by Wisconsin Department of Agriculture, Trade and Consumer Protection or Wisconsin Department of Natural Resources. Field studies did not involve any endangered or protected species.

Groundwater Contamination

Permanent groundwater monitoring wells, maintained by the WI DATCP-EQ, were used to measure neonicotinoid contamination of subsurface water resources as one component of an ongoing study documenting agrochemical (e.g., insecticides, herbicides, nutrients) impact on groundwater quality. Beginning in 2006, analytical water quality assessments for neonicotinoid contamination were conducted by the Wisconsin Department of Agriculture Trade and Consumer Protection-Bureau of Laboratory Services. Concentrations of acetamiprid, clothianidin, dinotefuran, imidacloprid, and thiamethoxam were monitored in 20–30 different monitoring well locations from 2006–2012. Presented are positive detections of those insecticides in different monitoring wells from 2006–2012 [17],[18]. Data provided by WI DATCP-EQ characterize the temporal and spatial profile of thiamethoxam and other neonicotinoid detections that occurred between 2008–2012. These data are presented in summary as a foundation for following objectives (Table 1).

Experimental Site and Design

In 2011 and 2012, leaching experiments were conducted 6 km west of Coloma, Wisconsin. Experiments were planted in two different fields approximately 0.5 km apart on 20 May 2011 and 11 May 2012. The soil at both sites consisted of Richford loamy sand (sandy, mixed, mesic, Typic Udipsamments) [19]. Soil composition was 7% clay, 82% sand, and 11% silt. Organic matter was 0.53 percent by weight. Study sites soils had a high infiltration rate (Hydrological Soil Group A), a high saturated hydraulic conductivity (K_{sat}) at 28 micrometers per second, and an available water capacity rating of 0.1 cm per cm [19]. No restrictive layer that would impede water movement through the soil has been documented [19]. Study site soil was formed in the bed of glacial Lake Wisconsin from parent material of glacial till overlain by glacial outwash [20]. Upper soil horizons (A and B) are sand with minimal structure. Subsurface soil (C horizon) had no structure. Irrigation pivots in sample fields withdrew water at a depth of 37 m and the water table depth (static water level) was approximately 6 m for both sites [21].

A randomized complete block design with four insecticide delivery treatments and an untreated control was established using the potato cultivar, 'Russet Burbank'. Plots were 0.067 ha in size and planted at a rate of one seed piece per 0.3 m with 0.76 m spacing between rows. Each year, experiments were nested within a different ~32 ha commercial potato field, and maintained under commercial management practices by the producer (e.g., nutrient application timing, chemical usage, tillage practices, etc.), with the exception of insecticide inputs. The decision to locate these experiments in commercial fields was, in part, based upon access to a center pivot irrigation system to best duplicate water inputs used to produce commercial potato in Wisconsin. All other inputs and production strategies (e.g. tillage, fumigation, fertility, and disease management) were conducted by the producer with equipment and products in a manner consistent with the best management practices for potato production in Wisconsin. Prior to planting in each season, a tension plate lysimeter (25.4×25.4×25.4 cm) was buried at a depth of 75 cm below the soil surface. Lysimeters were constructed of stainless steel with a porous stainless steel plate affixed to the top to allow water to flow into the collection basin over each sampling interval. Experimental blocks were connected with 9.5 mm copper tubing to a primary manifold and equipped with a vacuum gauge. A predefined, fixed suction was maintained under regulated vacuum at 107 ± 17 kPa (15.5 ± 2.5 lb per in^2) with a twin diaphragm vacuum pump (model UN035.3 TTP, KnF, Trenton, NJ) connected to a 76 L portable air tank. Each treatment block was equipped with a data-logging rain gauge (Spectrum Technologies, Inc. model # 3554WD1) recording daily water inputs at a five minute interval. Data was offloaded with Specware 9 Basic software (Spectrum Technologies, Inc., Plainfield, IL, USA) and aggregated into daily irrigation or rain event totals using the *aggregate* and *dcast* function in R (package: reshape2, [22]). Irrigation event records were obtained from the grower to identify days and estimated inputs of water application throughout the growing season.

Insecticides and Application

Thiamethoxam treatments (Platinum 75SG, 75% thiamethoxam per formulated unit, Syngenta, Greensboro, NC) were selected to represent a common, soil-applied insecticide in potato. A second formulation of thiamethoxam was selected to represent a common pre-plant insecticide seed treatment in potato (Cruiser 5FS, 47.6% thiamethoxam per formulated unit, Syngenta, Greensboro, NC). Each insecticide formulation is used to manage early season infestations of Colorado potato beetle, potato

Table 1. Positive (means±SD) neonicotinoid detections in groundwater from 2008–2012, State of Wisconsin Department of Agriculture Trade and Consumer Protection.

Year	County	Area potato (ha)[a]	Row crops (ha)[b]	Percent potato[c]	Well ID	N positive samples	Insecticide concentration (µg/L)[d]		
							clothianidin	imidacloprid	thiamethoxam
2008	Adams	2,617	21,385	10.9	6	2	-	-	4.34 (4.97)
	Grant	0	47,827	0.0	10	1	-	-	1.25
	Iowa	18	25,795	0.1	11,12,13	9	-	-	1.50 (0.67)
	Richland	29	9,582	0.3	16	1	-	-	0.69
	Sauk	30	31,931	0.1	17	2	-	-	2.41 (1.32)
	Waushara	2,630	29,447	8.2	20	2	-	-	0.67 (0.05)
2009	Adams	3,989	24,894	13.8	6	2	-	-	5.31 (5.12)
	Dane	22	101,527	0.0	9	1	-	-	1.61
	Iowa	343	33,375	1.0	11,12	3	-	-	1.31 (0.68)
	Richland	87	14,402	0.6	16	1	-	-	1.26
	Sauk	328	40,571	0.8	17	2	-	-	3.00 (0.94)
2010	Adams	4,188	24,871	14.4	6	4	3.43	-	2.97 (2.04)
	Brown	1	39,322	0.0	7	1	-	-	0.52
	Dane	34	110,979	0.0	8,9	4	0.54 (0.24)	0.54	1.08
	Grant	49	74,566	0.1	10	1	0.73	-	-
	Iowa	356	38,840	0.9	11,12,13	7	-	-	1.25 (1.02)
	Sauk	188	45,309	0.4	17	5	0.41	-	1.81 (0.88)
	Waushara	4,184	33,576	11.1	19,20	2	-	2.77 (0.81)	-
2011	Adams	4,066	27,693	12.8	2,5,6	9	0.63 (0.36)	0.33	0.63 (0.26)
	Brown	7	38,309	0.0	7	1	-	-	0.21
	Dane	33	107,214	0.0	8	2	0.62 (0.19)	-	-
	Grant	13	75,436	0.0	10	1	0.30	-	-
	Iowa	47	40,138	0.1	12	4	-	0.34 (0.09)	0.88 (0.23)
	Portage	7,364	45,324	14.0	15	1	-	-	0.32
	Sauk	213	46,686	0.5	17,18	5	0.54 (0.10)	-	1.92 (0.43)
	Waushara	4,536	36,676	11.0	19,20,21,23	23	0.25 (0.03)	0.78 (0.69)	1.40 (0.56)
2012	Adams	4,263	27,037	13.6	1,3,4,6	6	0.52 (0.30)	0.51 (0.26)	0.27
	Dane	11	115,501	0.0	8	1	0.67	-	-
	Grant	4	72,920	0.0	10	1	0.26	-	-
	Iowa	369	40,764	0.9	12	2	0.24	0.28	0.44
	Juneau	907	28,542	3.1	14	2	0.42 (0.18)	-	0.20
	Portage	7,622	46,337	14.1	15	2	-	0.47	0.47
	Waushara	5,904	38,999	13.1	21,22,23	13	-	0.68 (0.88)	1.51 (0.72)
	summary			N = 23		67	25	30	68

Table 1. Cont.

Year	County	Area potato (ha)[a]	Row crops (ha)[b]	Percent potato[c]	Well ID	N positive samples	Insecticide concentration (μg/L)[d]		
							clothianidin	imidacloprid	thiamethoxam
						Average	0.62 (0.63)	0.79 (0.83)	1.59 (1.51)
						Range	0.21–3.34	0.26–3.34	0.20–8.93

[a]Acreage estimates generated from USDA National Agricultural Statistics Service – Cropland Data Layer, 2008-2012 [26].
[b]Row crops class is the sum of the following crop areas (ha): maize, soy, small grains, wheat, peas, sweet corn, and miscellaneous vegetables and fruits.
[c]Percent potato calculated as the potato area grown annually divided by total arable row crop acreage (other row crops + potato).
[d]Positive neonicotinoid detections extracted from long-term, groundwater wells maintained by the WI-DATCP-EQ Program.

leafhopper, and colonizing aphid in Wisconsin potato crops. Commercially formulated insecticides were applied at maximum labeled rates for in-furrow (140 g thiamethoxam ha^{-1}) and seed treatment (112 g thiamethoxam ha^{-1} at planting density of 1,793 kg seed ha^{-1}) for potato [23]. A calibrated CO_2 pressurized, backpack sprayer with a single nozzle boom was used to deliver an application volume of 94 liters per hectare at 207 kPa through a single, extended range, flat-fan nozzle (TeeJet XR80015VS, Spraying Systems, Wheaton, IL) for in-furrow applications. Spray applications were directed onto seed pieces in the furrow at a speed of one meter per second and furrows were immediately closed following application. Seed treatments were applied using a calibrated CO_2 pressurized backpack sprayer with a single nozzle boom delivering an application volume of 102.2 L per hectare at 207 kPa through a single, extended range, flat-fan nozzle (TeeJet XR80015VS, Spraying Systems, Wheaton, IL) was used for delivery of thiamethoxam in water (130 mL) directly to suberized, cut seed pieces (23 kg) 24 hours prior to planting. Seed treatments were allowed to dry in the absence of light at 20°C during that pre-plant period. A novel soil application method, impregnated copolymer granules, was included as another treatment in an attempt to stabilize applied insecticide in the soil. Polyacrylamide horticultural copolymer granules (JCD-024SM, JRM Chemical, Cleveland, OH) were impregnated at an application rate of 16 kg per hectare. The polyacrylamide treatment was included as a novel delivery method to stabilize insecticide in the rooting zone and possibly reduce leaching in the early season. Thiamethoxam (0.834 g, Platinum 75SG) was initially diluted in 250 mL of deionized water and 100 μL of blue food coloring was incorporated into solution to ensure uniform mixing (brilliant blue FCF). Insecticide solutions were mixed with 75 g polyacrylamide then stirred until the liquid was absorbed and a uniform color was observed. Impregnated granules were vacuum dried in the absence of light for 24 hours at 20°C. Treated granules were divided into even quantities per row and evenly distributed into the four treatment rows for each polyacrylamide plot. A single untreated flanking row was planted between plots. All soil-applied insecticides were applied on 20 May 2011 and 11 May 2012 at the time of planting.

Two foliar applications of thiamethoxam (Actara 25WG, 25% thiamethoxam per formulated unit, Syngenta, Greensboro, NC) sprayed on the same plot were included as a fourth delivery treatment. Two successive neonicotinoid applications are recommended for foliar control of pests in potato [23]. Foliar thiamethoxam was applied using a calibrated CO_2 pressurized backpack sprayer delivering an application volume of 187.1 liters per hectare at 207 kPa through four, extended range flat-fan nozzles (TeeJet XR80015VS, Spraying Systems, Wheaton, IL) spaced at 45.2 cm. The first foliar application was followed approximately seven days later with a second equivalent rate of thiamethoxam to total the season-long maximum labeled rate (105 g thiamethoxam ha^{-1}) [23] and were timed to coincide with the appearance of 1st and 2nd instar larvae of native populations of *L. decemlineata*. Foliar applications of thiamethoxam were applied on 28 June and 5 July in 2011 and 15 and 22 June in 2012. Although total amounts of active ingredient differ by formulation, these rates are identical to registered label recommendations [23] and reflect the maximum amount of active ingredient used on an average hectare of cultivated potato. Specific chemical properties of formulated thiamethoxam that affect solubility and leaching potential in soil can be found in Gupta et al. [15] and the references therein (Fig. S1).

Chemical Extraction and Quantification

Lysimeter leachate was sampled twice monthly beginning on June 1 of each year and concluding in October of 2011 and November of 2012. Total leachate volume was recorded for each plot. A 500 mL subsample was taken from each plot into a 0.5 L glass vessel and immediately placed on ice and refrigerated at 4–6°C in the laboratory prior to analysis. Samples were homogenized into a 400 mL monthly (i.e., two samples per month) sample as percent volume per volume dependent on total catch measured in the field. Neonicotinoid residues from monthly water samples were extracted using automated solid phase extraction (AutoTrace SPE workstation, Zymark, Hopkinton, MA) with LiChrolut EN SPE columns (Merk KGaA, Darmstadt, Germany). If visual inspection of sample found excessive sediment contamination, samples were filtered through a 0.45 µm filter prior to extraction. Columns were conditioned prior to extraction with 3 mL of methanol (MeOH) and 3 mL of water. 210 mL of sample were loaded onto columns and rinsed with 10 mL of water then dried under flowing nitrogen for 15 minutes (N-evap, Organomation, Berlin, MA). Samples were eluted using a 50% ethyl acetate (EtOAc) and 50% methanol solution to collect a 2 mL sample fraction. Sample extract fractions were analyzed using a Waters 2690 HPLC/Micromass Quattro LC/MS/MS (Waters Corporation, Milford, MA). All thiamethoxam residues were identified, quantified, and confirmed using LC/MS/MS by the Wisconsin Department of Agriculture Trade and Consumer Protection-Bureau of Laboratory Services. The method detection limit (MDL) of the extraction procedure was $0.2 \ \mu g \ L^{-1}$. Specific conditions for all quantitative procedures follow WI-DATCP Standard Operating Procedure #1009 developed from Seccia et al. [24] and references therein.

Irrigation Use and Crop Area

To determine the extent of irrigated agriculture present within the watershed, we utilized current high capacity well pumping data and irrigated agriculture estimates derived from digital imagery. Publically available operator reporting data for high capacity agricultural pivots were obtained from the Wisconsin Department of Natural Resources Bureau of Drinking Water and Groundwater. Records included location information and pumping volume for the year 2012. High capacity wells service several irrigated fields and often these fields are further divided into individual crop management units each with unique irrigation requirements. We digitized the area watered by all identifiable center pivot, linear move, and traveling gun irrigation systems using digital aerial photography to measure the total number of management units present within the greater Central Wisconsin Water Management Unit watershed [25] (ArcGIS version 10.1, Redlands, CA). Fields were subdivided into management units using the consistent divisions in crop types with a sequence of National Agricultural Statistics Service Cropland Data Layer (NASS-CDL) [26] thematic data and aerial photography images [25] from 2010–2012.

To determine agronomic trends in the Central Sands vegetable production region of Wisconsin, we used a combination of publically available land use data and current neonicotinoid registration information. A geospatial watershed management boundary layer delineated by the Wisconsin Department of Natural Resources [27] was used to generally define the spatial extent where agriculture could be contributing to the detection of neonicotinoid insecticides in subsurface water. The Central Wisconsin Water Management Unit extent was used to estimate annual crop composition using the NASS-CDL [26] from 2006–2012 using ArcGIS. From these data, we selected major crops that

frequently receive either seed or in-furrow soil-applied neonicotinoid insecticide treatments. Application rates were identical for several similar crops (e.g. soybean and green bean), and so, we chose to aggregate crops based on insecticide rate and crop type into three primary groups: maize, beans, and potato [23],[28–30]. These crop groups comprise the majority of production area in the Central Wisconsin Water Management Unit extent. To our knowledge, limited information exists documenting the proportion of different soil-applied neonicotinoid active ingredients that are used on a per crop basis in the Central Wisconsin Water Management Unit. Based on this level of uncertainty, we chose not to extend tabulated crop areas to a direct calculation or estimate of neonicotinoid active ingredients applied.

Data Analysis

To determine the impact of different insecticide delivery treatments on thiamethoxam leachate detected over time, we reported the mean concentration over a period of several months. All lysimeter analyses included samples where neonicotinoid insecticides were not detected (i.e., zero detections). All data manipulation and statistical analyses of leachate concentrations were performed in R, version 2.15.2 [31] using the base distribution package. Functions used in the analysis are available in the base package of R unless otherwise noted. Observed concentration for time points in each year were subjected to a repeated-measures analysis of variance (ANOVA) using a linear mixed-effects model to determine significant delivery (i.e. treatment), date, and delivery×date effects ($P < 0.05$). Because the agronomic conditions differed between years and given that our comparison of interest was at the insecticide delivery treatment level, insecticide concentrations were analyzed separately for each year. Mixed-effects models (i.e., repeated-measures analysis of variance) were fit using the *lme* function (package nlme, [32]). Empirical autocorrelation plots from unstructured correlation model residuals were examined using the *ACF* function (package nlme, [32]). Correlation among within-group error terms were structured and examined in three ways: first, unstructured correlation, second, with compound symmetry using the function *corCompSymm* and third, with autoregressive order one covariance using the function *corAR1* (package nlme, [32][33]). Since models were not nested, fits of unstructured, compound symmetry, and autoregression order one covariance were compared using Akaike's information criterion statistic with the function *anova* (test = "F"). Data were transformed with natural logarithms before analysis to satisfy assumptions of normality, however untransformed means are graphically presented. In 2012, a single lysimeter in the polyacrylamide treatment of the leachate study malfunctioned and these observations were dropped from subsequent analyses leading to an unbalanced replicate number for that treatment (N = 3) in 2012. Water input data collected from tipping bucket samplers were averaged across block by day and aggregated as cumulative water inputs using the *cumsum* function. All summary statistics and model estimates were extracted using *aggregate*, *summary*, and *anova* functions.

Results and Discussion

Groundwater Detections

Neonicotinoid insecticides were detected at 23 different well monitoring well locations by WI-DATCP-EQ surveys between the years 2008 and 2012 (Table 1). These annual surveys, administered by WI-DATCP-EQ, occur at sensitive geologic or hydrogeologic locations that are at high risk of non-point source agrochemical leaching. Specifically, two agriculturally intensive

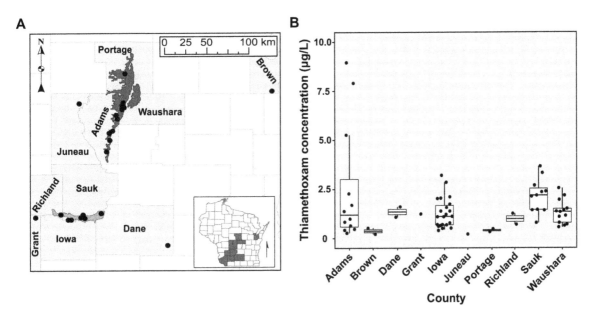

Figure 1. Positive thiamethoxam residue detections in groundwater 2008–2012. Points in the map (A) correspond to positive detection locations. Dark grey shaded region indicates the Central Sands potato production region. Light grey delimits the Lower Wisconsin River potato production region. Positive detections were obtained from established agrochemical monitoring wells collected by the Wisconsin Department of Agriculture, Trade and Consumer Protection (DATCP)-Environmental Quality division in collaboration with the Wisconsin DATCP Bureau of Laboratory Services. Boxplots (B) indicate average concentration detected from 2008–2012. Points show individual measured concentrations.

production regions of the state, the Central Sands and Lower Wisconsin River valley, are classified as high-risk areas for groundwater contamination and are frequently monitored for the presence of common agrochemicals (Fig. 1A). These regions have well-drained, sandy soils and easily accessible groundwater for irrigation that has driven agricultural intensification focused on vegetable production. Commercial potato is a key component in the agricultural production sequence, but is also rotated with many other specialty crops such as: carrots, onions, peas, pepper, processing cucumber, sweet corn, and snap beans. Unfortunately, the unique soil and water characteristics supporting a profitable specialty crop production system are also particularly vulnerable to groundwater contamination with water-soluble agricultural products [34–36]. Regulatory exceedences of nitrates and herbicide products (e.g. triazines, triazinones, and chloroacetamide) have been commonplace for several years [34–37], but recent detections of neonicotinoid contaminants have created new groundwater quality concerns. Beginning in the spring of 2008, two wells had detections of 1.25 and 1.47 μg L^{-1} thiamethoxam in Grant and Sauk Counties, WI (Fig. 1B, Table 1). Subsequent sampling later that season identified six additional locations for a total of 17 independent positive thiamethoxam detections that year. Since

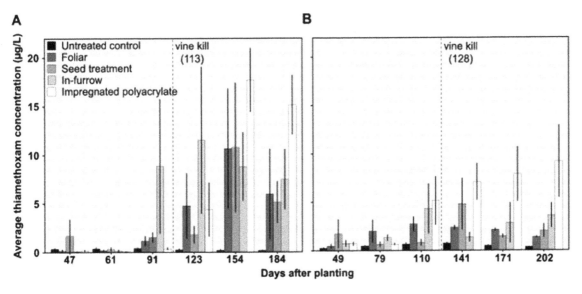

Figure 2. Thiamethoxam concentration in leachate from potato. Average thiamethoxam ($\pm$SD) recovered from in-furrow and foliar treatments in (A) 2011 an (B) 2012. Dotted lines indicate the date that the producer applied vine desiccant prior to harvest. Lysimeter studies continued in undisturbed soil following vine kill.

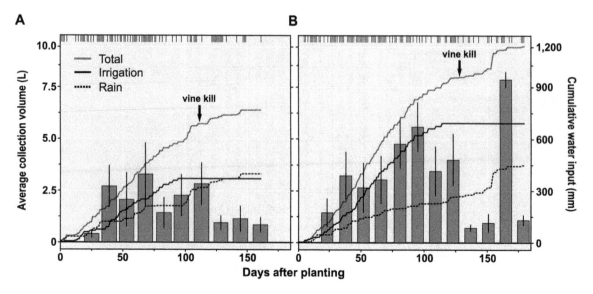

Figure 3. Water input volumes, 2011 and 2012. Water inputs and leachate volume collected in lysimeter studies in (A) 2011 and (B) 2012. Lines indicate cumulative water measured in tipping bucket rain gauges installed in plots each season. Bar plots indicate average leachate volume (±SD) collected in lysimeters on a bi-monthly sampling frequency. Hash marks at the top of each figure indicate days that overhead irrigation or rainfall occurred in each season.

these early detections, the WI-DATCP-EQ [17],[18] has repeatedly detected thiamethoxam, imidacloprid, and clothianadin residues at 23 different monitoring well locations over a five-year period (Table 1). Although the sampling effort was not uniformly distributed within the state, neonicotinoid detections often correspond to areas where intensive irrigated agricultural production occurs (Fig. 1A). As an indication of specialty crop production intensity, we used county-level potato abundance to better describe trends in historical neonicotinoid detections. Observed frequency and magnitude of neonicotinoid detections did not consistently correspond to potato abundance (Table 1). Although the contribution of potato production to the observed detections was not clear, regulatory agencies have continued to pursue this interaction by sampling where potato occurs at a high density, specifically the Central Sands and Wisconsin River Valley. Groundwater sampling strategies have provided a useful timeline of non-point source agrochemical pollution events in subsurface water resources. Identifying the origin of pollutants in the state is complicated by the diversity of neonicotinoid registrations, application methods and formulations; currently Wisconsin has 164 different registrations for field, forage, tree fruit, vegetable, turf, and ornamentals crops (6 acetamiprid, 18 clothianadin, 4 dinotefuran, 108 imidacloprid, 1 thiacloprid, 26 thiamethoxam) [38].

Neonicotinoid Losses and Concentrations in Leachate

The neonicotinoid insecticide thiamethoxam was included in field experiments to investigate the potential for leaching losses associated with different types of pesticide delivery. Specifically, formulations of thiamethoxam were applied as foliar and as at plant systemic treatments in commercial potato over two years and at two different irrigated fields. We hypothesized that thiamethoxam would be most vulnerable to leaching early in the season when plants were small and episodic heavy rains can be common. Interestingly, we observed the greatest insecticide losses following vine-killing operations which occurred more than 100 days after planting (Fig. 2). Detections of thiamethoxam in lysimeters varied between insecticide delivery treatments through time in 2011 (delivery×date interaction, $F = 2.1$; d.f. $= 20,88$; $P = 0.0131$) and again in 2012 (delivery×date interaction, $F = 1.8$; d.f. $= 20,87$; $P = 0.0384$). Moreover, the impregnated polyacrylamide delivery produced the greatest amount of thiamethoxam leachate late in each growing season (Fig. 2) when compared with other types of insecticide delivery.

Early season rainfall was not exceptionally heavy in either year of this experiment (Fig. 3). The accumulation of leachate detections in lysimeters likely is reflected by the steady application of irrigation water and rainfall. One clear exception to this pattern occurred in 2012 at 155–156 days after planting when 89 mm of

Table 2. Neonicotinoid concentration from irrigation water, 2011 and 2012.

Date	Days after planting	Insecticide concentration (µg/L)[a]	
		clothianidin	thiamethoxam
28 June 2011	39	-	0.310
1 September 2011	114	-	0.327
10 July 2012	60	-	0.533
15 August 2012	96	0.225	0.580

[a]Samples obtained from irrigation pivots while under operation in potato fields containing lysimeter experiments.

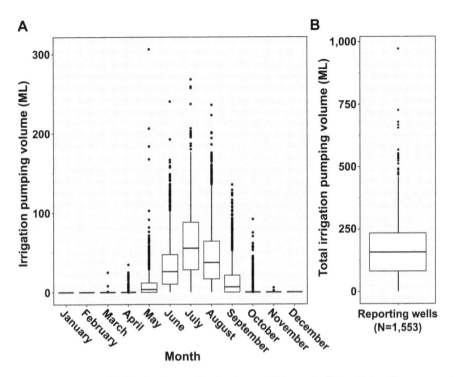

Figure 4. Reported irrigation inputs in the Central Wisconsin River Water Management Unit. Average reported agricultural pumping (megaliters, ML) in the Central Wisconsin River Water Management Unit for 2012. Monthly pumping records were reported by growers to the Wisconsin Department of Natural Resources Bureau of Drinking Water and Groundwater. Upper and lower whiskers extend to the values that are within 1.5*Inter-quartile range beyond the first (25%) and third (75%) percentiles. Data beyond the end of whiskers indicate outlier values and have been plotted as points.

rain fell within a 24-hour period. Peak detections of thiamethoxam in 2012 began to trend upward following this rain event, however the timing of similar detections across treatments in 2011 occurred at about the same time. One additional explanation may be that increased levels of pesticide losses are associated with plant death or senescence. In each year of this study, the largest proportion of pesticide detections in leachate occurred after vine killing with herbicide in the potato crop. Vine killing in commercial potato production is a common practice designed to aid the tubers in developing a periderm. Perhaps the rapid loss in root function following plant death permits excess pesticide to be solubilized and washed through the soil profile more quickly in root channels. In both seasons of this study, however, large episodic rain events did not occur early in the growing season. These results do appear, however, to document low to moderate levels of leaching losses that occur throughout the season even when the crop is managed at nominal evapo-transpirative need.

Untreated control plots also yielded low-level detections of thiamethoxam throughout both seasons. To better understand these insecticide detections in control plots, we sampled water directly from the center pivot irrigation system providing irrigation directly to the potato crop. Samples were taken while the systems were operational from lateral spigots mounted on the well casings. In both years, samples revealed low concentrations of thiamethoxam present in the groundwater at two time points in each sample season (Table 2) from which irrigation water was being drawn. Clothianidin was also present at a single time point in 2012 (Table 2). These positive detections of low-dose thiamethoxam were obviously being unintentionally applied directly to the crop through irrigation and this information is new to the producers in the Central Sands of Wisconsin. Although systemic neonicotinoids have recently been detected from surface water runoff and catch

basins associated with irrigated orchards [10], [39], to our knowledge no other study has documented the occurrence of neonicotinoids in subsurface groundwater being recycled through operating irrigation wells. Currently, the known exposure pathways for insecticide residues are most often associated with direct application or systemic movement of insecticides in floral structure and guttation water [8],[9],[40].

The implications for non-target effects resulting from these groundwater contaminants is currently unknown, but could be important considering the scale of irrigation ongoing in the Central Sands potato agroecosystem in Wisconsin (Fig. S2). Using a combination of aerial photography and NASS Cropland Data Layers, we identified 2,530 different irrigated field units distributed within the Central Wisconsin River Water Management Unit (Fig. S2). In all, 71,864 hectares of irrigated cropland were identified within the extent of the water management unit. Average irrigated field unit size was 28.4 ± 17.7 hectares (min. 1, max 138). Irrigation use patterns demonstrated clear increases in the summer months of the 2012 growing season (Fig. 4). Average annual pumping volume reported to the Wisconsin Department of Natural Resources in 2012 was 170.6 ± 115.6 megaliters (ML) of irrigation water (min. 0.00001, max 972.1) distributed over 1,553 reporting wells. Peak pumping volumes occurred in the month of July, averaging 61 ± 43.3 ML (min. 0, max 286.4). The timing of peak pumping correspond with crop demands for and reproductive phases of common open and closed pollination crops grown in the region.

While considerable attention has been focused on the positive attributes of the neonicotinoids [1–3], an increasing body of research suggests substantial negative impacts not only in terms of pest resistance development (e.g., Colorado potato beetle), but also impacts on non-target organisms and surrounding ecosystems

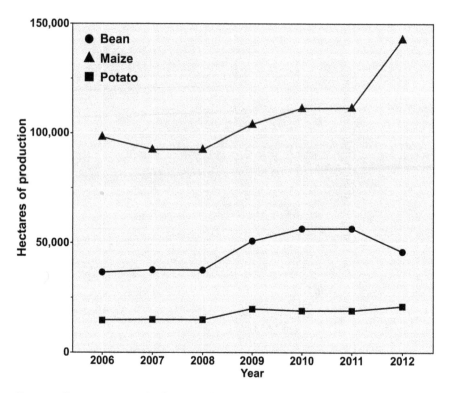

Figure 5. Crop area grown in the Central Wisconsin River Water Management Unit. Cropping trends in the Central Wisconsin River Water Management Unit from 2006–2012. Crop groups are often planted with a soil-applied neonicotinoid insecticide for insect pest management. Crop totals within the water management unit were tabulated from annual USDA-NASS Cropland Data Layers [26].

[8],[10],[41–44]. Recent studies have documented the negative influence of neonicotinoids on pollinator population health (both native and managed) which, in turn, created substantial concern about the long-term sustainability of these pesticides in agriculture [7],[11],[43],[45–49]. Exposures to pollinators reportedly occur through chronic, sub-lethal contact with low concentrations of neonicotinoid residues in pollen, nectar, waxes, and guttation drops of common crop plants [50–53]. Gill et al. [43] and Whitehorn et al. [54] found that low concentrations (≤ 10 μg L^{-1}) of imidacloprid significantly reduced colony-level health in bumblebee (*Bombus terrestris* L.). Imidacloprid residues measured by those authors are consistent with insecticide concentrations found in nectar and pollen of flowering crops, further supporting the direct crop-pollinator toxicological pathway hypothesis [47],[52],[54],[55]. Though they have received much less attention, many closed pollination crops also provide resources for pollinators (e.g., pollen, water)[56],[57]. These crops also rely on neonicotinoids and may have currently undescribed risks for non-target organisms through indirect contaminant pathways in the agroecosystem [51],[58].

Possible exposure related to a high frequency of irrigation could drive the exposure of non-target arthropods to low concentrations of neonicotinoid insecticides in irrigation water. Although such impacts have yet to be documented directly, new comprehensive reviews of neonicotinoid environmental impacts have demonstrated numerous unanticipated impacts occurring at the ecosystem scale [9],[58]. In the Wisconsin agroecosystem, neonicotinoids are used on a large proportion of crops grown with irrigation [28],[29]. Trends in production show increased maize production over the past six years in the Central Wisconsin River Water Management Unit (Fig. 5). As a result of common neonicotinoid seed treatment on maize, accelerating production may partially explain the increased frequency of neonicotinoid detection in

groundwater. Unfortunately, little crop-specific pesticide information exists for individual neonicotinoids at the watershed scale [26]. Although measurement of specific contributions of crops to measured insecticide contamination is currently not available, this study demonstrates a research approach to better understand leaching from different application methods. Improved understanding of crops and insecticide delivery that results in greater risk of insecticide leaching will inform targets to reduce aquifer contamination and recirculation of soil-applied insecticides. Area-wide application of neonicotinoid insecticides through irrigation water applications may have considerable unanticipated or undocumented environmental impacts for non-target organisms through chronic low-dose exposure to insecticides.

Conclusions

To gain a better understanding of the seasonal cycle of neonicotinoids moving from the potato system, this study used an experimental approach to document the leaching potential of common neonicotinoid application methods. Results presented here benefit both potato producers and regulators by identifying trends in leachate losses for these commonly used, water-soluble insecticides. Lysimeter experiments documented loss of thiamethoxam following the application of vine desiccants at the conclusion of the potato production season. Leachate losses did vary among the different delivery methods over time indicating some variability in the patterns of pesticide leachate throughout the season. Quantification of crops commonly using neonicotinoid soil applications in the Central Wisconsin Water Management Unit highlights the need to research leaching potential from soil-applied neonicotinoids in other commodities. Documentation of several neonicotinoids in irrigation water suggests a new candidate pathway for non-target environmental impacts of insecticides.

Supporting Information

Figure S1 Chemical structures and properties of common neonicotinoid insecticides. Chemical structures were drawn using ChemDraw (version 13, Perkin Elmer Inc., Waltham, MA). Properties of each active ingredient were accessed from the National Center for Biotechnology Information PubChem online interface.

Figure S2 Irrigated field locations in the Central Wisconsin River Water Management Unit. Distribution of fields irrigated with high capacity wells (n = 2530) in the Central Wisconsin River Water Management Unit [27]. Points indicate locations of individual irrigation units identified from aerial photography using ArcGIS.

Acknowledgments

We thank the cooperating growers for generously allowing us to conduct lysimeter studies on their farm. We thank Amy DeBaker, Rick Graham, Jeff Postle, Wendy Sax, Stan Senger, and Steve Sobek of Wisconsin DATCP for their support of this project. We thank Dave Johnson and Robert Smail of Wisconsin DNR-Water Bureau for providing 2012 irrigation use data. We thank Birl Lowery and Mack Naber for input on lysimeter design and installation. We thank Scott Chapman, Ken Frost, and David Lowenstein for their help installing lysimeters. We thank Claudio Gratton, George Kennedy, Jessica Petersen, and Wesley Stone for their insightful comments on earlier versions of this manuscript. We thank the Wisconsin Potato and Vegetable Growers Association for continued support of our research efforts.

Author Contributions

Conceived and designed the experiments: ASH RLG. Performed the experiments: ASH. Analyzed the data: ASH. Wrote the paper: ASH RLG

References

1. Jeschke P, Nauen R, Schindler M, Elbert A (2010) Overview of the status and global strategy for neonicotinoids. J Agric Food Chem 59: 2897–2908.
2. Elbert A, Haas M, Springer B, Thielert W, Nauen R (2008) Applied aspects of neonicotinoid uses in crop protection. Pest Manag Sci 64: 1099–1105.
3. Jeschke P, Nauen R (2008) Neonicotinoids–from zero to hero in insecticide chemistry. Pest Manag Sci 64: 1084–1098.
4. United States Environmental Protection Agency (2003) Imidacloprid; pesticide tolerances. Fed Regist 68: 35303–35315.
5. United States Environmental Protection Agency (2012) What is the conventional reduced risk pesticide program? Available: http://www.epa.gov/opprd001/workplan/reducedrisk.html. Accessed 2012 Oct 17.
6. Szendrei Z, Grafius E, Byrne A, Ziegler A (2012) Resistance to neonicotinoid insecticides in field populations of the Colorado potato beetle (Coleoptera: Chrysomelidae). Pest Manag Sci 68: 941–946.
7. Cresswell JE, Desneux N, vanEngelsdorp D (2012) Dietary traces of neonicotinoid pesticides as a cause of population declines in honey bees: An evaluation by Hill's epidemiological criteria. Pest Manag Sci 68: 819–827.
8. Blacquiere T, Smagghe G, Van Gestel CA, Mommaerts V (2012) Neonicotinoids in bees: A review on concentrations, side-effects and risk assessment. Ecotoxicology 21: 973–992.
9. Goulson D (2013) An overview of the environmental risks posed by neonicotinoid insecticides. J Appl Ecol 50: 977–987.
10. Starner K, Goh KS (2012) Detections of the neonicotinoid insecticide imidacloprid in surface waters of three agricultural regions of California, USA, 2010–2011. Bull Environ Contam Toxicol 88: 316–321.
11. Miranda GR, Raetano CG, Silva E, Daam MA, Cerejeira MJ (2011) Environmental fate of neonicotinoids and classification of their potential risks to hypogean, epygean, and surface water ecosystems in Brazil. Human and Ecological Risk Assessment: An International Journal 17: 981–995.
12. Gupta S, Gajbhiye V, Agnihotri N (2002) Leaching behavior of imidacloprid formulations in soil. Bull Environ Contam Toxicol 68: 502–508.
13. Papiernik SK, Koskinen WC, Cox L, Rice PJ, Clay SA, et al. (2006) Sorption-desorption of imidacloprid and its metabolites in soil and vadose zone materials. J Agric Food Chem 54: 8163–8170.
14. Chiovarou ED, Siewicki TC (2008) Comparison of storm intensity and application timing on modeled transport and fate of six contaminants. Sci Total Environ 389: 87–100.
15. Gupta S, Gajbhiye V, Gupta R (2008) Soil dissipation and leaching behavior of a neonicotinoid insecticide thiamethoxam. Bull Environ Contam Toxicol 80: 431–437.
16. Juraske R, Castells F, Vijay A, Muñoz P, Antón A (2009) Uptake and persistence of pesticides in plants: measurements and model estimates for imidacloprid after foliar and soil application. J Hazard Mater 165: 683–689.
17. Wisconsin Department of Agriculture, Trade and Consumer Protection (2010) Fifteen years of the DATCP exceedence well survey. WI-DATCP, Madison, WI.
18. Wisconsin Department of Agriculture, Trade and Consumer Protection (2011) Agrichemical Management Bureau annual report – 2011. Available: http://datcp.wi.gov/Environment/Water_Quality/ACM_Annual_Report/. Accessed 2012 Jul 10.
19. United States Department of Agriculture - Natural Resources Conservation Soil Service (2013) Web Soil Survey. USDA-NRCS, Washington, DC. Available: http://websoilsurvey.sc.egov.usda.gov. Accessed 2014 Jan 8.
20. Cooley ET, Lowery B, Kelling KA, Speth PE, Madison FW, et al. (2009) Surfactant use to improve soil water distribution and reduce nitrate leaching in potatoes. Soil Sci 174: 321–329.
21. Wisconsin Department of Natural Resources (2013) DNR drinking water system: high capacity wells. WI DNR, Madison, WI. Available: http://dnr.wi.gov/topic/wells/highcapacity.html. Accessed 2013 Aug 22.

22. Wickham H (2007) Reshaping data with the reshape package. Journal of Statistical Software 21: 1–20. Available: http://www.jstatsoft.org/v21/i12/. Accessed 2011 Jan 15.
23. Bussan A, Colquhoun J, Cullen E, Davis V, Gevens A, et al. (2012) Commercial vegetable production in Wisconsin. Publication A3422. University of Wisconsin-Extension, Madison WI.
24. Seccia S, Fidente P, Barbini DA, Morrica P (2005) Multiresidue determination of nicotinoid insecticide residues in drinking water by liquid chromatography with electrospray ionization mass spectrometry. Anal Chim Acta 553: 21–26.
25. United States Department of Agriculture - National Agricultural Imagery Program (2010) Wisconsin NAIP. USDA-NAIP, Washington, DC. Available: http://datagateway.nrcs.usda.gov/. Accessed 2011 Jan 15.
26. United States Department of Agriculture - National Agricultural Statistics Service Cropland Data Layer (2012) Wisconsin Cropland data layer. USDA-NASS, Washington, DC. Available: http://nassgeodata.gmu.edu/CropScape/. Accessed 2013 May 10.
27. Wisconsin Department of Natural Resources (2002) Wisconsin DNR 2003 watersheds. Wisconsin DNR, Madison, WI. Available: http://dnr.wi.gov/maps/gis/documents/dnr_watersheds.pdf. Accessed 2013 Jun 12.
28. Thelin GP, Stone WW (2013) Estimation of annual agricultural pesticide use for counties of the conterminous United States, 1992–2009. US Department of the Interior, US Geological Survey.
29. Stone WW (2013) Estimated annual agricultural pesticide use for counties of conterminous United States, 1992–2009. U.S. Geological Survey Data Series 752, 1-p. pamphlet, 14 tables.
30. Cullen EM, Davis VM, Jensen B, Nice GRW, Renz M (2013) Pest management in Wisconsin field crops. Publication A3646. University of Wisconsin-Extension, Madison WI. Available: http://learningstore.uwex.edu/pdf/A3646.PDF as of 08/18/2013. Accessed 2013 Aug 23.
31. Team R Core (2011) R: A language and environment for statistical computing (Version 2.15.2). Vienna, Austria: R foundation for statistical computing; 2012. Available: http://cran.r-project.org. Accessed 2012 Jun 15.
32. Pinheiro J, Bates D, DebRoy S, Sarkar D (2007) Linear and nonlinear mixed effects models. R package version 3.1–108.
33. Pinheiro J, Bates D (2000) Mixed-effects models in S and S-PLUS. New York: Springer.
34. Mossbarger Jr W, Yost R (1989) Effects of irrigated agriculture on groundwater quality in corn belt and lake states. J Irrig Drain Eng 115: 773–790.
35. Kraft GJ, Stites W, Mechenich D (1999) Impacts of irrigated vegetable agriculture on a humid North-Central US sand plain aquifer. Ground Water 37: 572–580.
36. Saad DA (2008) Agriculture-related trends in groundwater quality of the glacial deposits aquifer, central Wisconsin. J Environ Qual 37: 209–225.
37. Postle JK, Rheineck BD, Allen PE, Baldock JO, Cook CJ, et al. (2004) Chloroacetanilide herbicide metabolites in Wisconsin groundwater: 2001 survey results. Environ Sci Technol 38: 5339–5343.
38. Agrian Inc. (2013) Advanced product search. Available: http://www.agrian.com/labelcenter/results.cfm. Accessed 2013 Mar 21.
39. Hladik ML, Calhoun DL (2012) Analysis of the herbicide diuron, three diuron degradates, and six neonicotinoid insecticides in water–Method details and application to two Georgia streams: U.S. Geological Survey Scientific Investigations Report 2012–5206.
40. Hopwood J, Vaughan M, Shepherd M, Biddinger D, Mader E, et al. (2012) Are neonicotinoids killing bees? A review of research into the effects of neonicotinoid insecticides on bees, with recommendations for action. Xerces Society for Invertebrate Conservation, USA.
41. Casida JE (2012) The greening of pesticide–environment interactions: Some personal observations. Environ Health Perspect 120: 487–493.

42. Krupke CH, Hunt GJ, Eitzer BD, Andino G, Given K (2012) Multiple routes of pesticide exposure for honey bees living near agricultural fields. PLoS ONE 7: e29268.

43. Gill RJ, Ramos-Rodriguez O, Raine NE (2012) Combined pesticide exposure severely affects individual-and colony-level traits in bees. Nature 491: 105–108.

44. Seagraves MP, Lundgren JG (2012) Effects of neonicotinoid seed treatments on soybean aphid and its natural enemies. J Pest Sci 85: 125–132.

45. Cresswell JE, Page CJ, Uygun MB, Holmbergh M, Li Y, et al. (2012) Differential sensitivity of honey bees and bumble bees to a dietary insecticide (imidacloprid). Zoology 115: 365–371.

46. Henry M, Beguin M, Requier F, Rollin O, Odoux J, et al. (2012) A common pesticide decreases foraging success and survival in honey bees. Science 336: 348–350.

47. Stoner KA, Eitzer BD (2012) Movement of soil-applied imidacloprid and thiamethoxam into nectar and pollen of squash (Cucurbita pepo). PloS ONE 7: e39114.

48. Tapparo A, Marton D, Giorio C, Zanella A, Soldà L, et al. (2012) Assessment of the environmental exposure of honeybees to particulate matter containing neonicotinoid insecticides coming from corn coated seeds. Environ Sci Technol 46: 2592–2599.

49. Tomé HVV, Martins GF, Lima MAP, Campos LAO, Guedes RNC (2012) Imidacloprid-induced impairment of mushroom bodies and behavior of the native stingless bee Melipona quadrifasciata anthidioides. PloS ONE 7: e38406.

50. Chauzat M, Faucon J, Martel A, Lachaize J, Cougoule N, et al. (2006) A survey of pesticide residues in pollen loads collected by honey bees in France. J Econ Entomol 99: 253–262.

51. Girolami V, Mazzon L, Squartini A, Mori N, Marzaro M, et al. (2009) Translocation of neonicotinoid insecticides from coated seeds to seedling guttation drops: A novel way of intoxication for bees. J Econ Entomol 102: 1808–1815.

52. Laurent FM, Rathahao E (2003) Distribution of (14C) imidacloprid in sunflowers (Helianthus annuus L.) following seed treatment. J Agric Food Chem 51: 8005–8010.

53. Mullin CA, Frazier M, Frazier JL, Ashcraft S, Simonds R, et al. (2010) High levels of miticides and agrochemicals in North American apiaries: Implications for honey bee health. PLoS ONE 5: e9754.

54. Whitehorn PR, O'Conner S, Wackers FL, Goulson D (2012) Neonicotinoid pesticide reduces bumble bee colony growth and queen production. Science 336: 351–352.

55. Dively GP, Kamel A (2012) Insecticide residues in pollen and nectar of a cucurbit crop and their potential exposure to pollinators. J Agric Food Chem 60: 4449–4456.

56. Free JB (1993) Insect pollination of crops. London: Academic Press.

57. Klein A, Vaissière BE, Cane JH, Steffan-Dewenter I, Cunningham SA, et al. (2007) Importance of pollinators in changing landscapes for world crops. Proc R Soc B 274: 303–313.

58. Sánchez-Bayo F, Tennekes HA, Goka K (2013) Impact of systemic insecticides on organisms and ecosystems. In Stanislav T, editor, Insecticides - Development of Safer and More Effective Technologies. Rijeka: InTech. 367–416.

11

Affordable Nutrient Solutions for Improved Food Security as Evidenced by Crop Trials

Marijn van der Velde[1]*, Linda See[1], Liangzhi You[2,3], Juraj Balkovič[1], Steffen Fritz[1], Nikolay Khabarov[1], Michael Obersteiner[1], Stanley Wood[4]

1 International Institute for Applied Systems Analysis (IIASA), Ecosystem Services and Management Program, Laxenburg, Austria, 2 International Food Policy Research Institute (IFPRI), Washington D.C., United States of America, 3 College of Economics and Management, Huazhong Agricultural University, Wuhan, China, 4 Global Development Program, Bill & Melinda Gates Foundation, Seattle, Washington, United States of America

Abstract

The continuing depletion of nutrients from agricultural soils in Sub-Saharan African is accompanied by a lack of substantial progress in crop yield improvement. In this paper we investigate yield gaps for corn under two scenarios: a micro-dosing scenario with marginal increases in nitrogen (N) and phosphorus (P) of 10 kg ha^{-1} and a larger yet still conservative scenario with proposed N and P applications of 80 and 20 kg ha^{-1} respectively. The yield gaps are calculated from a database of historical FAO crop fertilizer trials at 1358 locations for Sub-Saharan Africa and South America. Our approach allows connecting experimental field scale data with continental policy recommendations. Two critical findings emerged from the analysis. The first is the degree to which P limits increases in corn yields. For example, under a micro-dosing scenario, in Africa, the addition of small amounts of N alone resulted in mean yield increases of 8% while the addition of only P increased mean yields by 26%, with implications for designing better balanced fertilizer distribution schemes. The second finding was the relatively large amount of yield increase possible for a small, yet affordable amount of fertilizer application. Using African and South American fertilizer prices we show that the level of investment needed to achieve these results is considerably less than 1% of Agricultural GDP for both a micro-dosing scenario and for the scenario involving higher yet still conservative fertilizer application rates. In the latter scenario realistic mean yield increases ranged between 28 to 85% in South America and 71 to 190% in Africa (mean plus one standard deviation). External investment in this low technology solution has the potential to kick start development and could complement other interventions such as better crop varieties and improved economic instruments to support farmers.

Editor: Luis Herrera-Estrella, Centro de Investigación y de Estudios Avanzados del IPN, Mexico

Funding: The Austrian Research Funding Agency (FFG) funded projects. The funders had no role in study design, data collection and analysis, decision to publish, or preparation of the manuscript. This research was supported by the Austrian Research Funding Agency (FFG) for the Project FarmSupport (No. 833421).

Competing Interests: The authors have declared that no competing interests exist.

* E-mail: velde@iiasa.ac.at

Introduction

Farming looks mighty easy when your plow is a pencil and you're a thousand miles from the corn field. –Dwight D. Eisenhower, 1956.

The increases in global population and food demand clearly indicate that current growth in agricultural productivity is not sufficient to sustain the 9 billion people that will inhabit the Earth by 2050 [1]. Feeding the world is a multifaceted and complex challenge and a number of solutions have been offered, where closing the yield gap is one of the most frequently cited recommendations [2;3;4;5;6]. The FAO [1] suggests that 70% of the required increase in crop production in developing countries should be realized through boosting the productivity of fields already under cultivation. Without this intensification, the inevitable cropland expansion will lead to deforestation, accelerate land degradation and threaten natural habitats and biodiversity [7]. A large part of this augmentation in crop production must therefore come from soils in tropical regions which are often highly weathered, have low levels of chemical soil fertility and will need additional inputs to improve crop productivity [8]. Global fertilizer use has already increased significantly since 1960 [9] and this increase has played an important role in the Green Revolution, benefitting many developing countries in South America and Asia. Yet substantial progress is still lacking in Africa [10]. From 1960 to 2000, yields of staple crops such as wheat, rice and corn increased in South America by over 180% while African yields did not improve substantially (see Figure 1). These contrasting trajectories reflect disparities in infrastructure development, primary crop types grown, agricultural R&D and extension capacities, socio-economic conditions as well as environmental differences [10;11;12].

In many smallholder fields, fertilizer and manure inputs have been too low for too long [8]. Agricultural soils cultivated without adequate nutrient replenishment cannot reach their full crop production potential and are at risk of irreversible degradation [13]. In large parts of Africa, this has led to seemingly perpetual low per capita food production [8]. Without maintaining adequate soil fertility levels, crop yields cannot be sustained, increase over time, or respond to improved agricultural management practices.

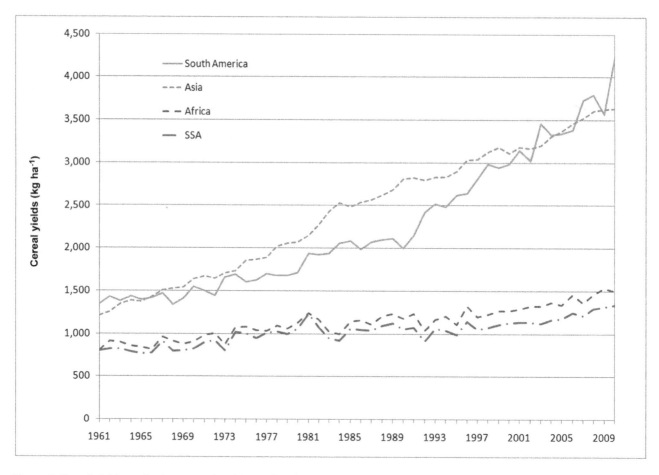

Figure 1. Cereal yield trends since 1960 in Africa, Sub-Sahara Africa (SSA), South America and Asia.

To improve nutrient input, smallholder farmers need actionable strategies such as micro-dosing: applications of small quantities of fertilizers. Field studies have shown that micro-dosing presents an attainable strategy for smallholder farmers in line with their financial means that can result in significant yield gains [14]. Importantly, previous higher fertilizer rate recommendations have ignored the sizeable but unlikely investment that would be required by poor and risk adverse smallholders.

Fertilizer prices in Africa are often higher relative to other developing countries. The small size of the fertilizer market, the high transportation and handling costs and the inefficient supply chain all contribute to the relatively high retail prices for fertilizers in Africa [15]. Fertilizer price also varies across regions, through different years and even among cropping seasons in the same year. In a landlocked country such as Uganda, the prices for urea in 2000 ranged from 600 shillings kg^{-1} (300 US\$ ton^{-1}) in the central districts to over 750 shillings kg^{-1} (375 US\$ ton^{-1}) in the eastern districts while prices for phosphate (Diammonium Phosphate, DAP) ranged from 560 shillings kg^{-1} (280 US\$ ton^{-1}) in the long rainy season to over 700 shillings kg^{-1} (350 US\$ ton^{-1}) in the short rainy season [16]. As a comparison, US farmers in 2000 paid from US\$80 to US\$120/ton for Urea, and US\$140 to US\$170/ton for DAP. Depending on the locations and seasons, the NPK (Nitrogen-Phosphorus-Potassium) compound fertilizer ranged from 700 shillings kg^{-1} (350 US\$ ton^{-1}) to over 1000 shillings kg^{-1} (500 US\$ ton^{-1}) for 1:1:2 NPK, the most common compound fertilizer in Uganda. The transportation cost in Uganda is over one third of the total fertilizer cost while it is

about a fifth in Tanzania [17] due to the major sea port of Dar Es Salaam. Moreover, there are global pressures that lead to price volatility in fertilizer prices. For example, there was a fourfold increase in the price of urea from 2000 to 2008, reaching over 500 US\$ ton^{-1}, falling to around 200 US\$ ton^{-1} in 2009 and which is currently at around 400 US\$ ton^{-1}.

To formulate more realistic sustainable intensification pathways, we need better estimates of smallholder yield gaps in tropical countries and to then align these with local fertilizer prices and associated investment costs. There are a number of yield gap approaches that estimate different types of attainable yield potentials across varying spatial and temporal scales [18]. Many global assessments of yield gaps use crop models or data that currently lack sufficiently detailed spatial information on soil characteristics, crop management practices and crop responses to fertilizers [19,20]. Mueller et al. [19] found that large crop production increases are possible, but will require considerable changes in nutrient and water management. Crop trials represent a valuable source of information for yield gap analysis and could be analyzed more comprehensively for this purpose, yet are rarely collected systematically. Furthermore, nutrient specific analysis of the relationships between fertilizers and crop yields has been limited, especially in the tropical and subtropical regions where crop yields are relatively low and must increase the most to meet growing demands [21].

In this paper we analyze historic data from FAO corn fertilizer trials carried out between 1969–1993 at 1358 locations in Africa and South America (Figure 2) where corn is the most commonly cultivated crop [18]. Mitscherlich-Baule crop response functions were fit to the crop trial data by optimizing the factors describing yield responses to elemental nitrogen (N) and phosphorus (P) inputs as well as an initial (residual) soil N and P. Nitrogen and Phosphorus specific fertilizer application rates from [22] were then used as inputs to the crop response functions, and the resulting yields were validated using sub-national yield statistics on corn from the International Food Policy Research Institute (IPFRI) [23] to estimate yield gaps in corn at the continental level. Only water-limited (i.e. rain fed) yield potential as opposed to irrigated yields potentials are considered here. Furthermore, we consider the importance of soil nutrient stoichiometry and the viability of micro-dosing [14,24] in order to further the African crop productivity discourse [25,26]. Yield increases associated with two different scenarios are considered here: 1) depicting a micro-dosing strategy and 2) a topping up to a conservative estimate of average nutrient fertilizer rates in the USA. Finally, the investment costs of scaling up these scenarios are calculated using the latest average fertilizer prices in Africa and South America and used to evaluate whether these approaches can function as part of an actionable development blueprint for Sub-Saharan Africa.

Materials and Methods

Crop Trials and Response Functions

Recently historic FAO crop fertilizer field trials have become publically accessible (http://www.fao.org/ag/agl/agll/nrdb/). These data were collected as part of FAO's Fertilizer Programme [27] that ran from 1969 until 1993. The purpose behind the programme was to undertake trials to determine suitable fertilizer application rates for locally grown crops and to demonstrate to as many farmers as possible, the positive effect of fertilizer application on crop yields and farm income. The information available from the trials includes crop yields (kg ha^{-1}), and application rates of the main nutrients (nitrogen (N), phosphorus (P) and potassium (K) and farmyard manure (kg ha^{-1})). Unfortunately, detailed soil or

meteorological information was not recorded. Nitrogen was mostly applied as urea, phosphorus (P) was mostly applied as superphosphate and potassium as part of NPK compound fertilizers. All the applied nutrients were recalculated to elemental application rates (with P calculated from P_2O_5 and K from K_2O). The application rates of farm yard manure were converted to N, P-P_2O_5 and K-K_2O application rates following [28]. Data on corn yields from trials with at least five N and P input combinations were selected for this analysis. The Mitscherlich-Baule crop response function was used to analyze relations between nitrogen (N) fertilizer input, phosphorus-phosphate (P) fertilizer input and corn yields y_{mb}:

$$y_{mb} = a_1[1 - \exp(-a_2(a_3 + N))] * [1 - \exp(-a_4(a_5 + p))] \quad (1)$$

The function allows for growth that plateaus with increasing fertilizer application, and accommodates cases of both near perfect factor substitution and near zero factor substitution, and performs superior to a quadratic and von Liebig type production function [29]. The growth plateau is represented by a_1, which was set equal to the maximum yield obtained in each field trial, while a_3 and a_5 represent the residual available nitrogen and phosphorus in the soil. Taking account of residual soil phosphorus is important; the cumulative cropland P surplus in certain countries in Western Europe has led to a buildup of residual soil P with expected future benefits to crop production, although lower and no effects are generally expected for Latin America and Africa [30]. The coefficients a_2 and a_4 describe the influence of the corresponding N or P fertilization on yield. The parameters a_2, a_3, a_4, and a_5 were obtained by minimizing the sum of squared errors for all applications in each experiment. The Nelder-Mead multidimensional unconstrained nonlinear minimization algorithm was used to minimize the objective function. Only those trials where a crop response function could be fit were used. This resulted in a total of 1358 unique experiments with at least five N and P input combinations; 752 in Africa and 606 in South America.

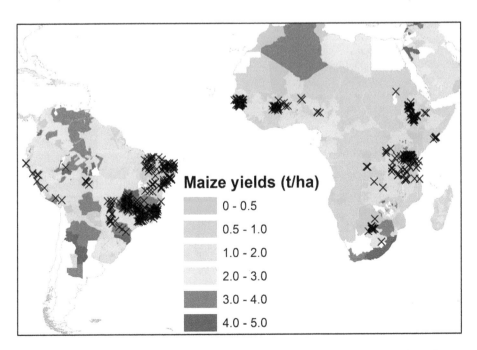

Figure 2. Crosses indicate locations of 1358 historic FAO corn field trials with at least five N and P input combinations in Africa and South America carried out between 1969 and 1993. Colors indicate (subnational) maize (corn) yields (ton/ha) as collected by [23].

Current Fertilizer Application Rates

The fertilizer dataset of [22] containing crop fertilizer rates was used to assign the current N and P inputs from chemical fertilizer (Nfer, Pfer) and manure (Nman, Pman; see Figures S1, S2, S3) at the trial locations and considered representative for corn fertilizer rates [31]. Data on corn yields collected by IFPRI [23] were used as a comparison with the yields obtained from the crop response functions (Figure S4). The average cereal area, production and yield as reported by FAOSTAT for 2008–2010 were calculated for each of the scenarios. In the area scenario a constant production was assumed and an increase in yield would reduce the requirement for cropland. In the people scenario an increase in yield would produce more on the same cropland area. We assume the average cereal calorie content to be 3000 kcal kg^{-1} and the average annual calorie need of a person to be 1 million kcal $year^{-1}$.

Costs

The cost of the two proposed scenarios was calculated using the Agricultural GDP in US dollars for 2009 [32]: South America (SA) $192 billion USD; Sub-Saharan Africa (SSA) $150 billion USD. Then the costs for each region were calculated by taking the total cost, dividing by the Agricultural GDP and multiplying by 100 to arrive at the percentage of Agricultural GDP that would be required to finance the scenario.

Results

An example of crop response trial data and the corresponding modeled Mitscherlich-Baule crop response function is shown in Figure 3. The median r^2 obtained by fitting the individual crop trials equaled 0.81; the 25th percentile equaled 0.66 and the 75th percentile 0.91. The resulting median, 25th and 75th percentiles for the a_1, a_2, a_3 and a_4 parameters obtained across all crop trials are presented in Table 1. Median values corresponded to 0.017 ton kg^{-1}, 68.4 kg N ha^{-1}, 0.29 ton kg^{-1}, 3.18 kg P ha^{-1} for parameters a_1, a_2, a_3 and a_4 respectively. This is in correspondence with the parameter values reported by [29] and [33]. The results from the individual crop trials indicate that out of the 1358 trials, there were 1037 trials (76%) that responded stronger to added phosphorus than nitrogen. Similarly, for 82% of the trials the pool of residual soil N was larger than the accessible residual P. Clearly, these site-specific analyses indicate that overall, phosphorus is the nutrient most limiting crop yield. Nevertheless, at the same time, the range of parameter values obtained highlights the variety of crop yield responses depending on site-specific conditions.

Overall, the yields modeled using Mitscherlich-Baule crop response functions show a good relationship ($r^2 = 0.94$, Figure 4).

Yield Gaps and Potentials

The fertilizer dataset of Potter et al. [22] was used to assign the current N and P inputs from chemical fertilizer (Nfer, Pfer) and manure (Nman, Pman) to the crop response functions; the average of these inputs and their distribution across Africa and South America as well as manure and fertilizer nutrient specific histograms are shown in Figures S1, S2, S3. Corn yield from the crop responses functions was compared to IFPRI reported data in Figure S4; median yields are comparable but there is a much larger variability in the yields derived from the crop response functions. This reflects both the coarser resolution of the IFPRI data [23] and the more realistic representation of the frequency distribution of yields that are attained at individual locations across both continents from the FAO crop trial data.

To indicate the potential for production increase, we calculated the average percentage yield increases resulting from an additional application of 10 kg N ha^{-1}, 10 kg P ha^{-1}, and both. Adding only N will lead to increases in crop production by ~4% and ~8%. Adding only P, on the other hand, will lead to substantially larger increases of ~12% and ~26% for South America and Africa respectively (Figure 5). This highlights the critical importance of P, of which many subsistence farmers may not be aware. The addition of both nutrients leads to increases of ~15% and ~35%, respectively, indicating that the effect of both nutrients is additive once P is applied. Thus P is clearly the limiting nutrient in improving crop yields.

In Africa, adding 10 kg ha^{-1} of N or P will result in mean and median percentage increases of respectively ~5.5 and ~5.7% and ~11.7 and ~16.3%; this would thus bring a significant proportion of farmers with the lowest yields closer towards attaining average yield levels and effectively shift a bulk of smallholders out of current marginal productivity. Since these are indicative for rainfed yields, additional water resources to attain these yield increases would not be required [6]. A final experiment considers the percentage yield increase that would be obtained if 80 kg ha^{-1} of N and 20 kg ha^{-1} of P - a relatively conservative estimate of average rates in the USA - were applied (Figure 6). In South America this would lead to average yield increases of 30%, up to a maximum of 90% while in Africa these increases would be considerably larger, i.e. average yield increases of 70%, up to a maximum of 190%.

Even though these yield increases are considerable, they are lower than the yield potentials generally estimated in other studies. For example, yield gaps of 180 to 540% for maize have been estimated for sub-Saharan Africa [34] while a yield gap of 118% was found by Tittonell et al. [35] for western Kenya. Since most of the crop trials were done more than 20 years ago, our results will provide conservative yield estimates. In soils that have become increasingly depleted, and with new and better crop varieties that have become available since then, the response to fertilizer may be even stronger than predicted here.

Implications

The implications for cropland expansion will be significant. Unless both N and P nutrient inputs are increased considerably and other complementary inputs and rural services such as seed, irrigation, market access and extension are available to improve crop yields, then necessary crop production gains will largely come from cropland expansion. This would have considerable negative impact on forest and grassland habitats and biodiversity [5]. In contrast, improving cereal yield by just 5% globally, over 33 million ha of forest or grassland would be saved. To put the benefits of these higher yields in context, our first scenario of applying an additional 10 kg N ha^{-1} and 10 kg P ha^{-1} (leading to a corn yield increase by 15% and 32% in South America and Africa) would save more than 4 million ha and 25 million ha of cropland conversion in South America and Africa respectively. Alternatively, if such yield improvement occurs in the currently cultivated cereal areas on these two continents, the improved productivity could feed an additional 64 million and 150 million people respectively.

Costs

Such a scenario would require a total investment of US$148 million in sub-Sahara Africa (using an average of fertilizer prices paid by farmers in 2012 of Urea ($620/ton or 0.29$ kg^{-1} N) and DAP ($950/ton or 0.22$ kg^{-1} P) and US$79 million in South America (using an average fertilizer price paid by farmers in 2012

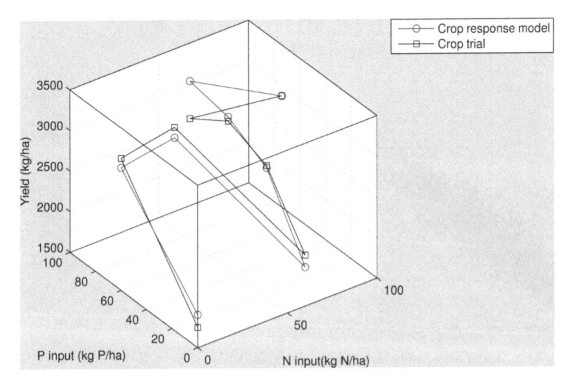

Figure 3. A typical example of a fitted crop response trial with experimental (blue line with circles) and modeled data (black line with squares) with eight N and P input combinations and resulting yields.

of Urea ($460/ton or 0.22$ kg^{-1} N) and DAP ($680/ton or 0.16$ kg^{-1} P)). The second scenario or larger nutrient inputs would amount to investments of US$798 million in sub-Sahara Africa and US$428 million in South America respectively. Maize yield increases range from 15% to over 70%, and such an investment would therefore bring considerable additional revenue to maize farmers. We acknowledge the fact that the prices of both fertilizer and maize vary with location so the actual profitability of fertilizer investment would vary spatially.

However, the direct investment in fertilizers is actually very small and is less than 1% of Agricultural GDP of both Sub-Saharan Africa and South America for the micro-dosing scenario. The calculation of the investment in terms of the percentage of Agricultural GDP that would be needed in SSA for scenario 1 equates to dividing 148 million USD by the Agricultural GDP of $150 billion USD multiplied by 100. The percentage of investment in terms of Agricultural GDP for the other scenario for SSA and the scenarios for SA were calculated in the same way (see Table 2).

Table 1. The median, 25th and 75th percentile values from the distributions of the Mitscherlich-Baule crop response function parameters (a_1, a_2, a_3 and a_4) fitted for the 1358 individual crop trials.

Parameter	Median	25th percentile	75th percentile
a1	0.017321	0.0077335	0.047077
a2	68.4297	21.6344	285.0599
a3	0.29219	0.047732	0.29219
a4	3.1803	0.99095	13.0806

Discussion

In reality a full development blueprint would need to have a broader scope and costs would be compounded with investments in roads, agricultural extension, (local) market access, etc. [5,6]. Nevertheless, external investment in this low technology solution has the potential to kick start development and could complement other interventions such as better crop varieties, improved economic instruments to support farmers as well as new technologies involving mobile phones, crowdsourcing and data mining of internet searches [36,37]. To improve our understanding of how best to target, design and support rural development, it is insightful to compare the costs calculated here with the costs involved in a project such as the Millennium Villages (MV). The MV is an integrated approach to eradicate poverty by involving an entire community in improving their livelihoods and health in a sustainable way (http://www.millenniumvillages.org). The focused investments calculated here are significantly lower per person per year compared to the costs in the MV, which vary between 35 to 100 USD per person per year [38]. However, we clearly acknowledge that the MV has a much broader scope, engages entire communities and contributes to many other aspects of well-being and improved livelihoods as set out in the Millennium Development Goals, which would not be part of the scenarios suggested here.

We have clearly demonstrated the importance of phosphorus for closing yield gaps in Africa. The phosphorus deficiency reflects soil P supply problems that are of widespread concern in highly weathered tropical soils notorious for low levels of available P and exhibit a strong P fixation capacity [39]. For instance, approximately 82% of the land area of the American tropics is deficient in P in its natural state [40]. Combined with the fact that P reserves are likely to become exhausted during the next 30–300 years [41], this paints a bleak picture indeed. Nevertheless, if recycling programs were put in place for animal and

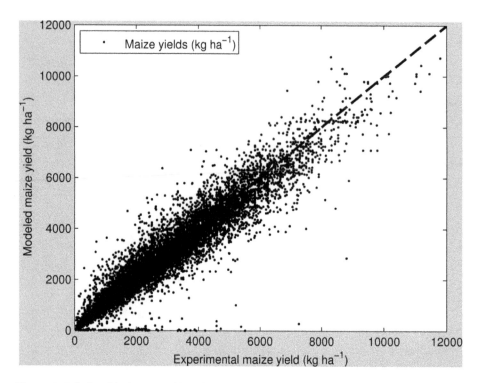

Figure 4. Relationship between historic FAO experimental corn field trials with at least five N and P input combinations and corn yields calculated with the Mitscherlich-Baule crop response function totaling 1358 unique nutrient-yield relations ($r^2 = 0.94$).

human excreta some of these effects might be mitigated, e.g. [42]. Raising awareness of the need to provide a more balanced stoichiometry is also a critical element in improving yields. If farmers continue to add increased supplies of N without P, they will soon reach a saturation point in yields and effectively waste valuable resources as well as contaminating groundwater due to the leaching of nitrogen. This finding also has implications for the current fertilizer subsidy programs in many developing countries. Most of these programs focus mainly on N and do not emphasize the importance of P sufficiently. Our results demonstrate that better balanced subsidy

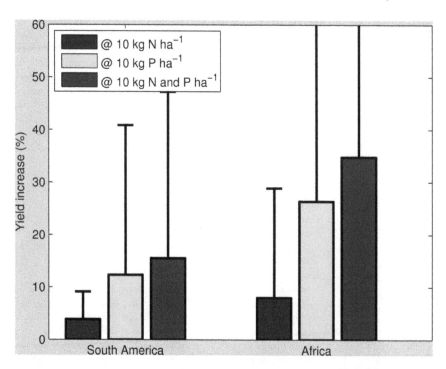

Figure 5. Mean corn yield increase (%) across trial sites at additional applications of 10 kg N ha^{-1}, 10 kg P ha^{-1} or 10 kg N and P ha^{-1} (error bars refer to the standard deviation of the obtained yield increases observed across all trials).

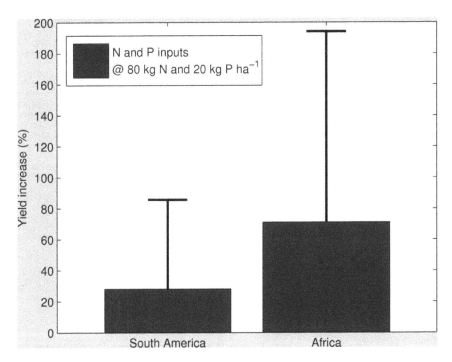

Figure 6. Mean corn yield increase (%) across trial sites at applications of 80 kg N ha^{-1} and 20 kg P ha^{-1} (error bars refer to the standard deviation of the obtained yield increases observed across all trials).

schemes taking account of both N and P would have a larger effect on crop yields.

Climate change is expected to generally have a negative impact on corn in Africa with estimates of lowering yield ranging from 3 to over 12% [43]. Corn will be the most heavily impacted crop in sub-Saharan Africa, where yield losses would occur in 65% of corn growing regions for a 1°C warming, increasing to 100% losses in areas subject to drought stress [44]. Although these pressures are considerable, and will require adaptation and fundamental changes to agricultural management, our results indicate that significant increases in yield are possible by improved nutrient management; especially during growing seasons when soil water availability is not constraining crop yields. Thus, achieving trend-growth in crop productivity through sufficient and balanced nutrient applications coupled with effective storage policies could partly offset negative climate change impacts on food security in Africa. The risk averseness of poor farmers that are prone to drought is one of the reasons why these farmers will be hesitant to invest in higher fertilizer applications. However, we have shown that a relatively small external investment would yield large improvements in both crop production and sustainable use of soils.

In contrast to previous crop productivity assessments for Africa and South America, our study allows farm input management interventions to be directly based on small scale on-the-ground observations accounting for both site-specific conditions, as well as reflecting variability in soil conditions and climates. Furthermore, the study provides yield gaps that are realistically attainable as the assessment is based on farmers' field trials and the costs associated with these interventions amount to less than 1% of Agricultural GDP in both Sub-Saharan Africa and South America. Crop field trials might be considered costly by some but provide essential and hard-won insights, and when analyzed comprehensively, have the potential - through better formulated policies and agreements - to reward global society with improved food security status for many.

Supporting Information

Figure S1 Current N and P inputs from chemical fertilizer (Nfer, Pfer) and manure (Nman, Pman) extracted and averaged from [22] for the 1358 trial locations.

Figure S2 Histograms of the N and P nutrient inputs from chemical fertilizer at the 1358 locations [22].

Figure S3 Histograms of the N and P nutrient inputs from manure at the 1358 locations [22].

Figure S4 Boxplots of regionally reported corn yields collected by IFPRI and corn yields obtained from the 1358 crop response functions (CRFs) with current N and P inputs from chemical fertilizer and manure (Nman, Pman) as reported by [22].

Table 2. Cost of the proposed scenarios expressed as the percentage of Agricultural GDP.

Region	Cost as a % of Agricultural GDP	
	Scenario 1	Scenario 2
SSA	0.10%	0.53%
SA	0.04%	0.22%

Author Contributions

Analyzed the data: MvdV LS LY. Wrote the paper: MvdV LS LY JB SF NK MO SW.

References

1. FAO (2009). How to Feed the World in 2050. Report from the High-Level Expert Forum. http://www.fao.org/fileadmin/templates/wsfs/docs/expert_paper/How_to_Feed_the_World_in_2050.pdf. Accessed 2013 Feb 28.

2. Rosegrant MW, Cline SA (2003) Global food security: challenges and policies. Science 302: 1917–1918.

3. Godfray HCJ, Beddington JR, Crute IR, Haddad L, Lawrence D, et al. (2010) Food security: the challenge of feeding 9 billion people. Science 327(5967): 812–818. doi: 10.1126/science.1185383.

4. Licker R, Johnston M, Foley JA, Barford C, Kucharik CJ, et al. (2010) Mind the gap: how do climate and agricultural management explain the 'yield gap' of croplands around the world? Global Ecology and Biogeography 19: 769–782.

5. Foley JA, Ramankutty N, Brauman KA, Cassidy ES, Gerber JS, et al. (2011) Solutions for a cultivated planet. Nature 478: 337–342.

6. Mueller ND, Gerber JS, Johnston M, Ray DK, Ramankutty N, et al. (2012) Closing yield gaps through nutrient and water management. Nature. Doi:10.1038/nature11420.

7. Maitima JM, Mugatha SM, Reid RS, Gachimbi LN, Majule A, et al. (2009) The linkages between land use change, land degradation and biodiversity across East Africa. African Journal of Environmental Science and Technology 3(10): 310–325.

8. Sanchez PA (2002) Soil fertility and hunger in Africa. Science 295: 2019–2020.

9. Bumb B, Baanante C (1996) World trends in fertilizer use and projections to 2020. 2020 Brief No. 38, International Food Policy Research Institute, Washington, DC, USA.

10. Ejeta G (2010) African Green Revolution needn't be a mirage. Science 327: 831–832.

11. Thurlow R, Kilman S (2009) Enough: Why the world's poorest starve in an age of plenty. New York: Perseus Books. 302 p.

12. Nin-Pratt A, Johnson M, Magalhaes E, You L, Diao X, et al. (2011) Yield gaps and potential growth in Western and Central Africa. IFPRI Research Monograph. International Food Policy Research Institute (IFPRI), Washington, USA. DOI: 10.2499/9780896291829.

13. Lal R (2010) Managing soils for a warming earth in a food-insecure and energy-starved world. J Plant Nutr Soil Sci 173: 4–15.

14. Tmowlow S, Rohrbach D, Dimes J, Rusike J, Mupangwa W, et al. (2010) Micro-dosing as a pathway to Africa's Green Revolution: evidence from broad-scale on-farm trials. Nutr Cycl Agroecosyst 88: 3–15.

15. The World Bank (2007) Africa fertilizer policy toolkit: Promoting fertilizer use in African agriculture: Lessons learned and good practice guidelines. http://www.worldbank.org/html/extdr/fertilizeruse/about.html. Accessed 2012 August 22.

16. UBOS (2002) Uganda national household survey 1999/2000. Report on the crop survey module. Entebbe, Uganda: Uganda Bureau of Statistics: 4–30.

17. Bumb B, Johnson M, Fuente PA (2011) Policy Options for Improving Regional Fertilizer Markets in West Africa. IFPRI Discussion Paper 01084, International Food Policy Research Institute, Washington, DC, USA.

18. Lobell DB, Cassman KG, Field CB (2009) Crop yield gaps: their magnitudes, and causes. Annu Rev Environ Resourc 34: 179–204.

19. Mueller ND, Gerber JS, Johnston J, Ray DK, Ramankutty N, et al. (2012) Closing yield gaps through nutrient and water management. Nature 490: 254–257.

20. Sanchez PA, Ahamed S, Carré F, Hartemink AE, Hempel J, et al. (2009) Digital soil map of the world. Science 325(5941): 680–681.

21. FAO (2011) The state of the world's land and water resources for food and agriculture. Managing systems at risk. FAO and Earthscan.

22. Potter P, Ramankutty N, Bennett EM, Donner SD (2010) Characterizing the spatial patterns of global fertilizer application and manure production. Earth Interac 14: 1–22.

23. IFPRI (2012) Global subnational crop database. IFPRI, Washington, DC, USA.

24. Van der Velde M, See L, Fritz S (2012) Soil remedies for small-scale farming. Nature 484: 318, doi:10.1038/484318c.

25. Gilbert N (2012) Africa agriculture: dirt poor. Nature 483: 525–527, doi:10.1038/483525a.

26. Bindraban PS, Van der Velde M, Ye L, Van den Berg M, Materechera S, et al. (2012) Assessing the impact of soil degradation on food production. Current Opinion in Environmental Sustainability, 4(5), 478–488, doi:10.1016/j.cosust.2012.09.015.

27. FAO (1989) Fertilizers and food production: Summary review of trial and demonstration results 1961–1986. FAO, Rome.

28. Van Averbeke W, Yoganathan S (1997) Using Kraal Manure as a Fertiliser. Department of Agriculture and Agricultural and Rural Development Research Institute, Fort Hare.

29. Frank MD, Beattie BR, Embleton ME (1990) A comparison of alternative crop response models. Am J Agric Econ 72(3): 597–603.

30. Sattari SZ, Bouwman AF, Giller KE, van Ittersum MK (2012) Residual soil phosphorus as the missing piece in the global phosphorus crisis puzzle. Proc Natl Acad Sci U S A doi: 10.1073/pnas.1113675109.

31. IFA (2002) Fertiliser use by crop. Rome, 2002, 45 pp.

32. The World Bank (2011) Data. http://data.worldbank.org/indicator. Accessed 2013 Feb 28.

33. Finger R, Hediger W (2008) The application of robust regression to a production function comparison. The Open Agriculture Journal 2: 90–98.

34. Pingali PL, Pandey S (2001) World maize needs meeting: technological opportunities and priorities for the public sector. In: PL Pingali (Ed.) CIMMYT 1999–2000 World Maize Facts and Trends. Meeting World Maize Needs: Technological Opportunities and Priorities for the Public Sector, 1–24. Mexico: CIMMYT.

35. Tittonell P, Vanlauwe B, Corbeels M, Giller KE (2008) Yield gaps, nutrient use efficiencies and response to fertilizers by maize across heterogeneous smallholder farms of western Kenya. Plant Soil 313: 19–27.

36. Fritz S, McCallum I, Schill C, Perger C, See L, et al. (2012) Geo-Wiki: An online platform for improving global land cover. Environmental Modelling & Software 31: 110–123, doi: 10.1016/j.envsof.2011.11.015.

37. Van der Velde M, See L, Fritz S, Verheijen FGA, Khabarov N, et al. (2012) Generating crop calendars with Web search data. Environmental Research Letters 7(2): 024022, doi:10.1088/1748-9326/7/2/024022.

38. Pronyk P (2012) The Costs and Benefits of the Millennium Villages: Correcting the Center for Global Development. http://www.millenniumvillages.org/field-notes/archive/2012/4. Accessed 2013 Feb 28.

39. Buresh RJ, Smithson PC, Hellums DT (1997) Building soil phosphorus capital in Africa. In Buresh RJ, Sanchez PA and Calhoun F (eds.) Replenishing soil fertility in Africa. SSSA/ASA. Madison, WI. 111–149.

40. Sanchez PA, Salinas JG (1981) Low-input technology for managing Oxisols and Ultisols in tropical America. Advances in Agronomy 34: 279–406.

41. Cordell D, Drangeert JO, White S (2009) The story of phosphorus: Global food security and food for thought. Glob Environ Change 19: 292–305.

42. Baker LA (2011) Can urban P conservation help to avoid the brown devolution. Chemosphere 84: 779–784.

43. Nelson GC, Rosegrant M, Palazzo A, Gray I, Ingersoll C, et al. (2010) Food security, farming, and climate change to 2050. IFPRI Research Monograph. International Food Policy Research Institute (IFPRI), Washington, USA. DOI: 10.2499/9780896291867.

44. Lobell DB, Banziger M, Magorokosho C, Vivek B (2011) Nonlinear heat effects on African corn as evidenced by historical yield trials. Nat Clim Chang DOI: 10.1038/NCLIMATE1043.Dafadf.

12

Determination of Critical Nitrogen Dilution Curve Based on Stem Dry Matter in Rice

Syed Tahir Ata-Ul-Karim, Xia Yao, Xiaojun Liu, Weixing Cao, Yan Zhu*

National Engineering and Technology Center for Information Agriculture, Jiangsu Key Laboratory for Information Agriculture, Nanjing Agricultural University, Nanjing, Jiangsu, P. R. China

Abstract

Plant analysis is a very promising diagnostic tool for assessment of crop nitrogen (N) requirements in perspectives of cost effective and environment friendly agriculture. Diagnosing N nutritional status of rice crop through plant analysis will give insights into optimizing N requirements of future crops. The present study was aimed to develop a new methodology for determining the critical nitrogen (N_c) dilution curve based on stem dry matter (S_{DM}) and to assess its suitability to estimate the level of N nutrition for rice (*Oryza sativa* L.) in east China. Three field experiments with varied N rates (0–360 kg N ha^{-1}) using three Japonica rice hybrids, Lingxiangyou-18, Wuxiangjing-14 and Wuyunjing were conducted in Jiangsu province of east China. S_{DM} and stem N concentration (SNC) were determined during vegetative stage for growth analysis. A N_c dilution curve based on S_{DM} was described by the equation ($N_c = 2.17W^{-0.27}$ with W being S_{DM} in t ha^{-1}), when S_{DM} ranged from 0.88 to 7.94 t ha^{-1}. However, for $S_{DM} < 0.88$ t ha^{-1}, the constant critical value $N_c = 1.76\%$ S_{DM} was applied. The curve was dually validated for N-limiting and non-N-limiting growth conditions. The N nutrition index (NNI) and accumulated N deficit (N_{and}) of stem ranged from 0.57 to 1.06 and 51.1 to −7.07 kg N ha^{-1}, respectively, during key growth stages under varied N rates in 2010 and 2011. The values of ΔN derived from either NNI or N_{and} could be used as references for N dressing management during rice growth. Our results demonstrated that the present curve well differentiated the conditions of limiting and non-limiting N nutrition in rice crop. The S_{DM} based N_c dilution curve can be adopted as an alternate and novel approach for evaluating plant N status to support N fertilization decision during the vegetative growth of Japonica rice in east China.

Editor: Guoping Zhang, Zhejiang University, China

Funding: This work was supported by grants from the National High-Tech Research and Development Program of China (863 Program) (2011AA100703), Special Program for Agriculture Science and Technology from Ministry of Agriculture in China (201303109), Priority Academic Program Development of Jiangsu Higher Education Institutions (PAPD), and Science and Technology Support Plan of Jiangsu Province (BE2011351, BE2012302). The funders had no role in study design, data collection and analysis, decision to publish, or preparation of the manuscript.

Competing Interests: The authors have declared that no competing interests exist.

* Email: yanzhu@njau.edu.cn

Introduction

Estimating nitrogen (N) nutritional status is a key to investigating, monitoring, and managing cropping systems [1]. Conventional farming has led to extensive use of N as a tool for ensuring profitability in the soils with uncertain fertility levels, which has raised the concerns about environmental sustainability. A reliable diagnosis of crop N requirement and nutritional status give insight into optimization of qualitative and quantitative aspects of crop production. It also improve N use efficiency and add to environmental protection [2]. Soil and plant-based strategies are two principle approaches, extensively used to derive information about the N nutrition status of crops, for satisfying their demand for N and to minimize N losses [3]. The former rarely describes the intensity of N release over a longer period, so the latter are widely accepted and adopted. Therefore, the present study investigates a plant-based strategy for an in-season assessment of N nutrition status for rice crop.

In plant-based approaches, the N nutrition status is generally monitored to determine the requirement for top dressing in crops [3]. For this purpose, several plant-based diagnostic tools, such as critical N concentration (N_c) approach, chlorophyll meter, hyper-

spectral reflectance and remote sensing, have been successfully used for in-season N management [4]. They differ in scope, in context of reference spatial scale, in terms of monetary and time resources, as well as skills and expertise required for their implementation at field [5]. Despite being simple, chlorophyll meter readings are affected by leaf thickness, abiotic stress and nutrient variability [6]. Canopy reflectance method's accuracy is affected by solar illumination, soil background effects and sensor viewing geometry [4]. However, the concept of N_c can be used as a potential alternate to these techniques, and it can give insight into relative N status of a crop. The present study utilizes this concept for an in-season N fertilizer management in rice crop.

The concept of N_c is crop specific, precise, simple and biologically sound, because it is based on actual crop growth. Whole plant dry matter based N_c approach was successfully applied for N management in winter wheat [7,8], corn [9] and spring wheat [10]. This approach was successfully applied for a Indica rice in tropics and Japonica rice in subtropical temperate region [11,12]. Dry matter partitioning among different plant organs affects the weight/N concentration relationship, and changes the shape of the dilution curve, thus limits its acceptance as a reliable method [13,14]. The concept of N_c for specific plant

Table 1. Changes of stem dry matter (S_{DM}) with time (days after transplantation) under different N rates in two rice cultivars in experiments conducted during 2010 and 2011.

Year	Cultivar	DAT	Sampling date	Stem dry matter/Applied N (kg ha^{-1})					F prob.	LSD
				0	80	160	240	320		
2010	LXY-18	16	07-Jul	0.23	0.27	0.38	0.48	0.49	*	0.028
	LXY-18	26	17-Jul	0.63	0.78	0.95	1.11	1.12	*	0.055
	LXY-18	36	27-Jul	1.04	1.28	1.55	1.77	1.81	*	0.075
	LXY-18	48	08-Aug	2.23	2.73	3.25	3.61	3.51	*	0.226
	LXY-18	60	20-Aug	3.87	4.47	4.94	5.23	5.29	*	0.146
	LXY-18	70	30-Aug	4.72	5.56	6.7	7.01	7.22	*	0.279
	WXJ-14	16	07-Jul	0.22	0.27	0.32	0.36	0.35	*	0.019
	WXJ-14	26	17-Jul	0.39	0.54	0.73	0.9	0.91	*	0.045
	WXJ-14	36	27-Aug	0.55	0.8	1.13	1.38	1.46	*	0.063
	WXJ-14	48	08-Aug	1.22	1.65	1.99	2.18	2.23	*	0.11
	WXJ-14	60	20-Aug	2.77	3.46	3.72	4.19	3.97	*	0.233
	WXJ-14	70	30-Sep	3.69	4.4	5.04	5.81	5.7	*	0.205
				Stem dry matter/Applied N (kg ha^{-1})						
				0	90	180	270	360		
2011	LXY-18	18	09-Jul	0.22	0.33	0.4	0.56	0.59	*	0.042
	LXY-18	30	21-Jul	0.67	0.76	0.92	1.19	1.19	*	0.053
	LXY-18	42	02-Aug	1.12	1.28	1.42	1.78	1.76	*	0.132
	LXY-18	54	15-Aug	2.24	2.41	2.76	3.43	3.48	*	0.127
	LXY-18	64	25-Aug	3.59	4.02	4.37	4.75	4.85	*	0.164
	LXY-18	74	04-Sep	5.68	5.91	6.41	7.84	8.04	*	0.172
	WXJ-14	18	09-Jul	0.19	0.28	0.32	0.34	0.36	*	0.02
	WXJ-14	30	21-Jul	0.37	0.6	0.71	0.86	0.9	*	0.06
	WXJ-14	42	02-Aug	0.54	0.96	1.2	1.37	1.47	*	0.128
	WXJ-14	54	15-Aug	1.51	1.84	2.24	2.52	2.68	*	0.137
	WXJ-14	64	25-Aug	2.49	3.07	3.36	4.06	4.04	*	0.169
	WXJ-14	74	04-Sep	4.41	5.04	5.75	6.2	6.27	*	0.178

*: F statistic significant at 0.01 probability level.

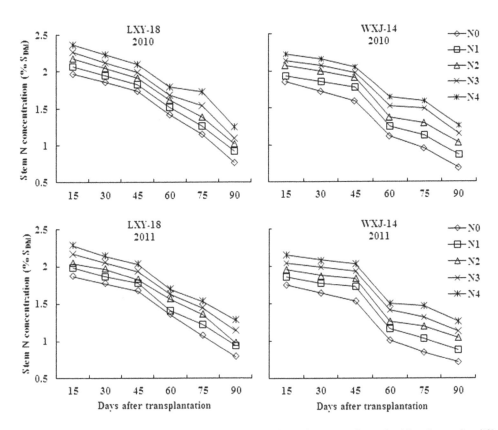

Figure 1. Changes of stem nitrogen concentration (% S_{DM}) with time (days after transplantation) for rice under different N rates in experiments conducted during 2010 and 2011.

organs (e.g., leaves and stem) is similar to that on whole plant basis. Leaf based diagnosis of N status in crops is affected by progressive shading by newer leaves, decline of leaf N concentration due to aging, pest attack, abiotic stresses and increase in the proportion of structural tissues [15]. Stem sap nitrate concentration is influenced by phenological phase, cultivar, temperature and solar radiation [16]. During vegetative phase, the contribution of stem dry matter (S_{DM}) towards total plant dry matter is significantly higher than that of leaf dry matter (L_{DM}), hence it is the most determining factor for N dilution of the whole plant [17]. Thus, the idea of using N_c curve based on S_{DM} over whole plant dry matter and L_{DM} based methods, can be used as an alternate approach for determination of N_c dilution curve.

The objectives of this work were to develop a N_c dilution curve based on S_{DM} and to assess the plausibility of this curve to estimate N nutrition status of Japonica rice. The estimation based on this approach will be more reliable than existing methods due to consistency at different growth stages.

Materials and Methods

Ethics statement

The experiments land is owned and managed by Nanjing Agricultural University, Nanjing, China. Nanjing Agricultural University permits and approvals obtained for the work and study. The field studies did not involve wildlife or any endangered or protected species.

Experimental details

Three field experiments with multiple N rates (0–360 kg N ha^{-1}) were conducted using three contrasting Japonica

rice hybrids, Lingxiangyou-18 (LXY-18), Wuxiangjing-14 (WXJ-14) and Wuyunjing (WYJ), at Yizheng (32°16′N, 119°10′E) and Jiangning (31°56′N, 118°59′E) located in lower Yangtze River Reaches of east China. The soil was clay loam and was classified as Ultisoles. The rice-wheat cropping system is practiced in the region. The applied N rates varied significantly among different farmers. The average rate of N fertilizer reached 387 kg ha^{-1} during the period of 2004–2008 [18].

The whole experimental area was ploughed and subsequently harrowed before transplanting. All bunds were compacted to prevent seepage into and from adjacent plots. A plastic lining was installed to a depth of 40 cm between drain and the bund of each plot to minimize seepage across the bunds towards the drains. To further minimize seepage of water from control plot (N$_0$), double bunds were constructed separating them and the adjacent plots. Experiments were arranged in a randomized complete block design with three replications. The size of each experimental plot was 8 m by 4.5 m, with planting density of approximately 22.2 hills per m^2. At site 1, soil pH, organic matter, total N, available phosphorous (P), and available potassium (K) were 6.2, 17.5 g kg^{-1}, 1.6 g kg^{-1} 43 mg kg^{-1}, 90 mg kg^{-1}, and 6.4, 15.5 g kg^{-1}, 1.3 g kg^{-1} 38 mg kg^{-1}, and 85 mg kg^{-1} in 2010 and 2011, respectively. The corresponding soil properties were 6.5, 13.5 g kg^{-1}, 1.13 g kg^{-1} 45 mg kg^{-1}, 91 mg kg^{-1} in 2007 at site 2. For experiments conducted at site 1 in 2010 and 2011, treatment consisted of five N rates as 0, 80, 160, 240, and 320 kg N ha^{-1}, and 0, 90, 180, 270, and 360 kg N ha^{-1}, respectively, while for experiment conducted at site 2 in 2007, treatment consisted of three N rates as 110, 220, and 330 kg N ha^{-1}. N in all experiments was distributed as 50% at pre planting, 10% at tillering, 20% at jointing, and 20% at booting, with urea as the N source. Aside from N fertilizer, phosphorus (135 kg ha^{-1}) and potassium (190 kg ha^{-1}) fertilizers were basally incorporated at the last harrowing and leveling in all plots before transplanting as monocalcium phosphate $Ca(H_2PO_4)_2$ and potassium chloride (KCl). Rice seedlings at five leaves stage were transplanted in experimental fields on June 20 (site 1) in 2010 and 2011, and on 29 June (site 2) in 2007, respectively. Pre-emergence herbicides were used to control weeds at early growth stages. Also plots were regularly hand-weeded until canopy was closed to prevent weed damage. Insecticides were used to prevent insect damage. All other agronomic practices were used according to local recommendations to avoid yield loss.

Sample collection and measurement

Rice plants were sampled from each plot at the intervals of 10–12 days from 0.23 m^2 area (5 hills) at active tillering, mid tillering, stem elongation, panicle initiation, booting and heading stages during the period of each experiment for growth analysis. The plants were manually severed at ground level on each sampling date. Fresh plants were divided into green leaf blades and culm plus sheath. Samples were oven-dried at 105°C for half an hour to rapidly stop metabolism and then at 70°C until constant weight to obtain stem dry matter (S_{DM}, t ha^{-1}). The dried stem samples were ground and analyzed for total stem N concentration (SNC, %) by Kjeldahl method. Stem N accumulation (SNA, kg N ha^{-1}) was obtained as summed product of the S_{DM} by the SNC. The SNC of whole-plant stem was calculated as SNA divided by S_{DM}.

Statistical analysis

The S_{DM} and SNC data for each sampling date, year and variety was separated and subjected to analysis of variance (ANOVA) using GLM procedures in SPSS-16 software package (SPSS Inc., Chicago. IL, USA). The differences among treatment means were measured by using the least significant difference (LSD) test at 90% level of significance, instead of classically used 95% in order to reduce the occurrence of Type II errors that could be high in such field experiments. For each measurement date, year and variety, the variation in the SNC versus S_{DM} across the different N levels was combined into a bilinear relation composed of a linear regression representing the joint increase in SNC and S_{DM} and a vertical line corresponding to an increase in SNC without significant variation in S_{DM}. The theoretical N_c points corresponds to the ordinate of the breakout of the bilinear regression. Regression analysis was performed using Microsoft Excel (Microsoft Cooperation, Redmond, WA, USA).

Construction and validation of critical, maximum and minimum N dilution curves

For determination of N_c dilution curve it is necessary to determine the N concentration that did not limit the S_{DM} production either by its excess or deficiency. The data used to construct the N_c dilution curve came from two experiments conducted in 2010 and 2011 by distinguishing the data points for N-limiting and non-N-limiting growth. The N-limiting growth treatment is defined as a treatment for which an additional N application leads to a significant increase in S_{DM}. The non-N-

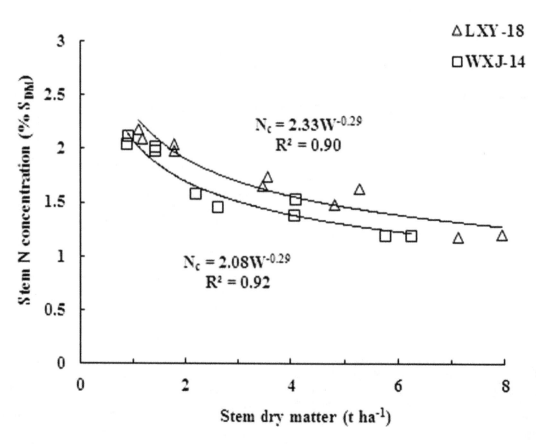

Figure 2. Critical nitrogen data points and N_c dilution curves in stem obtained by non-linear fitting for two rice cultivars (LXY-18, $N_c = 2.33W^{-0.29}$ and WXJ-14, $N_c = 2.08W^{-0.29}$) under different N rates in experiments conducted during 2010 and 2011.

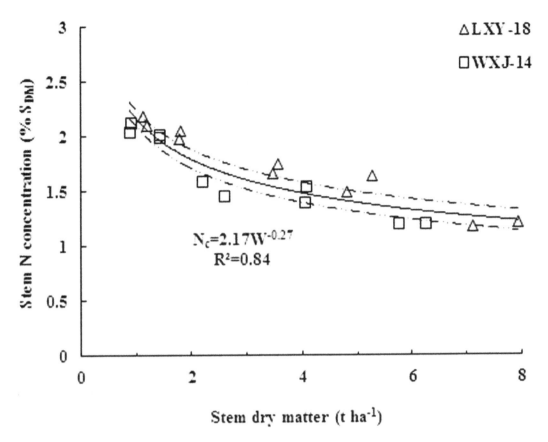

Figure 3. Critical nitrogen data points used to define the N_c dilution curve when data were pooled over for two rice cultivars (LXY-18 and WXJ-14). The solid line represents the N_c dilution curve ($N_c = 2.17W^{-0.27}$; $R^2 = 0.84$) describing the relationship between the N_c and stem dry matter of rice. The dotted lines represent the confidence band ($P = 0.95$).

limiting growth treatment is defined as a treatment, for which a supplement of N application does not lead to an increase in S_{DM} and, at the same time, exhibits a significant increase in SNC. If at the same measurement date, statistical analysis distinguished at least one set of N-limiting and non-N limiting data point, these data points were used either for construction of the N_c dilution curve or to validate it [7]. Consistent with earlier studies, an allometric function based on power regression (Freundlich model) was used to determine the relationship between the observed decreases in N_c with increasing S_{DM}. The N_c dilution curve was validated first by using the data points not retained for establishing the parameters of the allometric function in 2010 and 2011, and then with independent data set from experiment conducted in 2007.

The data points ($n = 13$) from most plethoric N treatments (N_4 plots) was assumed to represent the maximum N dilution curve (N_{max}) while the minimum N dilution curve (N_{min}) was determined by using the data points ($n = 13$) from the most N-limiting treatments for which N application was zero (N_0 check plots).

Calculation of critical N dilution curve based diagnostic tools

To identify the N status in the S_{DM} of rice during vegetative growth, the nitrogen nutrition index (NNI) and accumulated nitrogen deficit (N_{and}) were established for each sampling date, experiment and variety. The NNI value was obtained by dividing the total N concentration of S_{DM} by N_c value determined by critical dilution curve, [9]. The N_{and} value for rice crop on each

sampling date was obtained by subtracting the N accumulation under the N_c condition (N_{cna}) from actual N accumulation (N_{na}) under different N rates [12]. For in-season recommendation of supplemental N application, the difference value of NNI (ΔNNI), N_{and} (ΔN_{and}) and difference value of N application rate (ΔN between different N treatments was calculated according to the method proposed by Ata-Ul-Karim et al. [12].

Results

Stem dry matter and nitrogen concentration

The S_{DM} production was significantly affected by N fertilization during the growth period of rice. The increase in S_{DM} followed a continuous increasing trend along with sampling dates for both the varieties during each year with increasing N rates from N_0 to N_4; however, there was no significant difference between N_3 and N_4 in all the cases (Table 1). This increase in the S_{DM} production with N fertilization may be linked to a higher absorption of N fertilizer. S_{DM} ranged from minimum 0.22 t ha^{-1} and 0.19 t ha^{-1} (N_0) in WXJ-14 to a maximum of 7.22 t ha^{-1} and 8.04 t ha^{-1} (N_4) in LXY-18 during 2010 and 2011, respectively. The results showed that there was no positive correlation between S_{DM} and N rates, as the S_{DM} tend to decrease when N rate exceeded a critical level. During each experimental year, S_{DM} conferred with the following inequality under different N ratess.

$$S_{DM0} < S_{DM1} < S_{DM2} < S_{DM3} = S_{DM4} \qquad (1)$$

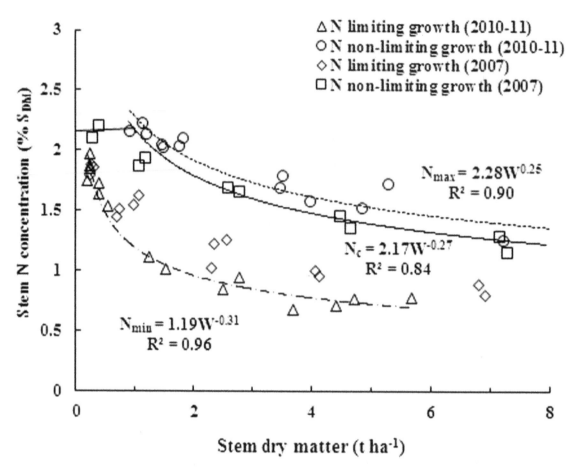

Figure 4. Comprehensive validation of N_c dilution curve using independent data set from experiment conducted in 2007. Data points ($\diamond$) represent N limiting growth conditions, while ($\square$) represent N non-limiting conditions. The solid line in the middle represents the N_c curve ($N_c = 2.17W^{-0.27}$) describing the relationship between the N_c and stem dry matter of rice. The data points (Δ) and ($\bigcirc$) not engaged for establishing the parameters of allometric function (2010 and 2011) were used to develop two boundary curves, (–•–•–•) minimum limit curve ($N_{min} = 1.19\,W^{-0.31}$) and (------) maximum limit curve ($N_{max} = 2.27W^{-0.25}$).

where S_{DM0}, S_{DM1}, S_{DM2}, S_{DM3} and S_{DM4} stands for S_{DM} of N_0, N_1, N_2, N_3 and N_4, respectively.

Stem N concentration response to N fertilizer rates was usually linear and a higher rate of N mostly resulted in a higher SNC, hitherto a decline in SNC was observed with increasing S_{DM} from active tillering to heading. Maximum variation in SNC of both cultivars was observed on 16 and 18 DAT, while minimum on 70 and 74 DAT, in years 2010 and 2011, respectively. The SNC ranged from 2.28 to 0.78 for LXY-18 and 2.16 to 0.71 for WXJ-14 during 2010, while 2.36 to 0.77 for LXY-18 and 2.23 to 0.68 for WXJ-14 during 2011 (Fig. 1).

Determination of critical nitrogen dilution curves based on stem dry matter

A set of twenty theoretical data points for both cultivar, obtained from two experiments (10 data points for each cultivar) from active tillering to heading were used to calculate the N_c for a given level of S_{DM}. The S_{DM} data that fit the statistical criteria for establishing N_c dilution curve varied from 0.88 t ha^{-1} to 7.94 t ha^{-1}. A power functions were fitted to the calculated N_c points as equations (2) and (3), the coefficient for which were 0.90 and 0.92 for LXY-18 and WXJ-14, respectively (Fig. 2).

$$N_c = 2.33\,W^{-0.29} \qquad (W \geq 0.88\ t\ ha^{-1}, R^2 = 0.90, n = 10) \quad (2)$$

$$N_c = 2.08\,W^{-0.29} \qquad (W \geq 0.88\ t\ ha^{-1}, R^2 = 0.92, n = 10) \quad (3)$$

where W is the S_{DM} expressed in t ha^{-1}; N_c is the critical N concentration in stem expressed in % S_{DM}; a and b are estimated parameters. The parameter a represents the N concentration in the S_{DM} when W = 1 t ha^{-1}, and b represents the coefficient of dilution describing the relationship between N concentration and S_{DM}.

The F-value (0.72) of two curves was less than the critical value of $F_{(1-18)} = 4.41$ at 5% probability level, showing non-significant difference between the curves [19], thus the data for the two varietal groups were united, and a unified dilution curve was determined as equation 4.

$$N_c = 2.17\,W^{-0.27} \qquad (W \geq 0.88\ t\ ha^{-1}, R^2 = 0.84, n = 20) \quad (4)$$

The model accounted for 84% of the total variance. At early growth stages of rice crop, the N_c varied between 2.24% S_{DM} to 2.10% S_{DM} (95% confidence interval) for a S_{DM} of 0.88 t ha^{-1} at the lower end while 7.94 t ha^{-1} at the higher end, respectively (Fig. 3).

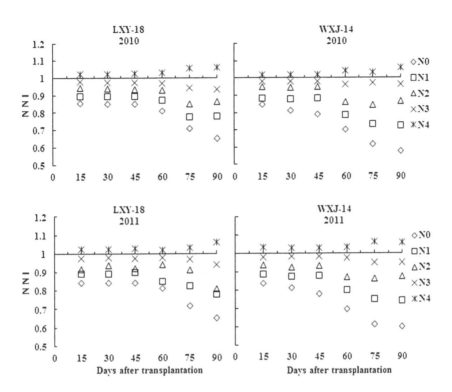

Figure 5. Changes of nitrogen nutrition index (NNI) with time (days after transplantation) for rice stem under different N rates in experiments conducted during 2010 and 2011.

For the S_{DM} range of 0.1 to 0.88 t ha^{-1}, corresponding to early growth stages, increasing N rates at sowing did not significantly affect S_{DM}, because N requirement is relatively low during these early stages. Therefore, the N_c dilution curve cannot be applied to the low S_{DM}<0.88 t ha^{-1} at early growth stages due to relatively smaller decline of N_c with increasing S_{DM}. For these S_{DM}, the N_c could not been determined by the same statistical method because the very high slope of the linear regression resulted in a highly variable estimate [7]. Hence, for the data points of S_{DM} ranging from 0.37 to 0.88 t ha^{-1} a constant N_c (1.76% S_{DM}) was calculated as the mean value between the minimum N concentration of non-limiting N points (2.26% S_{DM}) and the maximum N concentration of limiting N points (1.25% S_{DM}), based on extrapolation of equation 4.

The above S_{DM} based N_c dilution curve was dually validated for N-limiting and non-N-limiting situations within the range for which it was developed. First, the curve was partially validated by combining the data points not engaged for establishing the parameters of the allometric function. In addition, the comprehensive validation of the curve was performed by using the data points from an independent experiment conducted in 2007. The results revealed that the N concentration data that led to the highest significant yields in S_{DM} were positioned close to or above the N_c dilution curve and considered to be non-N-limiting concentrations, whereas the data for the lowest significant S_{DM} yields, were positioned close to or under the N_c dilution curve and classified as N limiting values (Fig. 4). To determine N_{max}, data points were selected only from non-N-limiting treatments (n = 13), and for N_{min}, data points were selected from the treatment without N application (n = 13). Thus, the present N_c dilution curve could well discriminate the N limiting and non-N-limiting growing conditions in this study

Changes of NNI and N_{and}

Nitrogen nutrition index and N_{and} are helpful in determining the crop nutrition status i.e. deficient, optimal or excess of N nutrition. N nutrition is considered as optimum when NNI = 1 and N_{and} = 0, while NNI>1 and N_{and}<0 indicates luxury consumption of N nutrition, values of NNI<1 and N_{and}>0 represents N shortage. NNI and N_{and} can be used to quantify the intensity of the N stress after the onset of N deficiency. Our results of significant differences in NNI and N_{and} across the growing seasons, N rates, and phenological stages in rice are in agreement with earlier reports for maize and wheat [10]. As seen in Figure 5 and 6, during 2010 and 2011 the NNI ranged from 0.65 to 1.06 for LXY-18 and 0.57 to 1.06 for WXJ-14, while the N_{and} ranged from 51.1 kg ha^{-1} to −7.07 kg ha^{-1} for LXY-18 and 43.3 kg ha^{-1} to −4.5 kg ha^{-1} for WXJ-14. The results showed that NNI amplified while N_{and} declined with increasing N rates, while both intensified steadily with growth of rice crop and reached to peaks at heading stage for N_0, N_1, N_2 and N_3 (N limiting treatments), nevertheless, for N_3 this intensification was minor. In contrast, surplus N nutrition existed till heading stage for N_4 (non-N-limiting treatment). The estimates based on NNI and N_{and} can be used to identify the N nutritional status at any stage of rice growth, allowing us to assess whether the N fertilizer dosage was ample enough to obtain higher yield in practice. These results confirmed the plausibility of using NNI and N_{and} to assess the status of N nutrition in rice plants growing under various conditions and stages.

Figure 7 and 8 showed that ΔN had a positive correlation with ΔNNI and ΔN_{and}. The simple linear regression equation showed non-significant differences between two varieties, although noticeable differences were observed among different phenological stages. Therefore, ΔN during growth period for both varieties could be derived from ΔNNI and ΔN_{and}, respectively, according

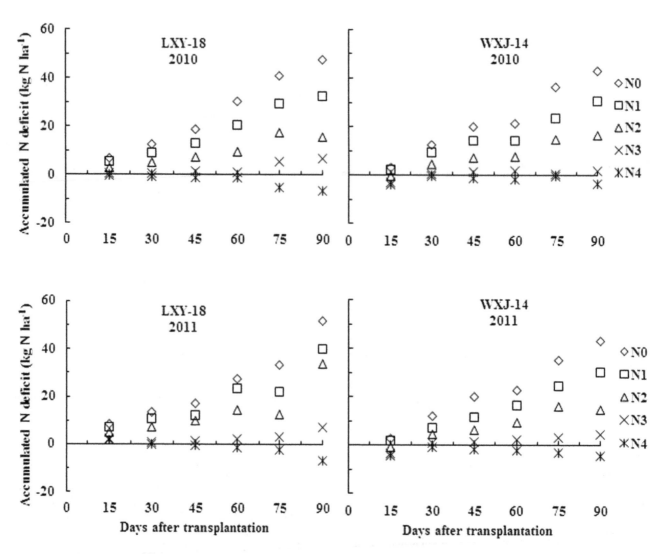

Figure 6. Changes of accumulated N deficit (N_{and}) with time (days after transplantation) for rice stem under different N rates in experiments conducted during 2010 and 2011.

to the equations 5 & 6 as follows:

$$\Delta N = A \times \Delta NNI + B \qquad (5)$$

$$\Delta N = C \times \Delta N_{and} + D \qquad (6)$$

The parameters A, B, C, and D could be calculated from days after transplanting (DAT) using the equations as:

$$A = -16.60 \times DAT + 2101 \qquad (R^2 = 0.95) \quad (7)$$

$$B = -0.024 \times DAT^2 + 2.57 \times DAT - 40.07 \qquad (R^2 = 0.62) \quad (8)$$

$$C = 18.97 ln(DAT) - 89.22 \qquad (R^2 = 0.98) \quad (9)$$

$$D = -9.98 ln(DAT) + 52.76 \qquad (R^2 = 0.19) \quad (10)$$

The ΔN obtained on the basis of relationship between ΔNNI, ΔN_{and} and ΔN, allowed us to make corrective decisions of N dressing recommendation for the precise N management during or even before the period of highest demand of the rice crop.

Discussion

Application of N fertilizer for crop production is an economically viable option in terms of low cost as compared to the value of the marketable agricultural products themselves; however, N usage cannot assure a significant increase in crop productivity due to diminishing returns after certain levels. There is an increasing demand by strategy makers for simple-to-use, technically established and economically viable N indicators, which may allow monitoring and assessment of policy measures and offer tools for farm N management. With the advent of technology, more emphasis should be put on plant-based indicators, which simultaneously reflect the interactions between the plant and the

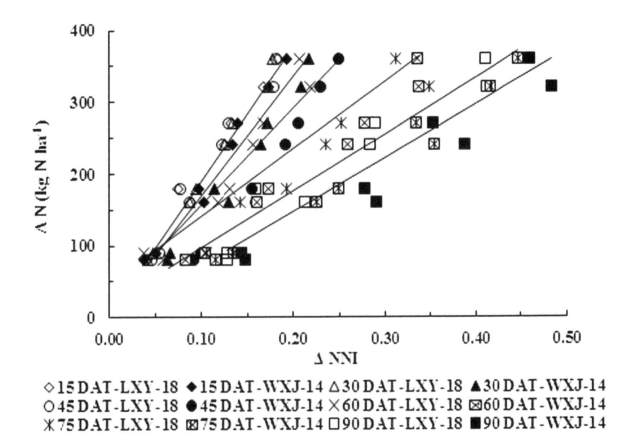

Figure 7. Relationship between changes of nitrogen nutrition index (ΔNNI) and changes of nitrogen application rates (ΔN, kg N ha⁻¹) at different growth stages in experiments conducted during 2010 and 2011. The open symbols represent different growth stages for LXY-18 while filled symbols represent different growth stages for WXJ-14. (ΔN = A×ΔNNI+B; A = −16.60×DAT+2101, R^2 = 0.95; B = −0.024×DAT²+ 2.57×DAT−40.07, R^2 = 0.62).

soil. So far, there have been several reports on estimating the N_c concentration on whole plant dry matter basis in various crops, including rice [11,12], and on L_{DM} basis in rice [20], yet no attempt was made to determine the N_c dilution curve on S_{DM} basis for any crop including rice. The current study has developed a S_{DM} based N_c dilution curve for rice in east China, thus providing a new approach for diagnosing and regulating N in crop species.

Minimum and maximum nitrogen dilution curves

An obvious variability in SNC for a given range of S_{DM} was observed when all the data from three year experiments were analyzed for interpretation. This variability in SNC towards maturity of rice crop in present study was in agreement with earlier studies on winter wheat [7] and Japonica rice [12], and this variability could be attributed to a decline in the fraction of total plant N associated with photosynthesis [21], change in leaf/shoot ratio and self-shading of leaves [8].

Two boundary curves for N maximum (N_{max}) and minimum (N_{min}) have been determined by using maximum and minimum N concentration in S_{DM} and can be represented as equations:

$$N_{max} = 2.28 W^{-0.25} \qquad (11)$$

$$N_{min} = 1.19 W^{-0.31} \qquad (12)$$

The N_{max} curve corresponding to the maximum N uptake in the S_{DM} without interfering with productivity and it can be considered as the first assessment of a maximum N dilution on S_{DM} basis in crops, and can be obtained with increasing N rates for maximum growth and N accumulation. This curve is an estimate of the maximum N accumulation capacity of stem which is regulated by mechanism associated with the growth and availability of soil N directly or indirectly via N metabolism [22]. The N_{max} curve in the present study shows a luxury consumption of N under N_4 treatment, when N concentration exceeds N_c dilution curve and S_{DM} does not increase with increasing N rate. In contrast, the N_{min} curve is considered as a lower limit at which the N metabolism would soon stop to function. It corresponds to the minimum N taken up by rice plants under N_0 treatment in present study. Thus, the N_{min} were used as the threshold concentration for proper metabolic functionality of the plant.

Moreover, the value of parameter b for the N_{max} was not significantly different from that of N_c dilution curve, which indicate that the partitioning of dry matter remains relatively constant when N uptake exceeds the N_c dilution curve. This is consistent with the concept of N_c dilution curve, which represents the lowest N at which maximum dry matter accumulation occurs. This implies that under luxury consumption of N, when N exceeds N_c dilution curve, dry matter accumulations does not increase with N and hence, dry matter partitioning will have similar value of parameter b. In contrast, for N_{min} curve under N stress, the value for parameter b tended to be slightly lower than the dilution curve.

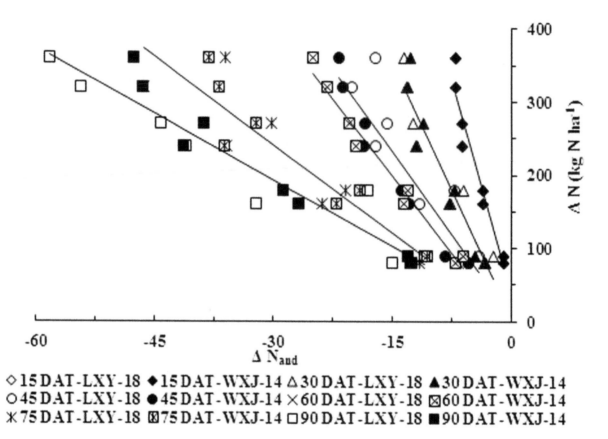

$$\diamondsuit\,15\,DAT\text{-}LXY\text{-}18 \quad \blacklozenge\,15\,DAT\text{-}WXJ\text{-}14 \quad \triangle\,30\,DAT\text{-}LXY\text{-}18 \quad \blacktriangle\,30\,DAT\text{-}WXJ\text{-}14$$
$$\bigcirc\,45\,DAT\text{-}LXY\text{-}18 \quad \bullet\,45\,DAT\text{-}WXJ\text{-}14 \quad \times\,60\,DAT\text{-}LXY\text{-}18 \quad \boxtimes\,60\,DAT\text{-}WXJ\text{-}14$$
$$\maltese\,75\,DAT\text{-}LXY\text{-}18 \quad \boxtimes\,75\,DAT\text{-}WXJ\text{-}14 \quad \square\,90\,DAT\text{-}LXY\text{-}18 \quad \blacksquare\,90\,DAT\text{-}WXJ\text{-}14$$

Figure 8. Relationship between changes of accumulated N deficit (ΔN_{and}) and changes of nitrogen application rates (ΔN, kg N ha^{-1}) at different growth stages in experiments conducted during 2010 and 2011. The open symbols represent different growth stages for LXY-18 while filled symbols represent different growth stages for WXJ-14 ($\Delta N = C \times \Delta N_{and} + D$; $C = 18.97$ ln(DAT)-89.22, $R^2 = 0.98$ and $D = -9.98$ ln(DAT)+ 52.76, $R^2 = 0.19$), respectively.

The relatively low value for b was associated with a change in dry matter partitioning.

Comparison with other critical nitrogen dilution curves

The concept of N_c dilution curve on whole plant dry matter and L_{DM} basis have already been successfully implicated for several crops including rice, yet no attempt was made to construct a S_{DM} based N_c dilution curve in any crop including rice. Figure 9 showed that the parameter a of N_c dilution curve on S_{DM} basis with Japonica rice developed in present study (2.17) was lower than the reference curve on whole plant dry matter basis of Indica rice in tropics (5.20) by Sheehy et al. [11] as well as lower than the curves developed with Japonica rice on whole plant dry matter basis (3.53) by Ata-Ul-Karim et al. [12], and L_{DM} basis (3.76) by Yao et al. [20].

The differences observed between the parameter a of dilution curve developed in present study and the curves on whole plant dry matter basis [11,12] were due to morphological aggregation of structural components, which relates to the weight/N concentration in the whole plant [13]. Stress responses may cause differences in the partitioning of dry matter among various plant organs, and thereby affect the shape of the dilution curves. Moreover, dissimilarities in climatic conditions and genetic differences of Indica and Japonica rice contributed to the differences between the curves. The ability of Indica to hold higher plant N content and total N uptake [23–25] and faster growth rate [26], compared with those of Japonica rice, also lead to the differences between N_c dilution curve of Sheehy et al. [11] and that described in the present study. The differences of S_{DM} based curve with that of L_{DM} based one [20] are mainly attributed to leaf/stem ratio, because decrease in stem N during vegetative phase is related to decline in the metabolic biomass with high N contents, and increase in proportion of structural and non-photosynthetic biomass with low N contents [8]. Thus, higher proportion of structural biomass in stem than in leaves is responsible for the differences between the L_{DM} and S_{DM} based curves of Japonica rice.

The parameter b of the dilution curve indicates the dilution intensity of N during growth and the higher values of b indicate lower N dilutions [17]. The coefficients b were (-0.50, -0.28, -0.22 and -0.274) for N_c dilution curve of Indica rice and for Japonica rice based on whole plant dry matter, L_{DM}, and on S_{DM}, respectively. The observed differences between the coefficients b of Indica rice and current S_{DM} based dilution curve might be explained by the differences in duration of vegetative phase in tropical and subtropical climates, while the differences between coefficients b of the curves of Japonica rice based on L_{DM} and S_{DM} were directly related to the distribution of dry matter between green leaves and the stem [17]. In contrast, the differences between coefficients b of the curves of Japonica rice on whole plant dry matter basis compared with that of S_{DM} basis, are negligible due to the reason that stem have a dilution effect on the N in the above ground tissues, because of their higher weight percentage in the total dry matter [27]. Therefore, the S_{DM} based dilution curve can be used as a potential alternative for in-season estimation of

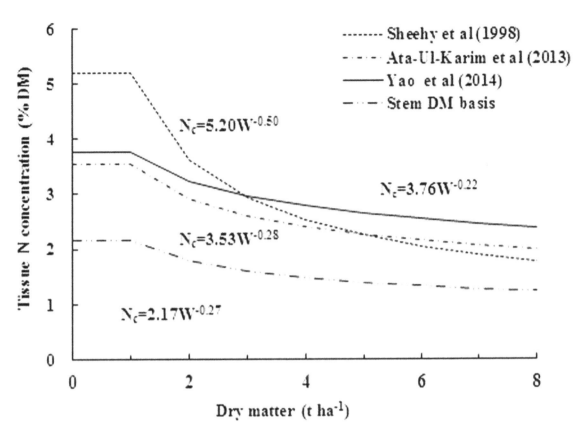

Figure 9. Comparison of different N_c dilution curves. The (------) represents the N_c dilution curve of Sheehy et al. (1998) ($N_c = 5.20W^{-0.50}$) on plant dry matter basis in Indica rice under tropic environment. The (-●-●-●) represents the N_c dilution curve of Ata-Ul-Karim et al. (2013) ($N_c = 3.53W^{-0.28}$) on plant dry matter basis in Japonica rice in Yangtze River Reaches. The (——) line represents N_c dilution curve of Yao et al. (2014) ($N_c = 3.76W^{-0.22}$) on leaf dry matter basis in Japonica rice in Yangtze River Reaches, and the (-●●-●●-) line represents N_c dilution curve on stem dry matter basis in present study ($N_c = 2.17W^{-0.27}$).

plant N nutrition status, instead of existing whole plant dry matter and L_{DM} based approaches.

Implication for nitrogen diagnosis

The application of the present N_c dilution curve as a diagnostic tool for accurate N management to make corrective decisions of N dressing recommendation during rice production is very interesting. The N_c dilution curve can be used for a priori analysis intended to optimize fertilizer N management or for a posteriori diagnosis intended to detect N limiting nutrition for rice within experimental trials or fields in production. The a priori diagnosis of plant N status consists of timely detection of plant N deficiency during the crop growth cycle to determine the necessity of applying additional N fertilizer. Present study showed that the N_c dilution curve, resulting NNI and N_{and} effectively distinguished conditions of deficient, optimal and surplus N nutrition in rice. The values of ΔN in present study obtained on the basis of relationship between ΔNNI, ΔN_{and} and ΔN, permitted us to make corrective decisions of N dressing recommendation for precise N management during or even before the period of peak demand of the rice crop. The main limitation in using the present NNI and N_{and} directly as diagnostic tools is the need to determine the actual dry matter and N concentration, which can be monitored by the non-destructive means including remote sensing [28–30]. Moreover, a good correlation between these analytical tools and chlorophyll meter readings was previously reported by [9]. These indirect methods could possibly be a substitute for assessing NNI

and N_{and} and portray crops and environments in conditions where they cannot be measured directly [31]. Thus, the models of NNI and N_{and}, based on N_c dilution curve in relation to actual growth status, can be exploited directly for the estimation of crop N status to recommend the necessities of further N application during plant growth. These novel algorithms can also be combined into crop growth and management models to forecast crop N status and quantify N dressing plan. Although, NNI and N_{and} calculated in present study distinguished well the N-limiting and non-N-limiting growth conditions, a more comprehensive validation using different N management practices, N availabilities and cultivars is mandatory to robustly confirm the reliability of NNI and N_{and} usage as an investigative indicators for different ecological regions and rice production systems.

Conclusions

In conclusion, we found that N fertilization endorses increase in the S_{DM}, which was influenced by variations in SNC. A higher rate of N fertilizer generally increased SNC in Japonica rice; however, towards advancing maturity this increase followed a declining trend under different N levels, sampling dates and growing seasons. S_{DM} during vegetative growth period ranged from minimum value of 0.19 (N_0) in WXJ-14 to a maximum value of 8.04 (N_4) in LXY-18, whereas SNC varied from 0.68% in WXJ-14 to 2.36% in LXY-18 on S_{DM} basis under different N rates and growth stages. A new N_c dilution curve on S_{DM} basis for Japonica rice grown in east China was developed and can be described by equation, $N_c = 2.17W^{-0.274}$, when S_{DM} ranges from 0.88 and 7.94 t ha^{-1}, however for $S_{DM} < 0.88$ t ha^{-1}, the constant critical value $N_c = 1.76\%$ S_{DM} was applied, which was independent of S_{DM}. Additionally, the values of NNI and N_{and} at different sampling dates for N limiting condition were generally <1 and >0, while >1 and <0, respectively for non-N-limiting supply. The values of ΔN derived on the basis of relationship between ΔNNI, ΔN_{and} and ΔN, can be used to make corrective decisions of N dressing recommendation for precise N management, prior to or on the onset of the period of highest demand of the rice crop. We conclude that the S_{DM} based dilution curve developed in the present study offers a new vision into plant N status and can possibly be adopted as an alternate practical tool for reliable diagnosis of plant N status to correct N fertilization decision during the vegetative growth of rice in east China.

Author Contributions

Conceived and designed the experiments: ST AUK XY XL WC YZ. Performed the experiments: ST AUK XY XL. Analyzed the data: ST AUK XY YZ. Wrote the paper: ST AUK YZ.

References

1. Jaggard K, Qi A, Armstrong M (2009) A meta-analysis of sugarbeet yield responses to nitrogen fertilizer measured in England since 1980. J Agric Sci-(Camb) 147: 287–301.
2. Ghosh M, Mandal B, Mandal B, Lodh S, Dash A (2004) The effect of planting date and nitrogen management on yield and quality of aromatic rice (*Oryza sativa*). J Agric Sci-(Camb) 142: 183–191.
3. Cabangon R, Castillo E, Tuong T (2011) Chlorophyll meter-based nitrogen management of rice grown under alternate wetting and drying irrigation. Field Crops Res 121: 136–146.
4. Lin FF, Qiu LF, Deng JS, Shi YY, Chen LS, et al. (2010) Investigation of SPAD meter-based indices for estimating rice nitrogen status. Compu Electron Agric 71: S60–S65.
5. Confalonieri R, Debellini C, Pirondini M, Possenti P, Bergamini L, et al. (2011) A new approach for determining rice critical nitrogen concentration. J Agric Sci-(Camb) 149: 633–638.
6. Smeal D, Zhang H (1994) Chlorophyll meter evaluation for nitrogen management in corn. Commun Soil Sci Plant Anal 25: 1495–1503.
7. Justes E, Mary B, Meynard JM, Machet JM, Thelier-Huche L (1994) Determination of a critical nitrogen dilution curve for winter wheat crops. Ann Bot 74: 397–407.
8. Yue S, Meng Q, Zhao R, Li F, Chen X, et al. (2012) Critical nitrogen dilution curve for optimizing nitrogen management of winter wheat production in the North China Plain. Agron J 104: 523–529.
9. Ziadi N, Brassard M, Bélanger G, Cambouris AN, Tremblay N, et al. (2008) Critical nitrogen curve and nitrogen nutrition index for corn in eastern Canada. Agron J 100: 271–276.
10. Ziadi N, Belanger G, Claessens A, Lefebvre L, Cambouris AN, et al. (2010) Determination of a critical nitrogen dilution curve for spring wheat. Agron J 102: 241–250.
11. Sheehy JE, Dionora MJA, Mitchell PL, Peng S, Cassman KG, et al. (1998) Critical nitrogen concentrations: implications for high-yielding rice (*Oryza sativa* L.) cultivars in the tropics. Field Crops Res 59: 31–41.
12. Ata-Ul-Karim ST, Yao X, Liu X, Cao W, Zhu Y (2013) Development of critical nitrogen dilution curve of Japonica rice in Yangtze River Reaches. Field Crops Res 149: 149–158.
13. Kage H, Alt C, Stützel H (2002) Nitrogen concentration of cauliflower organs as determined by organ size, N supply, and radiation environment. Plant Soil 246: 201–209.
14. Vouillot MO, Huet P, Boissard P (1998) Early detection of N deficiency in a wheat crop using physiological and radiometric methods. Agronomie 18: 117–130.
15. Ziadi N, Bélanger G, Gastal F, Claessens A, Lemaire G, et al. (2009) Leaf nitrogen concentration as an indicator of corn nitrogen status. Agron J 101: 947–957.

16. Lemaire G, Jeuffroy MH, Gastal F (2008) Diagnosis tool for plant and crop N status in vegetative stage: Theory and practices for crop N management. Eur J Agron 28: 614–624.
17. Oliveira ECAd, de Castro Gava GJ, Trivelin PCO, Otto R, Franco HCJ (2013) Determining a critical nitrogen dilution curve for sugarcane. J Plant Nutr Soil Sci 176: 712–723.
18. Chen J, Huang Y, Tang Y (2011) Quantifying economically and ecologically optimum nitrogen rates for rice production in south-eastern China. Agric Ecosyst Environ 142: 195–204.
19. Hahn WS (1997) Statistical Methods for Agriculture and Life Science. Seol: Free Academy Publishing Co. 747 p.
20. Yao X, Ata-Ul-Karim ST, Zhu Y, Tian Y, Liu X, et al. (2014) Development of critical nitrogen dilution curve in rice based on leaf dry matter. Eur J Agron 55: 20–28.
21. Bélanger G, Richards JE (2000) Dynamics of biomass and N accumulation of alfalfa under three N fertilization rates. Plant Soil 219: 177–185.
22. Gayler S, Wang E, Priesack E, Schaaf T, Maidl FX (2002) Modeling biomass growth, N-uptake and phenological development of potato crop. Geoderma 105: 367–383.
23. Islam M, Islam M, Sarker A (2008) Effect of phosphorus on nutrient uptake of Japonica and Indica rice. J Agric Rural Dev 6: 7–12.
24. Shan Y, Wang Y, Yamamoto Y, Huang J, Yang L, et al. (2001) Study on the differences of nitrogen uptake and use efficiency in different types of rice. J Yangzhou Univ (Nat Sci Ed) 4: 42.
25. Yoshida H, Horie T, Shiraiwa T (2006) A model explaining genotypic and environmental variation of rice spikelet number per unit area measured by cross-locational experiments in Asia. Field Crops Res 97: 337–343.
26. Ying J, Peng S, He Q, Yang H, Yang C, et al. (1998) Comparison of high-yield rice in tropical and subtropical environments: I. Determinants of grain and dry matter yields. Field Crops Res 57: 71–84.
27. Oliveira ECAd, Freire FJ, Oliveira RId, Freire M, Simoes Neto DE, et al. (2010) Extração e exportação de nutrientes por variedades de cana-de-açúcar cultivadas sob irrigação plena. Rev Bras de Ciênc Solo 34: 1343–1352.
28. Wang W, Yao X, Tian Y, Liu X, Ni J, et al. (2012) Common spectral bands and optimum vegetation indices for monitoring leaf nitrogen accumulation in rice and wheat. J Integr Agric 11: 2001–2012.
29. Zhao B, Yao X, Tian Y, Liu X, Ata-Ul-Karim ST, et al. (2014) New critical nitrogen curve based on leaf area index for winter wheat. Agron J 106: 379–389.
30. Ata-Ul-Karim ST, Zhu Y, Yao X, Cao W (2014) Determination of critical nitrogen dilution curve based on leaf area index in rice. Field Crops Res: In press.
31. Debaeke P, Rouet P, Justes E (2006) Relationship between the normalized SPAD index and the nitrogen nutrition index: Application to durum wheat. J Plant Nutr 29: 75–92.

13

Soil Carbon and Nitrogen Fractions and Crop Yields Affected by Residue Placement and Crop Types

Jun Wang[1], Upendra M. Sainju[2]*

1 College of Urban and Environmental Sciences, Northwest University, Xian, Shaanxi Province, China, **2** U.S. Department of Agriculture, Agricultural Research Service, Northern Plains Agricultural Research Laboratory, Sidney, Montana, United States of America

Abstract

Soil labile C and N fractions can change rapidly in response to management practices compared to non-labile fractions. High variability in soil properties in the field, however, results in nonresponse to management practices on these parameters. We evaluated the effects of residue placement (surface application [or simulated no-tillage] and incorporation into the soil [or simulated conventional tillage]) and crop types (spring wheat [*Triticum aestivum* L.], pea [*Pisum sativum* L.], and fallow) on crop yields and soil C and N fractions at the 0–20 cm depth within a crop growing season in the greenhouse and the field. Soil C and N fractions were soil organic C (SOC), total N (STN), particulate organic C and N (POC and PON), microbial biomass C and N (MBC and MBN), potential C and N mineralization (PCM and PNM), NH_4-N, and NO_3-N concentrations. Yields of both wheat and pea varied with residue placement in the greenhouse as well as in the field. In the greenhouse, SOC, PCM, STN, MBN, and NH_4-N concentrations were greater in surface placement than incorporation of residue and greater under wheat than pea or fallow. In the field, MBN and NH_4-N concentrations were greater in no-tillage than conventional tillage, but the trend reversed for NO_3-N. The PNM was greater under pea or fallow than wheat in the greenhouse and the field. Average SOC, POC, MBC, PON, PNM, MBN, and NO_3-N concentrations across treatments were higher, but STN, PCM and NH_4-N concentrations were lower in the greenhouse than the field. The coefficient of variation for soil parameters ranged from 2.6 to 15.9% in the greenhouse and 8.0 to 36.7% in the field. Although crop yields varied, most soil C and N fractions were greater in surface placement than incorporation of residue and greater under wheat than pea or fallow in the greenhouse than the field within a crop growing season. Short-term management effect on soil C and N fractions were readily obtained with reduced variability under controlled soil and environmental conditions in the greenhouse compared to the field. Changes occurred more in soil labile than non-labile C and N fractions in the greenhouse than the field.

Editor: Raffaella Balestrini, Institute for Sustainable Plant Protection, C.N.R., Italy

Funding: Funding came from the U.S. Department of Agriculture-Agricultural Research Service, Sidney, MT, USA and the National Natural Science Foundation of China (No. 31270484). The funders had no role in study design, data collection and analysis, decision to publish, or preparation of the manuscript.

* Email: upendra.sainju@ars.usda.gov

Introduction

Soil organic matter, as indicated by C and N levels, is an important component of soil quality and productivity. Increasing soil organic matter through enhanced C and N sequestration can also reduce the potentials for global warming by mitigating greenhouse gas emissions and N leaching by increasing N storage in the soil [1,2]. Carbon and N sequestration usually occur when non-harvested crop residues, such as stems, leaves, and roots, are placed at the soil surface due to no-tillage [3,4,5]. Carbon and N sequestration rates, however, depend on the balance between the amounts of plant residue C and N inputs and rates of C and N mineralized in the nonmanured soil [6,7]. Other benefits of increasing C and N storage include enhancement of soil structure and soil water-nutrient-crop productivity relationships [8].

Soil and crop management practices can alter the quantity, quality, and placement of crop residues in the soil, thereby influencing soil C and N storage, microbial biomass and activity, and N mineralization–immobilization [9,10]. Residue placement in the soil under different tillage systems can influence C and N levels by affecting soil aggregation, aeration, and C and N

mineralization [9,11]. Crop types can affect the quantity and quality (C/N ratio) of crop residue returned to the soil and therefore on soil C and N levels [9,12]. Legumes, such as pea, because of its higher N concentration and lower C/N ratio, decompose more rapidly in the soil and supply greater amount of N to succeeding crops than nonlegumes [12,13]. As a result, N fertilization rates to crops following pea can be reduced to sustain yields [14,15].

Because of large pool sizes and inherent spatial variability, soil organic C (SOC) and total N (STN) (slow or non-labile fractions) change slowly with management practices [16]. Therefore, measurements of SOC and STN alone may not adequately reflect changes in soil quality and nutrient status [16,17]. Active (or labile) C and N fractions, such as potential C and N mineralization (PCM and PNM) that indicate microbial activity and N mineralization, and microbial biomass C and N (MBC and MBN) that refer to microbial biomass and N immobilization, change seasonally [16,18]. Similarly, particulate organic C and N (POC and PON) that represent coarse organic matter and considered as intermediate C and N levels between slow and active fractions, provide substrates for microbes and influence soil aggregation [19,20].

Available N fractions that influence plant growth and N losses due to leaching, denitrification, or volatilization are NH_4-N and NO_3-N [10,12].

Although active C and N fractions in the soil can change more rapidly than the other fractions, these fractions sometime may not be readily changed within a crop growing season due to high variability in soil properties within a short distance in the field or in regions with limited precipitation, cold weather, and a short growing season [10,12,15]. Under controlled soil and environmental conditions, such as in the greenhouse, it may be possible to detect changes in these fractions more rapidly as affected by management practices than in the field. We hypothesized that surface placement of crop residue (a simulation of no-tillage in the field) under spring wheat can increase soil labile and non-labile C and N fractions and sustain crop yields compared to residue incorporation into the soil (a simulation of conventional tillage) under pea or fallow more in the greenhouse than in the field. Our objectives were to: (1) evaluate the effects of residue placement and crop types on crop yields, residue C and N losses, and soil labile and non-labile C and N fractions within a growing season in the greenhouse and the field and (2) determine if soil C and N fractions change more readily in the greenhouse than the field within a growing season.

Materials and Methods

Greenhouse experiment

The experiment was conducted under controlled soil and environmental conditions in the greenhouse with air temperatures of 25°C in the day and 15°C in the night. Soil samples were collected manually from an area of 5 m^2 using a shovel to a depth of 20 cm under a mixture of crested wheatgrass [*Agropyron cristatum* (L.) Gaertn] and western wheatgrass [*Pascopyrum smithii* (Rydb.) A. Love] from a dryland farm site, 11 km east of Sidney, Montana, USA. The research farm site where soil samples were collected is under the management of USDA, Agricultural Research Service, Sidney, Montana and no endangered or protected species were involved or was negatively impacted by this research. The soil was a Williams loam (fine-loamy, mixed, frigid, Typic Argiborolls [International classification: Luvisols]) with 350 g kg^{-1} sand, 325 g kg^{-1} silt, 325 g kg^{-1} clay, 1.42 Mg m^{-3} bulk density, and 7.2 pH at the 0–20 cm depth. Soil C and N fractions in the sample before the initiation of the experiment are shown in Table 1. Soil was air-dried and sieved to 4.75 mm after discarding coarse organic materials and rock fragments. Eight kilograms of soil was placed in a plastic pot, 25 cm high by 25 cm diameter, above 3 cm of gravel at the bottom.

Treatments consisted of two residue placements (surface placement vs. incorporation into the soil) and three crop types (spring wheat, pea, and fallow [or no crop]) arranged in a completely randomized design with three replications. In order to match the residue and crop type, spring wheat residue was placed under spring wheat and fallow and pea residue under pea. Residues included nine-week old spring wheat and pea plants collected from the field without grains, chopped to 2 cm, and oven-dried at 60°C for 3 d. Fifteen grams of residues per pot (corresponding to 2.6 Mg ha^{-1} of residue found in the field) were either placed uniformly at the soil surface or incorporated into the soil by mixing the residue with the soil by hand. The surface placement of residue corresponded to the simulated no-tillage system in the field, although the soil was disturbed during collection, and incorporated residue to the simulated conventional tillage system. Spring wheat received 0.96 g N pot^{-1} as urea, similar to the recommended N fertilization rate (80 kg N ha^{-1}) in

the field, while pea received 0.11 g N pot^{-1} (or 9 kg N ha^{-1}) while applying monoammonium phosphate as the P fertilizer. Half of 0.96 g N pot^{-1} was applied at planting and other half at four weeks later. Both spring wheat and pea also received P fertilizer (monoammonium phosphate) at 0.25 g P pot^{-1} (or 27 kg P ha^{-1}) and K fertilizer (muriate of potash) at 0.50 g K pot^{-1} (or 29 kg K ha^{-1}). No fertilizers were applied to the fallow treatment.

In July 2012, five spring wheat (cultivar Reeder) and pea (cultivar Majoret) seeds were planted per pot, except in the fallow treatment. At a height of 3 cm, seedlings were thinned to two plants per pot. In order to compensate for the water received as rainfall in the field, water was applied to all treatments in the greenhouse experiment to field capacity (0.25 m^3 m^{-3}) [21] at 300 to 500 mL pot^{-1}. Water was applied at planting and at 3 to 7 d intervals thereafter, depending on soil water content (as determined by a soil water probe [TDR 300, Spectrum Technologies Inc., Aurora, IL] installed to a depth of 15 cm). Since measured amount of water was applied according to soil water content and crop demand, only a negligible amount of water was leached below the pot that was not determined. Herbicides and pesticides were applied to plants as needed. At 105 d after planting, shoot biomass including grains was harvested from the pot, washed with water, oven-dried at 60°C for 3 to7 d, and dry matter yield was determined. Because of the small amount of grain production, grains were also included in the shoot biomass. After crop harvest, soil from the entire pot was sieved to 2 mm to separate coarse residue and root fragments, which were picked by hand, washed with water, and oven-dried at 60°C for 3 to7 d to determine dry matter yields. A portion (100 g) of residue and root-free soil sample visible to the naked eye was collected from each pot, air-dried, and used for determinations of C and N fractions. The remaining soil samples were further washed in a nest of 1.0 and 0.5 mm sieves under a continuous stream of water to separate fine roots. Roots left in the sieves were picked using a tweezers, oven-dried at 60°C for 3 to7 d, and dry matter yield was determined. Total root biomass was determined by adding biomass of coarse and fine roots.

Shoot and root biomass and crop residues added to the soil at the initiation of the experiment and those (>2.00 mm) recovered from the soil at the end were ground to 1 mm and C and N concentrations (g kg^{-1}) were determined with a high induction furnace C and N analyzer (LECO, St. Joseph, MI). Amounts of C and N in the residue added and recovered from the soil were determined by multiplying C and N concentrations by the weight of the soil in the pot. Carbon and N losses from the residue were determined as: Residue C and N losses (g kg^{-1}) = (Residue C and N added – Residue C and N recovered)×1000/Residue C and N added. While determining the amount of C and N recovered in the residue, it was assumed that fine residue (<2.00 mm) was a part of soil organic matter.

Field experiment

The field experiment was conducted using identical treatments, design, and replications as in the greenhouse from April to August 2012 near the place where soil samples were collected for the greenhouse experiment. As a result, soils were similar in both field and greenhouse experiments. The field site has mean monthly air temperature ranging from −8°C in January to 23°C in July and August. The mean annual precipitation (105-yr average) is 340 mm, 80% of which occurs during the crop growing season (April-October). Equivalent amounts of crop residues and fertilizers using the same treatments as in the greenhouse were applied to spring wheat, pea, and fallow in the field. Because the amount of residue applied was similar, the amounts of C and N

Table 1. Average soil organic C (SOC), total N (STN), particulate organic C and N (POC and PON), potential C and N mineralization (PCM and PNM), microbial biomass C and N (MBC and MBN), and NH_4-N and NO_3-N concentrations at the start of the experiment (n = 4).

Parameter	Concentration
SOC (g C kg^{-1})	11.80
POC (g C kg^{-1})	3.18
PCM (mg C kg^{-1})	9.25
MBC (mg C kg^{-1})	117.6
STN (g N kg^{-1})	1.29
PON (g N kg^{-1})	0.34
PNM (mg N kg^{-1})	8.95
MBN (mg N kg^{-1})	69.0
NH$_4$-N (mg N kg^{-1})	2.86
NO$_3$-N (mg N kg^{-1})	5.04

added in residue to the soil were also identical in the greenhouse and field. Residues and fertilizers were placed at the soil surface in the no-till system and incorporated to a depth of 10 cm using tillage with a field cultivator in the conventional tillage system. Plot size was 12.2×6.1 m.

Spring wheat and pea were planted in April with a no-till drill at a spacing of 20.3 cm. Growing season weeds were controlled with selective post emergence herbicides appropriate for each crop. Contact herbicides were applied at postharvest and preplanting. Crops were grown under dryland condition receiving only precipitation without irrigation. In August, biomass yield of spring wheat and pea was determined from two 0.5 m^2 areas outside yield rows within each plot and grain yield was determined by harvesting grains from a swath of 1.5 m×12.0 m using a combine harvester. Carbon and nitrogen concentrations in the grain and biomass were determined after oven drying subsamples at 55°C and using the C and N analyzer as above. Carbon and N contents (Mg C or N ha^{-1}) in grain and biomass were determined by multiplying C and N concentrations by grain and biomass yields, respectively. Total aboveground biomass and C and N contents were determined by adding yields and C and N contents of grain and biomass.

Soil samples were collected from five random locations in central rows of the plot to a depth of 20 cm using a truck-mounted hydraulic probe (3.5 cm inside diameter). Samples were composited within a plot, air-dried, ground, and sieved to 2 mm for determining C and N concentrations. No attempts were made to collect the surface residue at soil sampling because of residue loss and contamination with soil and residue from one plot to another due to actions of wind and water. Therefore, residue C and N losses were not determined in the field.

Soil carbon and nitrogen fractions measurements

The SOC concentration in the greenhouse and field soils were determined with a high induction furnace C and N analyzer as above after pretreating the soil with 5% H_2SO_3 to remove inorganic C [22]. The STN concentration was determined by using the analyzer without pretreating the soil with the acid. For determining POC and PON concentrations, 10 g soil sample was dispersed with 30 mL of 5 g L^{-1} sodium hexametaphosphate by shaking for 16 h and the solution was poured through a 0.053 mm sieve [19]. The solution and particles that passed through the sieve and contained mineral-associated and water-soluble C and N were

dried at 50°C for 3 to 4 d and SOC and STN concentrations were determined by using the analyzer as above. The POC and PON concentrations were determined by the difference between SOC and STN in the whole-soil and that in the particles that passed through the sieve after correcting for the sand content.

The PCM and PNM concentrations in air-dried soils were determined by the modified method of Haney et al. [23]. Two 10 g soil subsamples were moistened with water at 50% field capacity [21] and placed in a 1 L jar containing beakers with 4 mL of 0.5 mol L^{-1} NaOH to trap evolved CO_2 and 20 mL of water to maintain high humidity. Soils were incubated in the jar at 21°C for 10 d. At 10 d, the beaker containing NaOH was removed from the jar and PCM was determined by measuring CO_2 absorbed in NaOH, which was back-titrated with 1.5 mol L^{-1} $BaCl_2$ and 0.1 mol L^{-1} HCl. One beaker containing soil was removed from the jar and extracted with 100 mL of 2 mol L^{-1} KCl for 1 h. The NH_4-N and NO_3-N concentrations in the extract were determined by using the autoanalyzer (Lachat Instrument, Loveland, CO). The PNM was calculated as the difference between the sum of NH_4-N and NO_3-N concentrations in the soil before and after incubation.

The other beaker containing moist soil and incubated for 10 d (used for PCM determination above) was used for determining MBC and MBN concentrations by the modified fumigation–incubation method for air-dried soils [24]. The moist soil was fumigated with ethanol-free chloroform for 24 h and placed in a 1 L jar containing beakers with 2 mL of 0.5 mol L^{-1} NaOH and 20 mL water. As with PCM, fumigated moist soil was incubated for 10 d and CO_2 absorbed in NaOH was back-titrated with $BaCl_2$ and HCl. The MBC was calculated by dividing the amount of CO_2–C absorbed in NaOH by a factor of 0.41 [25] without subtracting the values from the nonfumigated control [24]. For MBN, the fumigated–incubated sample at 10 d was extracted with 100 mL of 2 mol L^{-1} KCl for 1 h and NH_4-N and NO_3-N concentrations were determined by using the autoanalyzer as above. The MBN was calculated by the difference between the sum of NH_4-N and NO_3-N concentrations in the sample before and after fumigation–incubation and divided by a factor of 0.41 [25,26]. The NH_4-N and NO_3-N concentrations determined in the nonfumigated–nonincubated samples were used as available fractions of N.

Data analysis

Data for C and N contents in crop biomass and residue and soil C and N fractions were analyzed by using the MIXED model of SAS [27]. Treatment was considered as the fixed effect and replication as the random effect. Means were separated by using the least square means test when treatments and interactions were significant [27]. Statistical significance was evaluated at $P \leq 0.05$, unless otherwise stated.

Results

Greenhouse experiment

Shoot and root biomass yields and carbon and nitrogen contents. Shoot and root biomass yields and C and N contents varied among residue placements and crop types (Table 2). Interaction between residue placement and crop types on these parameters was not significant. Shoot and root biomass yields and C and N contents were greater in surface placement than incorporation of residue into the soil. Shoot biomass yield and C and N contents were also greater in wheat than in pea. Because of the negligible amount of roots, root biomass yield and C and N contents in pea were not determined. Absence of plants in the fallow also resulted in non-existence of crop data in this treatment. The coefficient of variation (CV) for crop parameters ranged from 38.2 to 62.5%.

Residue carbon and nitrogen losses. Total amounts of C and N added through residue application and leaf fall and those recovered in coarse fractions (>2 mm) after crop harvest varied with residue placements and crop types, with the significant residue placement × crop type interaction for C and N recovered in the residue (Table 3). Although the amount of residue applied was similar in all treatments (15 g of wheat or pea residue pot^{-1}), differences in C and N concentrations between residues and those added through leaf fall during crop growth varied residue C and N additions among treatments. Averaged across crop types, residue C addition was greater in surface placement than incorporation of residue into the soil. Averaged across residue placements, residue C addition was greater under wheat than pea or fallow, but residue N addition was greater under pea than wheat or fallow. Residue C recovery was greater in surface placement under wheat and fallow than surface placement under pea and incorporation under fallow. Residue N recovery was also greater in surface placement under wheat and fallow than surface placement under pea and incorporation under fallow and wheat. Averaged across crop types, residue N recovery was greater in surface placement than incorporation of residue into the soil. Averaged across residue placements, residue C recovery was greater under wheat than pea. The coefficient of variation for residue C and N addition and recovery varied from 7.2 to 17.8%.

Residue C and N losses also varied with residue placements and crop species, with the significant residue placement × crop species interaction (Table 3). Residue C loss was greater in surface placement under pea and incorporation under fallow than surface placement under fallow. Residue N loss was in the order: surface placement and incorporation under pea > incorporation under wheat and fallow > surface placement under wheat > surface placement under fallow. Averaged across crop types, residue N loss was greater in residue incorporation than surface placement. Averaged across residue placements, residue N loss was greater under pea than under fallow and wheat. The coefficient of variation for residue C and N losses varied from 14.3 to 31.6%.

Soil carbon and nitrogen fractions. The SOC, POC, and PCM concentrations varied among residue placements and crop types (Table 4). Averaged across crop types, SOC and PCM were greater in surface placement than incorporation of the residue into the soil. Averaged across residue placements, SOC was greater under wheat than pea and fallow and POC was greater under wheat than pea. The MBC was not influenced by treatments. The coefficient of variation for soil C fractions ranged from 2.6 to 14.3%.

The STN, PNM, MBN, NH_4-N, and NO_3-N concentrations also varied among residue placements and crop types (Table 4). Averaged across crop types, PNM and NH_4-N were greater in surface placement than incorporation of residue into the soil. Averaged across residue placements, STN was greater under wheat than pea and MBN was greater under wheat than fallow. In contrast, PNM was greater under pea and fallow than wheat and NO_3-N was greater under fallow than wheat. The PON was not influenced by treatments. The coefficient of variation for soil N fractions ranged from 4.6 to 15.9%.

Field experiment

Aboveground total crop biomass yield and C and N contents varied with crop types (Table 5). Averaged across tillage practices, crop biomass yield and C content were greater in wheat than pea, but the trend reversed for N content. Tillage and its interaction with crop type were not significant for crop biomass yield and C and N contents. The coefficient of variation for crop biomass yield and C and N contents ranged from 28.1 to 41.9%.

Soil MBN, NH_4-N, and NO_3-N concentrations varied with tillage practices and MBC and PNM varied with crop types (Table 5). Averaged across crop types, MBN and NH_4-N were greater in no-tillage than conventional tillage, but NO_3-N was greater in conventional tillage than no-tillage. Averaged across tillage practices, MBC was greater under wheat than fallow and PNM was greater under pea than wheat and fallow. Tillage, crop type, and their interaction were not significant for SOC, POC, PCM, STN, and PON. The coefficient of variations for soil C and N fractions ranged from 8.0 to 36.7%.

Discussion

Enhanced soil water conservation due to mulch action of the residue at the soil surface [28] may have increased shoot and root biomass yields and C and N contents in surface placement compared to incorporation of residue into the soil in the greenhouse (Table 2). It has been reported that surface placement of residue in the no-till system increased spring wheat yield compared to residue incorporation in the conventional till system [3,15]. In our field experiment, crop biomass yield and C and N contents, however, were not influenced by tillage (Table 5). It may be possible that wheat and pea residues applied by hand at the soil surface were more uniformly distributed in the greenhouse than in the field where residues were distributed by a machine sprayer. As a result, soil water was probably conserved more, resulting in increased crop yield and C and N contents with the surface placement than incorporation of residue in the greenhouse compared to the field.

Differences in the amount of N fertilizer applied and N fixation capacity may have resulted in variation in crop biomass yields and C contents among crop species in the greenhouse and the field (Tables 2 and 5). Higher amount of N fertilizer application may have increased biomass yield and C and N contents in wheat than pea in the greenhouse. Higher amount of N fertilizer application also may have increased biomass yield and C content in wheat and pea, but greater N fixation may have increased N content in pea than wheat in the field [14,28]. Grain and biomass yields are usually greater in wheat which receives N fertilizer than pea which

Table 2. Effects of residue placement and crop type on crop shoot (grains+leaves+stems) and root biomass C and N contents in the greenhouse.

Residue placement	Crop type	Shoot biomass	Root biomass	Shoot biomass C	Root biomass C	Shoot biomass N	Root biomass N	Total biomass C	Total biomass N
		——— g pot⁻¹ ———		——— g C pot⁻¹ ———		——— g Npot⁻¹ ———		g C pot⁻¹	g N pot⁻¹
Incorporated		4.51b[a]	2.37b	1.83b	0.74b	0.14b	0.05b	2.91b	0.25b
Surface		7.41a	6.14a	3.07a	1.74a	0.24a	0.11a	5.55a	0.44a
	Pea	4.55b	———[b]	1.91b	———	0.12b	———	1.91b	0.12b
	Wheat	7.36a	4.25	3.00a	1.24	0.26a	0.08	4.23a	0.34a
CV (%)[c]		42.4	61.9	42.4	56.4	55.0	62.5	40.4	38.2
Significance									
Residue placement (R)		*	———	*	*	**	*	*	*
Crop species (C)		*	*	*	———	**	———	*	———
R×C		NS[d]	NS	NS	———	NS	———	———	———

*Significant at P=0.05.
**Significant at P=0.01.
[a]Numbers followed by different letters within a column in a set are significantly different at P≤0.05 by the least square means test.
[b]Non-measurable values due to negligible amount of root biomass.
[c]Coefficient of variation.
[d]Not significant.

Table 3. Effects of residue placement and crop type on residue C and N addition, recovered in coarse fragments (>2 mm), and losses during the crop growing period in the greenhouse.

Residue placement	Crop type	Crop residue					
		C added[a] g C pot^{-1}	N added[a] g N pot^{-1}	C recovered g C pot^{-1}	N recovered g N pot^{-1}	C loss g kg^{-1}	N loss g kg^{-1}
Incorporated	Fallow	7.80a[b]	0.40a	3.84b	0.23b	508a	438b
	Pea	7.80a	0.64a	4.24ab	0.26ab	457ab	588a
	Wheat	8.76a	0.44a	4.89ab	0.24b	430ab	456b
Surface	Fallow	7.80a	0.40a	5.16a	0.31a	339b	213d
	Pea	7.80a	0.64a	3.82b	0.23b	511a	640a
	Wheat	8.92a	0.46a	4.95a	0.31a	446ab	325c
CV (%)[c]		7.2	17.8	15.4	16.5	143	316
Means							
Incorporated		8.07b	0.49a	4.32a	0.24b	465a	494a
Surface		8.17a	0.50a	4.64a	0.28a	432a	393b
	Fallow	7.80b	0.40b	4.50ab	0.27a	423a	326b
	Pea	7.80b	0.64a	4.02b	0.25a	484a	614a
	Wheat	8.76a	0.45b	4.92a	0.28a	438a	390b
Significance							
Residue placement (R)		*	NS[d]	NS	*	NS	**
Crop species (C)		***	***	*	NS	*	***
R×C		NS	NS	*	*	*	**

*Significant at $P=0.05$.
**Significant at $P=0.01$.
*** Significant at $P=0.001$.
[a]Includes C and N added from the residue application and leaf fall.
[b]Numbers followed by different letters within a column in a set are significantly different at $P\leq0.05$ by the least square means test.
[c]Coefficient of variation.
[d]Not significant.

Soil Carbon and Nitrogen Fractions and Crop Yields Affected by Residue Placement and Crop Types

Table 4. Effects of residue placement and crop type on soil organic C (SOC), total N (STN), particulate organic C and N (POC and PON), potential C and N mineralization (PCM and PNM), microbial biomass C and N (MBC and MBN), and NH_4-N and NO_3-N concentrations in the greenhouse.

Residue placement	Crop type	SOC	POC	PCM	MBC	STN	PON	PNM	MBN	NH_4-N	NO_3-N
		g C kg^{-1}		mg C kg^{-1}		g N kg^{-1}		mg N kg^{-1}		mg N kg^{-1}	
Incorporated		12.0b[a]	3.28a	11.6b	128.0a	1.29a	0.33a	5.80b	50.8a	1.14b	11.8a
Surface		12.3a	3.28a	14.4a	143.8a	1.31a	0.31a	9.56a	57.5a	1.65a	16.5a
	Fallow	12.0b	3.26ab	11.4a	118.9a	1.30ab	0.32a	9.36a	46.1b	1.44a	23.7a
	Pea	12.0b	3.18b	13.1a	144.0a	1.26b	0.29a	9.68a	56.3ab	1.38a	12.8ab
	Wheat	12.4a	3.40a	14.5a	145.1a	1.34a	0.34a	5.98b	97.6a	1.36a	6.1b
CV (%)[b]		2.6	11.8	14.3	13.3	4.6	15.9	14.2	14.3	14.8	13.6
Significance											
Residue placement (R)		**	NS[c]	*	NS	NS	NS	*	NS	*	NS
Crop species (C)		**	*	NS	NS	*	NS	*	*	NS	*
R×C		NS	NS	NS	NS	NS	NS	NS	NS	NS	NS

Soil samples were collected at the 0–20 cm depth in the field and used for the greenhouse experiment.
*Significant at $P=0.05$.
**Significant at $P=0.01$.
[a]Numbers followed by different letters within a column in a set are significantly different at $P \leq 0.05$ by the least square means test.
[b]Coefficient of variation.
[c]Not significant.

Table 5. Effects of residue placement and crop type on crop aboveground biomass (grains+stems+leaves) yield, C and N contents, and soil organic C (SOC), total N (STN), particulate organic C and N (POC and PON), potential C and N mineralization (PCM and PNM), microbial biomass C and N (MBC and MBN), and NH_4-N and NO_3-N concentrations at the 0–20 cm depth in the field.

Tillage[a]	Crop type	Crop biomass yield	Crop C content	Crop N content	SOC	POC	PCM	MBC	STN	PON	PNM	MBN	NH_4-N	NO_3-N
		Mg ha^{-1}	Mg C ha^{-1}	kg N ha^{-1}	─── g C kg^{-1} ───		─── mg C kg^{-1} ───		─── g N kg^{-1} ───		──── mg N kg^{-1} ────			
CT		4.91a[b]	2.06a	59.6a	11.0a	2.56a	45.1a	114.4a	1.33a	0.24a	3.21a	12.6b	3.05b	4.54a
NT		5.08a	2.11a	65.3a	11.0a	2.56a	56.8a	122.9a	1.37a	0.28a	4.28a	19.6a	3.82a	2.08b
	Fallow	────[c]	─────	─────	10.6a	2.55a	45.61a	111.4b	1.30a	0.25a	2.85b	14.2a	2.93a	3.36a
	Pea	4.68b	1.87b	72.3a	11.4a	2.56a	51.0a	118.6ab	1.42a	0.27a	5.56a	15.5a	3.43a	2.87a
	Wheat	5.31a	2.18a	52.6b	11.0a	2.57a	56.0a	126.0a	1.34a	0.26a	2.83b	18.6a	3.93a	3.71a
CV (%)		28.1	30.0	41.9	8.0	18.7	27.1	28.9	9.3	18.4	36.4	36.7	27.9	25.0
Significance														
Tillage (T)		NS[d]	NS	NS	NS	NS	NS	NS	NS	NS	NS	*	*	*
Crop species (C)		*	*	**	NS	NS	NS	*	NS	NS	*	NS	NS	NS
T×C		NS	NS	NS	NS	NS	NS	NS	NS	NS	NS	NS	NS	NS

*Significant at $P = 0.05$.
**Significant at $P = 0.01$.
[a]Tillage are CT, conventional tillage; and NT, no-tillage.
[b]Numbers followed by different letters within a column in a set are significantly different at $P \leq 0.05$ by the least square means test.
[c]Crop absent in the fallow.
[d]Not significant.

receives no N fertilizer due to increased water-use efficiency, but higher N concentration due to increased atmospheric N fixation can increase N content in pea than wheat [10,15,28]. The fact that different trends in N content in pea vs. wheat occurred in the field and the greenhouse was probably related to root growing soil volume. It may be possible that roots exploited greater soil volume that resulted in increased N fixation by pea and therefore increased its N content in the field compared to the greenhouse where plants were grown in a limited soil volume in the pot.

Greater residue input due to higher biomass yield may have increased residue C addition in surface placement than incorporation of residue into the soil or increased under wheat than pea or fallow in the greenhouse (Tables 2 and 3). In contrast, higher N concentration may have increased residue N addition under pea than wheat or fallow. Greater C and N recovered in the residue placed at the soil surface under wheat and fallow were probably due to reduced mineralization of wheat residue as a result of its higher C/N ratio than pea residue. While surface placement of residue reduces its contact with soil microorganisms that result in reduced mineralization [29,30], increased mineralization of pea residue due to its lower C/N ratio may have resulted in reduced C and N recovery in the residue placed at the soil surface under pea. Residues of legumes, such as pea with lower C/N ratio, decompose more rapidly than those of nonlegumes, such as wheat with higher C/N ratio [12]. When incorporated into the soil, residue C and N recovery were lower under fallow and wheat. As a result, C and N losses were higher in surface placement of residue under pea or residue incorporation under pea and fallow than the other treatments. It may be possible that some of C and N lost from the residue converted into soil C and N fractions, as discussed below.

Reduced mineralization of residue may have increased SOC, PCM, PNM, and NH_4-N concentrations in surface placement than incorporation of residue into the soil in the greenhouse (Table 4). Similar increases in MBN and NH_4-N concentrations in no-tillage compared to conventional tillage were found in the field (Table 5). Several researchers [5,16,31,32,33] have reported greater SOC, POC, MBC, PCM, PNM, and MBN in surface residue placement in the no-tillage system than residue incorporation into the soil in the conventional tillage system. Increased N mineralization due to residue incorporation, however, may have increased NO_3-N concentration in conventional tillage than no-tillage in the field.

Higher C and N substrate availability due to increased yield probably increased SOC, POC, STN, and MBN under wheat than under pea or fallow in the greenhouse (Table 4) or increased MBC under wheat than fallow in the field (Table 5). Root biomass C, residue C addition (Tables 2 and 3), and amount of applied N fertilizer were greater in wheat than pea or fallow. Similar results probably occurred in the field, since treatments were identical in the greenhouse and the field and crop biomass C was higher in wheat than pea in the field (Table 5). Rhizodeposit C released by roots can increase microbial biomass and activity and soil C storage [34]. Liebig et al. [35] also found higher MBC under spring wheat than under fallow. In contrast, greater PNM and NO_3-N under pea and fallow than wheat in the greenhouse were probably either due to increased mineralization of pea residue as a result of its lower C/N ratio than wheat residue [12] or to greater mineralization of soil and wheat residue as a result of enhanced microbial activity from higher soil temperature and water content and absence of plants to uptake N under fallow [11,13,36]. Since residue N loss was greater under pea than wheat and fallow (Table 3), part of N from pea residue may have contributed to increased PNM and NO_3-N concentrations under pea. Similar

result of increased PNM under pea than wheat and fallow was also found in the field, since crop biomass N was greater in pea than wheat (Table 5).

Comparison of soil C and N fractions at the beginning and end of the experiment due to residue placement (Tables 1 and 4) showed that SOC increased by 4.2%, PCM by 55.7%, and PNM by 6.1% with surface residue placement in the greenhouse. Corresponding values in SOC, PCM, and PNM with residue incorporation were 1.7, 25.4, and −35.1%, respectively. In the field, MBN reduced by 71.5% in no-tillage and 81.7% in conventional tillage from the beginning to the end of the experiment. This shows that residue placement at the surface either increased soil C and N fractions in the greenhouse or reduced their losses in the field within a crop growing season compared to residue incorporation. Since soil NH_4-N and NO_3-N concentrations vary seasonally due to N mineralization from crop residue and soil, N fertilization, crop N uptake, and N losses due to leaching, volatilization, and denitrification [10,15], variations in their levels from the beginning to the end of the experiment were not taken into account.

Among crop types, SOC increased by 5.1%, POC by 6.9%, STN by 3.9%, and MBN by 41.4%, but PNM decreased by 33.1% under wheat from the beginning to the end of the experiment in the greenhouse. In the field, MBC increased by 7.1%, but PNM decreased by 68.3% under wheat during this period. The corresponding increases in SOC, POC, STN, and MBN or decrease in PNM during this period were lower under pea and fallow. This suggests that wheat increased more soil C and N fractions, except PNM, than pea or fallow due to increased substrate availability from root and rhizodeposition and/or to slow decomposition of wheat than pea residue due to differences in residue quality (e.g. C/N ratio). The greater PNM under pea than wheat or fallow was due to increased N contribution from its residue (Table 5).

When the greenhouse and field experiments were compared, trends in changes in soil C and N fractions due to treatments within a crop growing season were similar. However, greater changes in labile than nonlabile C and N fractions occurred more in the greenhouse than in the field. Furthermore, the coefficient of variations in soil C and N fractions were lower in the greenhouse (2.6 to 15.9%) than in the field (8.0 to 36.7%) (Tables 4 and 5). This indicates that soil C and N fractions changed more readily but with lower variability with management practices within a crop growing season when soil and environmental conditions are controlled in the greenhouse than in field where soil heterogeneity often results in non-significant differences among treatments in these fractions [16,18,30]. Use of disturbed soil in the greenhouse vs. undisturbed (especially in the no-till system) in the field also may have an influence on differences in changes in soil C and N fractions between the two experiments. The greater changes in labile than nonlabile C and N fractions as influenced by management practices within a short period in the greenhouse and the field suggests that labile C and N fractions are better indicators of changes in soil organic matter quality than nonlabile fractions, a case similar to that reported by various researchers [10,11,13,16,30]. The fact that more changes in labile than nonlabile C and N fractions occurred in the greenhouse than in the field suggests that better measurements of changes in soil organic matter due to management practices within a short period can be observed when soil and environmental conditions are controlled. Greater levels of most soil C and N fractions in the greenhouse than in field was probably a result of increased turnover rate plant C and N into soil C and N, because disturbed soil was used in the greenhouse and environmental condition for microbial transformation was more favorable in the greenhouse than the field.

Greenhouse study provided more information on plant and residue parameters, such as measurement of root biomass and C and N contents and residue C and N losses, which cannot be measured easily in the field. This resulted in the measurement of turnover rate of plant C and N into soil C and N in the greenhouse, a fact that was absent in the field. Because of greater changes in soil C and N fractions, greenhouse study provided a more robust method of evaluating C and N cycling and soil quality within a short period of time as affected by management practices than the field experiment. Such changes can also be measured in the field but it may take longer time. While all results from the greenhouse study may not be readily applied in the field, some information, such as root biomass and residue C and N losses, measured in the greenhouse can be extrapolated to the field condition. The effects of short-term study in the greenhouse can be useful to predict the long-term impact of management practices on soil C and N fractions in the field.

Conclusions

Crop yields, residue C and N losses, and soil C and N fractions varied with residue placement and crop types in the greenhouse and the field. Surface placement of residue increased crop yields, residue C and N losses, and enhanced SOC, PCM, MBN, and NH_4-N concentrations, but residue incorporation increased PNM and NO_3-N concentrations. Similarly, spring wheat had higher yield and increased SOC, POC, MBC, STN, and MBN than pea or fallow, but pea had higher N content and increased PNM than wheat or fallow. Placing nonlegume residue at the soil surface

using no-tillage can increase soil C and N sequestration and microbial biomass and activity that can improve soil health and quality. Using this practice, producers can claim for C credit. Incorporation of legume and nonlegume residues into the soil using conventional tillage can increase N mineralization and availability which can reduce N fertilization rate to succeeding crops, but can degrade soil quality due to reduced organic matter and increased erosion. Although soil labile C and N fractions changed more readily than nonlabile fractions within a crop growing season both in the greenhouse and field, greater changes in labile than nonlabile fractions occurred with reduced variability more in the greenhouse than in the field. Results suggest that greenhouse study provided a more robust measurement of crop growth and changes in soil C and N fractions within a short period as influenced by management practices than the field experiment. Longer time will be probably needed in the field to obtain results similar to those in the greenhouse. Additional information, such as root growth, residue C and N losses, turnover of plant C and N to soil C and N, and results of short-term study on soil C and N fractions as influenced by management practices in the greenhouse can be used to predict the long-term impact in the field.

Acknowledgments

We appreciate the excellent support of Joy Barsotti and Thecan Caesar-TonThat for analyzing soil and plant samples in the laboratory.

Author Contributions

Conceived and designed the experiments: UMS. Performed the experiments: JW. Analyzed the data: UMS. Contributed reagents/materials/analysis tools: UMS. Wrote the paper: UMS.

References

1. Lal R, Kimble JM, Stewart BA (1995) World soils as a source or sink for radiatively-active gases. In: Lal R, editor, Soil management and greenhouse effect. Advances in soil science. CRC Press, Boca Raton, FL, pp. 1–8

2. Paustian K, Robertson GP, Elliott ET (1995) Management impacts on carbon storage and gas fluxes in mid-latitudes cropland. In: Lal R, editor, Soils and global climate change. Advances in soil science. CRC Press, Boca Raton, FL, USA, pp. 69–83.

3. Halvorson AD, Peterson GA, Reule CA (2002a) Tillage system and crop rotation effects on dryland crop yields and soil carbon in the central Great Plains. Agron J 94:1429–1436.

4. Sherrod LA, Peterson GA, Westfall DG, Ahuja LR (2003) Cropping intensity enhances soil organic carbon and nitrogen in a no-till agroecosystem. Soil Sci Soc Am J 67:1533–1543.

5. Sainju UM, Caesar-TonThat T, Lenssen AW, Evans RG, Kolberg R (2007) Long-term tillage and cropping sequence effects on dryland residue and soil carbon fractions. Soil Sci Soc Am J 71:1730–1739.

6. Rasmussen PE, Allmaras RR, Rhoade CR, Roager NC Jr (1980) Crop residue influences on soil carbon and nitrogen in a wheat-fallow system. Soil Sci Soc Am J 44:596–600.

7. Peterson GA, Halvorson AD, Havlin JL, Jones OR, Lyon DG, Tanaka DL (1998) Reduced tillage and increasing cropping intensity in the Great Plains conserve soil carbon. Soil Tillage Res 47:207–218.

8. Bauer A, Black AL (1994) Quantification of the effect of soil organic matter content on soil productivity. Soil Sci Soc Am J 58:185–193.

9. Ghidey F, Alberts EE (1993) Residue type and placement effects on decomposition: Field study and model evaluation. Trans ASAE 36:1611–1617.

10. Sainju UM, Lenssen AW, Caesar-Tonthat T, Waddell J (2006b) Tillage and crop rotation effects on dryland soil and residue carbon and nitrogen. Soil Sci Soc Am J 70:668–678.

11. Halvorson AD, Wienhold BJ, Black AL (2002b) Tillage, nitrogen, and cropping system effects on soil carbon sequestration. Soil Sci Soc Am J 66:906–912.

12. Kuo S, Sainju UM, Jellum EJ (1997) Winter cover cropping influence on nitrogen in soil. Soil Sci Soc Am J 61:1392–1399.

13. Sainju UM, Lenssen AW, Caesar-TonThat T, Waddell J (2006a) Carbon sequestration in dryland soils and plant residue as influenced by tillage and crop rotation. J Environ Qual 35:1341–1349.

14. Miller PR, McConkey B, Clayton GW, Brandt SA, Staricka JA, Johnston AM, Lafond GP, Schatz BG, Baltensperger DD, Neill KE (2002) Pulse crop adaptation in the northern Great Plains. Agron J 94:261–272.

15. Sainju UM, Lenssen AW, Caesar-TonThat T, Evans RG (2009) Dryland crop yields and soil organic matter as influenced by long-term tillage and cropping sequence. Agron J 101:243–251.

16. Franzluebbers AJ, Hons FM, Zuberer DA (1995) Soil organic carbon, microbial biomass, and mineralizable carbon and nitrogen in sorghum. Soil Sci Soc Am J 59:460–466.

17. Bezdicek DF, Papendick DF, Lal R (1996) Introduction: Importance of soil quality to health and sustainable land management. In: Doran JW, Jones AJ, editors, Methods of assessing soil quality, Spec. Publ. 49, Soil Science Society of America, Madison, USA, pp. 1–18.

18. Franzluebbers AJ, Arshad MA (1997) Soil microbial biomass and mineralizable carbon of water-stable aggregates. Soil Sci Soc Am J 67:1090–1097.

19. Cambardella CA, Elliott ET (1992) Particulate soil organic matter changes across a grassland cultivation sequence. Soil Sci Soc Am J 56:777–783.

20. Six J, Elliott ET, Paustian K (1999) Aggregate and soil organic matter dynamics under conventional and no-tillage systems. Soil Sci Soc Am J 63:1350–1358.

21. Pikul JL Jr, Aase JK (2003) Water infiltration and storage affected by subsoiling and subsequent tillage. Soil Sci Soc Am J 67:859–866.

22. Nelson DW, Sommers LE (1996) Total carbon, organic carbon, and organic matter. In: Sparks DL, editor, Methods of soil analysis. Part 3. Chemical method. SSSA Book Ser. 5. Soil Science Society of America, Madison, pp. 961–1010.

23. Haney RL, Franzluebbers AJ, Porter EB, Hons FM, Zuberer DA (2004) Soil carbon and nitrogen mineralization: Influence of drying temperature. Soil Sci Soc Am J 68:489–492.

24. Franzluebbers AJ, Haney RL, Hons FM, Zuberer DA (1996) Determination of microbial biomass and nitrogen mineralization following rewetting of dried soil. Soil Sci Soc Am J 60:1133–1139.

25. Voroney RP, Paul EA (1984) Determination of k_C and k_N *in situ* for calibration of the chloroform fumigation-incubation method. Soil Biol Biochem 16:9–14.

26. Brookes PC, Landman A, Pruden G, Jenkinson DJ (1985) Chloroform fumigation and the release of soil nitrogen: A rapid direct-extraction method to measure microbial biomass nitrogen in soil. Soil Biol Biochem 17:937–942.

27. Littell RC, Milliken GA, Stroup WW, Wolfinger RR (1996) SAS system for mixed models. SAS Institute Inc., Cary, NC, USA.

28. Lenssen AW, Johnson GD, Carlson GR (2007) Cropping sequence and tillage system influences annual crop production and water use in semiarid Montana. Field Crops Res 100:32–43.

29. Coppens F, Garnier P, de Gryze P, Merckx R, Recous S (2006) Soil moisture, carbon and nitrogen dynamics following incorporation and surface application of labelled crop residues in soil columns. Europ J Soil Sci 57:894–905.

30. Giacomini SJ, Recous S, Mary B, Aita C (2007) Simulating the effects of nitrogen availability, straw particle size and location in soil on carbon and nitrogen mineralization. Plant Soil 301: 289–301.

31. Malhi SS, Lemke R (2007) Tillage, crop residue and nitrogen fertilizer effects on crop yield, nutrient uptake, soil quality and greenhouse gas emissions in the second 4-yr rotation cycle. Soil Tillage Res 96:269–283.

32. Wright AL, Hons FM, Lemon RG, MacFarland ML, Nichols RL (2008) Microbial activity and soil carbon sequestration for reduced and conventional tillage cotton. Appl Soil Ecol 38:168–173.

33. Lupwayi NZ, Lafond GP, Ziadi N, Grant CA (2012) Soil microbial response to nitrogen fertilizer and tillage in barley and corn. Soil Tillage Res 118:139–146.

34. Lu Y, Watanabe A, Kimura M (2002) Contribution of plant-derived carbon to soil microbial biomass dynamics in a paddy rice microcosm. Biol Fertil Soils 36:136–142.

35. Liebig MA, Tanaka DL, Wienhold BJ (2004) Tillage and cropping effects on soil quality indicators in the northern Great Plains. Soil Tillage Res 78:131–141.

36. Kuzyakov Y, Domanski G (2000) Carbon input by plants into the soil: Review. J. Plant Nutri Soil Sci 163:421–431.

14

Adjacent Habitat Influence on Stink Bug (Hemiptera: Pentatomidae) Densities and the Associated Damage at Field Corn and Soybean Edges

P. Dilip Venugopal*, Peter L. Coffey, Galen P. Dively, William O. Lamp

Department of Entomology, University of Maryland, College Park, Maryland, United States of America

Abstract

The local dispersal of polyphagous, mobile insects within agricultural systems impacts pest management. In the mid-Atlantic region of the United States, stink bugs, especially the invasive *Halyomorpha halys* (Stål 1855), contribute to economic losses across a range of cropping systems. Here, we characterized the density of stink bugs along the field edges of field corn and soybean at different study sites. Specifically, we examined the influence of adjacent managed and natural habitats on the density of stink bugs in corn and soybean fields at different distances along transects from the field edge. We also quantified damage to corn grain, and to soybean pods and seeds, and measured yield in relation to the observed stink bug densities at different distances from field edge. Highest density of stink bugs was limited to the edge of both corn and soybean fields. Fields adjacent to wooded, crop and building habitats harbored higher densities of stink bugs than those adjacent to open habitats. Damage to corn kernels and to soybean pods and seeds increased with stink bug density in plots and was highest at the field edges. Stink bug density was also negatively associated with yield per plant in soybean. The spatial pattern of stink bugs in both corn and soybeans, with significant edge effects, suggests the use of pest management strategies for crop placement in the landscape, as well as spatially targeted pest suppression within fields.

Editor: Youjun Zhang, Institute of Vegetables and Flowers, Chinese Academy of Agricultural Science, China

Funding: This work is supported by the United Soybean Board, http://unitedsoybean.org/ (GPD PDV); Maryland Soybean Board, http://mdsoy.com/ (GPD); Maryland Grain Producers Utilization Board, http://www.marylandgrain.com/ (GPD); Cosmos Club Foundation, http://www.cosmosclubfoundation.org/ (PDV); Northeast Sustainable Agriculture Research and Education - Graduate Student grant - GNE-12-047, http://www.nesare.org/ (PDV, WOL, GPD); Arthur B. Gahan Fellowship, Dept. of Entomology, Univ. of Maryland, http://entomology.umd.edu/academics/graduate/awardopportunities/gahanfellowshipfund (PDV); and United States Department of Agriculture - National Institute of Food & Agriculture Hatch Project #MD-ENTM-1016, http://www.csrees.usda.gov/ (WOL). Funding for Open Access provided by the UMD Libraries Open Access Publishing Fund. The funders had no role in study design, data collection and analysis, decision to publish, or preparation of the manuscript.

Competing Interests: The authors confirm that funding from the commercial source professional association.

* Email: venugopal.dilip@gmail.com

Introduction

Agricultural fields are components within a heterogeneous landscape that strongly connect to and interact with surrounding habitats [1]. The movement of insects between natural and agricultural habitats has important implications for agricultural ecosystem function [2]. The movement of pest insects among seasonal crop resources is often non-random and directional as pest species disperse and colonize crops [3]. This movement may result in insect pests immigrating into an agricultural field in an aggregated manner in specific areas within the field [1]. In such cases, species-specific characteristics such as host range, vagility, chemical ecology, and host developmental status influence the spatial pattern of a pest population within crop fields, often resulting in the aggregation of pests along field edges as they disperse between habitats. The seasonal availability and suitability of source and recipient habitats in relation to the life stages of a mobile, polyphagous insect pest influence the dispersal dynamics of pests from sources to recipient habitats [2,4,5]. Knowledge about insect pest movement among habitats in the landscape, and

the subsequent colonization of plants within crop fields, may inform risk of infestation by an insect pest prior to their subsequent population increase as well as provide opportunities for pest management [1].

Stink bugs in the family Pentatomidae are major pests of economically important crops globally [6], and are considered important pests in soybean *Glycine max* (L.) Merr. producing areas of the world [7]. While stink bugs caused economic losses in the southern United States, they were not considered serious pests of crops in the mid-Atlantic region until recently. The most common stink bugs in agricultural fields in the mid-Atlantic are *Chinavia hilaris* (Say 1832) and *Euschistus servus* (Say 1832), but these species have had little economic impact in the region [8,9]. The recent explosion in populations of the invasive brown marmorated stink bug, *Halyomorpha halys* (Stål 1855) however, has led to significant economic and ecological impacts.

Since its discovery near Allentown, Pennsylvania, USA, *H. halys* has been detected in 41 states within United States, and local populations and detections from Europe (Switzerland, France, Canada, Germany, Italy, and Liechtenstein) have also been reported [10]. This polyphagous stink bug has a wide range of host plants including tree fruits, vegetables, field crops, ornamental plants, and native vegetation in its native and invaded ranges. Since 2010, serious crop losses have been reported for apples, peaches, sweet corn, peppers, tomatoes, and row crops such as field corn and soybeans in the mid-Atlantic region [11]. *Halyomorpha halys* is also a nuisance pest in human dwellings. In this context, information on the movement of stink bugs into crops from adjacent managed and natural habitats, and the associated spatially-explicit crop damage, has direct implications for integrated pest management.

The dispersal and movement of various stink bug species between crops and other habitats has been addressed by many studies in the context of dispersal between habitats, adjacent habitat influences on densities at field edges, and their relationship to crop damage [12–23]. However, these studies mainly pertain to stink bug communities in crops in the southern U.S. Currently only anecdotal reports of high *H. halys* abundance in the edges of fields adjacent to woodlots [11] are available. These reports are consistent with the documented use of woody trees as a food source and as overwintering sites by *H. halys* [24]. This factor coupled with use of buildings as overwintering sites suggests that these habitats, which support high local stink bug populations that eventually invade soybean later in the season, are likely to play an important role in stink bug densities.

Many stink bug species cause significant seed quality and yield losses in field corn, *Zea mays* L., and soybean [19,23–29], and stink bugs are also associated with the transmission of plant diseases [32–34]. However, few studies quantify spatially-explicit field crop damage in relation to patterns of stink bug densities [35]. Soybean is one of the preferred hosts for *H. halys* [36], and both field corn and soybean constitute a very high proportion of overall crop area in the mid-Atlantic region and throughout the United States [37]. Research efforts aimed at determining the role of adjacent habitat in influencing stink bug dispersal, population density, and pattern of settlement into crop, contribute to the development of management strategies of *H. halys* in row crops.

In this study, our objectives were a) document the species composition and within-field distribution of stink bugs in field corn and soybean, b) examine the influence of adjacent managed and natural habitats on the density of stink bugs, and c) relate stink bug density to seed quality in field corn and soybean, and pod development and yield in soybean. We expected *H. halys* to be the most abundant stink bug in our study based on previous reports of stink species composition in mid-Atlantic row crops [8]. We predicted higher density of stink bugs along wooded habitats and buildings than open habitats as they provide host plants and over-wintering refuge [24]. We also predicted high density of stink bugs at the field edge, reducing with distance into the field interior as observed by anecdotal reports for *H. halys* [11].

Methods

Ethics statement

The field studies were conducted at facilities provided by the University of Maryland and the USDA Beltsville Agricultural Research Center where we had permission to conduct research. Stink bugs were enumerated *in situ* and were not collected, and no endangered or protected species were involved.

Field selection and stink bug sampling strategy

The study was conducted at the USDA Beltsville Agricultural Research Center at Beltsville, MD (39°01′50N; 76°53′28W) and University of Maryland Research and Education Center facilities

at Beltsville (39°00′45N; 76°49′32W), Clarksville (39°15′18N; 76°55′51W) and Keedysville, MD (39°30′37N; 77°43′57W). At these sites, field corn (76.2 cm row spacing) and full season soybean (17.8 cm row spacing) fields with a portion of their perimeter directly adjacent to wooded areas (henceforth wooded habitats), buildings (buildings, houses and barns; henceforth building habitats), crops (alfalfa, sorghum, and vegetable crops; henceforth crop habitats) and open, non-crop areas that were grassy, untilled land (henceforth open habitats) were sampled in 2012 and 2013. Corn fields were chosen as one of the adjacent habitat types *in lieu* of crops for several soybean fields. In each field, the sampling layout included four transects spaced 20 m apart. Each transect was marked for sampling plots at distances 0, 1.5, 3, 4.5, 6, 9, 12, and 15 m from the edge to field interior. Stink bugs were enumerated at each plot by carefully examining 10 consecutive corn plants and later converted to densities based on planting density, or all plants within a semicircular area of 0.5 m radius (1.57 m^2) in soybean.

Visual counts of stink bug adults, nymphs, and egg masses of *H. halys*, *E. servus*, *C. hilaris*, *Murgantia histrionica* (Hahn 1834), and *Thyanta custator* (Fabricius 1803) were recorded and converted to densities. Fields and plots were sampled each week between mid-July and mid-August in field corn, and mid-August and late-September in soybean. Sampling coincided with the kernel development stages of corn (R2–R5; blister – dent; [38]) and the seed development stages of soybean (R4–R7; full pod to physiological maturity; [39]), which are associated with high *H. halys* and other stink bug species densities [8,11,40]. Details on the number of corn and soybean field edges with different adjacent habitats, and the sampling dates during 2012 and 2013 are provided in Table 1. Including the repeated sampling of fields, a total of 4,315 field corn plots in 31 fields, and 2,968 soybean plots in 26 fields across all sites were sampled for stink bugs during 2012 and 2013.

Assessing seed quality in field corn and soybean

To relate stink bug density to ear damage in corn, eight fields with the highest observed count of stink bugs in 2013 and with different adjacent habitats were selected. Of these fields, 3, 3, 1 and 1 were adjacent to wooded, building, crops and open habitats, respectively. In each field, 6–10 consecutive corn ears were collected at each sampling plot prior to harvest maturity and stored in cloth bags for drying. Planting details of the fields used for assessing corn damage are provided in Table 2. For each ear, the following data were recorded once at the end of the season: 1) number of kernels damaged by stink bugs (identified by a characteristic puncture scar typically surrounded by a discolored cloudy marking); 2) number of collapsed kernels due to stink bug damage (this type of damage was carefully examined to distinguish between kernels damaged by stink bugs and dusky sap beetles, *Carcophilus lugubris* Murray); 3) number of kernel rows around the ear; 4) length of one kernel row (mm); and 5) average width of individual kernels (mm). With the individual ear measurements, the total number of kernels was derived by dividing the kernel row length by the width of a kernel times the number of rows. Data were then summed across all ears of a plot and stink bug damage was expressed as the percentage of damaged and collapsed kernels in relation to total number of kernels per sample. A total of 2,326 ears of corn from 252 sampling plots across 8 fields were assessed for stink bug damage.

To relate stink bug density to soybean pod development prior to harvest, samples of 10 consecutive plants at each plot of seven soybean fields in 2012–2013 were examined once at the end of the season *in situ* to count the total numbers of pods with 3 or more

Table 1. Details on field corn and soybean field edges with different adjacent habitats sampled for stink bugs and the sampling occasions at each field in Maryland, USA during 2012–2013.

Crop	Year	Site	Adjacent habitats (number of field edges)	Sampling dates (frequency)
Field Corn	2012	Beltsville	woods (4), buildings (3), crops (1), open (4)	10 July–15 Aug (7–10 days)
		Clarksville	woods (1), buildings (1), crops (3)	10 July–15 Aug (7 days)
	2013	Clarksville	woods (3), buildings (2), crops (3), open (1)	18 July–22 Aug (7 days)
		Keedysville	woods (1), buildings (1), crops (1), open (2)	16 July–20 Aug (7 days)
		Overall	woods (9), buildings (7), crops (8), open (7)	10 July–22 Aug (7–10 days)
Soybean	2012	Beltsville	woods (2), buildings (3), corn (1), open (2)	23 Aug–20 Sept (7–10 days)
		Keedysville	woods (2), buildings (1), corn (2), open (1)	30 Aug–26 Sept (7–10 days)
	2013	Beltsville	woods (1), buildings (1), corn (1), open (1)	13 Aug–06 Sept (7 days)
		Clarksville	buildings (1), corn (1), open (1)	16 Aug–12 Sept (5–7 days)
		Keedysville	woods (2), buildings (1), corn (1), open (1)	15 Aug–18 Sept (5–7 days)
		Overall	woods (7), buildings (7), corn (6), open (6)	13 Aug–26 Sept (5–10 days)

fully developed seeds (full pods), pods with fewer than 3 seeds (half pods), and flat, immature pods (flat pods). For standardization, the proportions for each pod type were calculated for each sample. Pod quality data were collected from 64 plots in 2 fields adjacent to wooded habitat at Keedysville. Planting details for the fields used for assessing soybean seed damage are provided in Table 2. Of the 7 fields sampled for seed quality data, 2 were adjacent to buildings, and 5 fields were adjacent to wooded habitats, and all had the highest counts of stink bugs observed in each year. Just prior to harvest, 20 plants each from 154 plots in the fields were collected, stored in mesh bags, and allowed further drying for optimal thrashing. Seeds were removed from pods for each sample by a stationary motor-driven thrasher (RL Brownfield-Swanson Machine Company; Model B1). Dirt, chaff, or un-thrashed pods were removed, and the remaining seed samples were then weighed to measure yield (measured as grams/20 plants).

To assess seed quality, subsamples of 200–300 seeds were removed from each sample, counted, and weighed to calculate test weight (expressed as weight per 100 seeds). Seed samples in 2012 were sieved to separate smaller, immature seeds (<0.3 cm), whereas these smaller seeds were not removed from subsamples in 2013. In both years, seeds were individually examined and categorized into six groups as follows: 1) stink bug damaged seed, distinguished by a puncture scar and often surrounded by a discolored cloudy area; 2) moldy seed, characterized by having milky white or grayish crusty growth on surface, sometimes with cracks and fissures; 3) shriveled seed that appeared wrinkled and often undersized; 4) purple seed recognized as purple or pink areas on the seed coat due to the fungus *Cercospora kikuchii* Matsumoto & Tomoy 1925 [41]; 5) green seed showing discolored green tissue in cross section, rather than the normal yellow; and 6) normal, undamaged seed. Seeds with characteristics of more than one

Table 2. Details on the field corn and soybean fields used for analyzing grain and seed damage in Maryland, USA during 2012–2013.

Crop	Year	Site	Field ID	Variety	Planting Date	Density/acre
Field Corn	2013	Clarksville	Corn1	P1319HR (113)	2May13	26,000
			Corn2	DKC61-21 (111)	15May13	26,000
			Corn3	P1319HR (113)	2May13	26,000
			Corn4	P1319HR (113)	2May13	26,000
			Corn5	DKC61-21 (111)	16May13	26,000
			Corn6	NK74R3000GT (114)	16May13	26,000
		Keedysville	Corn7	Doebler's 633HXR (110)	23Apr13	26,000
			Corn8	Doebler's 633HXR (110)	23Apr13	26,000
Soybean	2012	Beltsville	Soy1	AG3030 (3.0)	11May12	155,555
			Soy2	AG3030 (3.0)	11May12	155,555
		Keedysville	Soy3	Doebler 3809RR (3.8)	26May12	180,000
			Soy4	Doebler 3809RR (3.8)	4Jun12	180,000
			Soy5	Doebler 3809RR (3.8)	4Jun12	180,000
	2013		Soy6	SCS9360RR (3.5)	22May13	180,000
			Soy7	Doebler 3809RR (3.8)	27May13	180,000

For each of the corn and soybean varieties, the corn relative maturity in days and the soybean maturity groups respectively, have been provided in parenthesis.

damage category were assigned to whichever category seemed dominant, except for stink bug damaged seeds which were all assigned to that category. To standardize across samples, the percentage of seeds in each category was calculated. Soybean seed quality data were collected for a total of 154 plots from 4 fields in 2012 and 2 fields in 2013.

Statistical analyses

Adjacent habitat and distance from edge influences.   The influence of adjacent habitat and distance from field edge on the density and distribution of stink bugs was analyzed by Generalized Linear Mixed Models (GLMMs) based on Laplace approximation, with a Poisson-lognormal error distribution and log link function [42]. All analyses were performed with fields as replicates and the sampling plots along transects within the field as subsamples. For corn and soybean data, analyses were performed on –data pooled over all stink bug life stages. For each of these datasets, GLMMs were performed on the data pooled across species, years, and study sites, and on data from each study site pooled across years. Sampling points within each field edge was treated as a random factor to control for repeated measurement [43]; adjacent habitat, distance from edge, and their interaction were the fixed effects, and stink bug density was the response variable. For the overall data models, study site and year were also treated as random effects.

Model building and selection procedures for the mixed effects modeling followed the procedures used by Zuur and others [44]. First, several candidate models, each with different random effects but identical fixed effects, were tested to choose the optimal random effect model using a combination of AIC and BIC values for selection criteria [36]. For all optimum fixed effect models, an initial full model analysis including individual and interactive effects of adjacent habitat (4 levels - woods, buildings, mix crop/corn, and open) and distance from edge was performed. Specifically we designed a model matrix that directly estimated the intercepts and the slopes for each level of adjacent habitat (extension of means parameterization; see pages 61–64 in [45]) in relation to distance from field edge. The significance of the fixed effects was determined by Wald χ^2 tests. If a significant interaction was found these estimates, directly interpreted as the intercept and slope of the regression of stink bug density on distance for each of the adjacent habitat types, were then simultaneously compared through post-hoc comparisons with a Bonferroni correction. Models were evaluated for assumption appropriateness by testing for over-dispersion and correlations among random effect terms, and by visualizing variances in a location-scale plot with superimposed loess fit [42].

Relating stink bug density and seed damage.   Influence of stink bug density on damage to corn kernels was assessed using generalized linear models (GLMs) with Poisson or quasi-Poisson error distribution and log link function [46]. Percentage of collapsed and stink bug damaged seeds were used as the response variables and mean stink bug density, the explanatory variable. For significant results, the coefficient of determination was calculated by Nagalkerke's pseudo r^2 statistic [47].

Linear regression was used to assess the influence of stink bug density on soybean pod development. To meet normality assumptions, response variables (% flat or full pods) was square root transformed prior to analysis. Influence of stink bug density on soybean seed quality was assessed by linear mixed models (LMMs) with year as a random effect to account for minor differences in grading seed size protocols. LMMs related stink bug density to the percentage of seeds in each category of seed quality. Influence of stink bug density on soybean yield per plot was assessed by LMMs with field as a random effect to account for differences in soybean variety and other field conditions among sites. Response variables (% seeds in various damage categories, or yield) were log or square root transformed to meet normality requirements and the significance of the fixed effect was determined by Wald t-tests. Diagnostic plots of the models visualizing within-group residuals (standardized residuals vs fitted values, normal Q-Q plots, and histograms of residuals) and estimated random effects (normal Q-Q plots and pairs-scatter plot matrix) were used to assess model appropriateness. The coefficients of determination for the LMMs, based on the likelihood-ratio test, were calculated using Nagalkerke's pseudo r^2 statistic [47]. Patterns in damage to corn kernels, soybean pods, and seeds at different distances from the edge, in relation to stink bug density, were visualized by plotting average values of damage and stink bug density aggregated by distance.

All statistical analyses were performed in R program [48] and associated statistical packages. GLMMs were performed with package 'lme4' [49] and LMMs with package 'nlme' [50]. Multiple comparisons of means for GLMMs were computed with R packages 'contrast' [51] and 'multcomp' [52]. GLMMs and LMMs estimated coefficients were extracted through package 'effects' [53], converted to densities/m^2 and plotted using 'ggplot2' [54]. Coefficient of determination (pseudo r^2) for the GLM was calculated with package 'rms' [55], and with package 'MuMin' [56] for LMMs.

Results

Species composition and density

A total of 9440 individuals (66% nymphs; 34% adults) of four phytophagous stink bug species (*E servus*, *H. halys*, *C. hilaris*, and *M. histrionica*) were recorded in field corn, of which *H. halys* accounted for 97% of the total. Species composition varied among study sites and crop systems. *Halyomorpha halys* comprised 57% of the sampled populations in corn at Beltsville, followed by *E. servus* (35%), whereas *H. halys* accounted for ~97% of all stink bugs at Clarksville and Keedysville. In soybean, a total of 9867 individuals (68% nymphs; 32% adults) of five phytophagous stink bug species (*E. servus*, *H. halys*, *C. hilaris*, *M. histrionic*, *T. custator*) were recorded, of which *H. halys* accounted for 93% of the total. *Halyomorpha halys* comprised 83–85% of the stink bug numbers in soybean at Beltsville, while greater than 92% were *H. halys* at Clarksville and Keedysville. Results obtained from the statistical analyses hence pertain mainly to patterns of *H. halys* density, since this species constituted ~95% of all observed stink bugs in both field corn and soybean.

Adjacent habitat and distance from field edge influences

The trends for density among adjacent habitats at different distances from the edge observed in adult and nymph data sets were broadly similar and were hence pooled for the analyses, and we report only these results.

Field corn.   For the analysis of overall stink bug data from field corn edges, the random effects used for the GLMM included the field, study site and year. Results showed significant interactive influences of adjacent habitat and distance from edge on stink bug density (Wald $\chi^2 = 492.6$, DF $= 8$, $P<0.001$). Simultaneous comparison of intercepts and slopes among adjacent habitats accounting for the influence of distance from edge revealed significant differences. Overall, the density of stink bugs at different distance from the edge was significantly higher in fields adjacent to wooded habitats compared to density in fields adjacent to buildings and open habitats (Fig. 1A). Density between wooded and crop habitats did not significantly differ, although mean densities of stink bugs were consistently higher adjacent to wooded habitats (see Fig. S1 for variance around estimated means). Similarly stink bug density did not differ significantly between crop habitats and buildings although higher mean density was observed along buildings.

For site level GLMMs, significant interactive influences of adjacent habitat and distance from edge on stink bug density were observed for all the sites - Clarksville (Wald $\chi^2 = 426.9$, DF = 8, $P<0.001$), Keedysville (Wald $\chi^2 = 367.9$, DF = 8, $P<0.001$) and Beltsville (Wald $\chi^2 = 43.1$, DF = 3, $P<0.001$). Multiple comparisons of means of stink bug density at Clarksville showed similar trends to the results of analyses of data pooled over all study sites, with density in fields adjacent to wooded habitats higher than density adjacent to buildings and open habitats (Fig. 1C). Similar to overall data, at Clarksville, stink bug density did not differ significantly between wooded and crop habitats, although mean densities of stink bugs were consistently higher adjacent to wooded habitats. Similarly stink bug density did not differ significantly between crop habitats and buildings although higher mean density

was observed along buildings. At Keedysville however, stink bug density was highest along fields adjacent to crops, and the slopes and intercepts were significantly different among adjacent habitats (Fig. 1B). Additionally, corn fields at Keedysville had higher stink bug density along the outside rows (0 m) adjacent to crops than adjacent to woods. At Beltsville, with the least stink bug density, multiple comparisons did not reveal significant differences in stink bug density among adjacent habitats, after accounting for the distance from edge influence (Fig. 1D; also see Fig. S1 for variance around estimated means for each site). The raw means for each site and year (Fig. S2) are provided as supplementary information.

Soybean. The GLMM analysis of overall stink bug data treated field as the only random effect and showed significant interactive influences of adjacent habitat and distance on stink bug density (Wald $\chi^2 = 717.2$, DF = 8, $P<0.001$). Multiple comparisons of means showed that density of stink bugs was significantly higher at all distances in fields adjacent to wooded habitats compared to density in fields adjacent to buildings and open habitats (Fig. 2A). Stink bug density adjacent to wooded habitats was also consistently higher than density in fields adjacent to corn

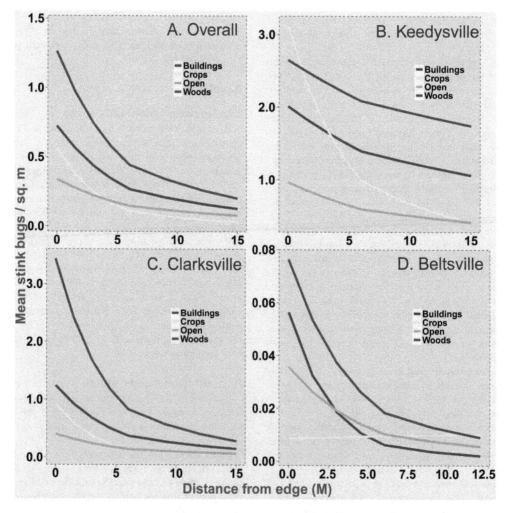

Figure 1. Mean stink bug density in field corn in relation to different adjacent habitats and distance from the field edge. Estimates derived from Poisson-lognormal GLMMs are plotted for overall stink bug data pooled over all study sites (A), Keedysville (B), Clarksville (C) and Beltsville (D). Values presented here have been back transformed from their original link function estimated model coefficients. Multiple comparison of means with a bonferroni correction ($\alpha = 0.05$) showed significant differences in: overall (A) - wooded habitats and buildings and, wooded and open habitats; Keedysville (B) – all adjacent habitats significantly different from each other; and Clarksville (C) - wooded habitats and buildings and, wooded and open habitats. For Beltsville (D), all multiple means comparisons were non-significant.

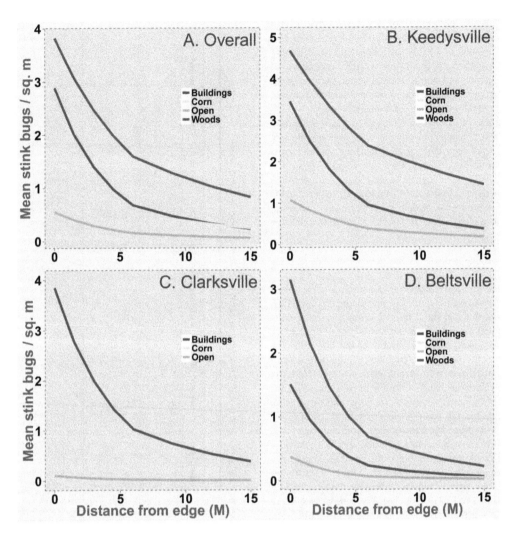

Figure 2. Mean stink bug density in soybean field edges in relation to different adjacent habitats and distance from field edge.
Estimates derived from Poisson-lognormal GLMMs are plotted for overall stink bug data pooled over all study sites (A), Keedysville (B), Clarksville (C) and Beltsville (D). Values presented here have been back transformed from its original link function estimated model coefficients. Multiple comparison of means with a bonferroni correction ($\alpha = 0.05$) showed significant differences in: overall (A) - wooded habitats and buildings, wooded and open habitats, buildings and corn habitats, buildings and open habitats; Keedysville (B) – wooded habitats and buildings, wooded and open habitats, buildings and corn habitats, buildings and open habitat; Clarksville (C) - buildings and open habitats, and corn and open habitats; Beltsville (D) – buildings and open habitats, buildings and wooded habitats, and wooded and corn habitats.

field habitats, but differences were not significant (see Fig. S3 for variances around estimated means). Across all the adjacent habitat types the highest density of stink bugs was recorded at the immediate field edge (0 m), and was lowest at 15 m from edge ($< 1/m^2$).

GLMMs performed by study site on overall stink bug data showed significant interactive influence of adjacent habitat and distance from edge on density at all the sites - Keedysville (Wald $\chi^2 = 485.4$, DF = 8, $P<0.001$), Clarksville (Wald $\chi^2 = 216.63$, DF = 5, $P<0.001$) and Beltsville (Wald $\chi^2 = 279.2$, DF = 8, $P< 0.001$). Multiple means comparisons for Keedysville data showed significantly higher densities adjacent to wooded habitats buildings and open habitats (Fig. 2B). The intercept and slopes for buildings and corn, and buildings and open areas were also significantly different. At Clarksville, buildings and corn habitats had significantly higher density of stink bugs than open areas (Fig. 2C). Model estimated stink bug densities for buildings and corn habitats were not significantly different. At Beltsville, wooded habitats and buildings had significantly higher stink bug density than corn as

adjacent habitat (Fig. 2D). The slope and intercept were significantly different for wooded habitats and buildings. Unlike the other sites where stink bugs were fewest adjacent to open habitats, at Beltsville, corn as adjacent habitat harbored the lowest density of stink bugs. The variance around the estimated density of stink bugs in soybean fields for each site (Fig. S3), and the raw means for each site and year (Fig. S4) are provided as supplementary information.

Corn and soybean seed damage

For field corn, results from the quasi-Poisson GLM showed a significant positive association between percentage of stink bug damaged kernels and mean stink bug density ($y = 0.57+0.15x$, n = 252, $P<0.001$, pseudo $r^2 = 0.47$). A Poisson GLM showed that the percentage of collapsed kernels was not significantly associated with mean stink bug density ($y = -6.75+0.14x$, n = 252, $P = 0.50$, pseudo $r^2 = 0.17$). For soybean pod development data, regression analysis revealed that the percentage of full pods was negatively influenced by mean stink bug density ($y = 5.9-0.17x$, n = 63, $P<$

0.001, $r^2 = 0.51$), while percentage of flat pods (square root transformed) was positively influenced ($y = 2.18+0.26x$, $n = 63$, $P<$ 0.001, $r^2 = 0.63$). Results of LMMs analyzing each seed quality category (Table 3) showed a significant positive association of mean stink bug density and purple stained seeds, percentage of stink bug damaged seeds, percentage of immature, shriveled and moldy seeds, and overall percentage of damaged seeds.

Overall percentage of normal, undamaged soybean seeds and yield had a significant negative relationship with stink bug density (Table 3). The overall seed damage by stink bugs in both corn (Fig. 3A) and soybean (Fig. 3C), and their impact on soybean pod development (Fig. 3D), were highest at field edge and declined gradually towards the field interior. Furthermore, soybean yields (grams/20 plants) were lowest at field edge, gradually increasing from field edge to highest yields at 12 and 15 m from the edge (Fig. 3B).

Discussion

Seasonal availability and suitability of source and recipient habitats influence dispersal dynamics of pest populations [2,4,5]. Aggregation of stink bugs at the field edge of target crops results from their directional movement among a sequence of hosts in an area in response to deteriorating suitability of the host crops or native vegetation and increasing suitability of the target crop plants [57]. Our results provide strong evidence that *H. halys* density exhibits a clear edge effect in both field corn and soybean. Across all adjacent habitats, density of *H. halys* and other stink bugs was highest within the first 3 meters, reaching lowest levels between 9–15 m from the edge, least at 15 m from the edge ($<1/m^2$). In addition to host suitability, species specific behaviors contribute to the resulting aggregations of stink bugs along crop edges. The strong edge effect exhibited by *H. halys* is similar to the within-field infestation pattern reported for other native stink bug species in U.S. crops [12,15,17–19,23,58]. We demonstrated that the aggregations are associated with specific kinds of damage to both corn grain and soybean seeds, thus suggesting that sitenspecific management at the field and farmscape levels may improve crop protection from stink bugs.

The timing of infestations during mid to late July in corn, and then later colonization of soybean fields in August, suggests that the majority of *H. halys* adults were offspring of the generation which developed on host plants earlier in the season (June/July). *Halyomorpha halys* is known to feed on a wide range of cultivated and wild hosts (up to 170 species) [59], of which many tree and shrub species are likely present in wooded habitats. Particularly high density of *H. halys* was observed (personal field observations, Dilip Venugopal) in soybean fields bordering wooded habitats with tree of heaven (*Ailanthus altissima* Swingle), princess tree (*Paulownia tomentosa* Baill.), and black cherry (*Prunus serotina* Ehrhart), all which support high population densities of reproducing *H. halys* [59,60]. Similarly, earlier reports show that *C. hilaris* populations in soybean are greater adjacent to wooded borders with black cherry and elderberry (*Sambucus canadensis* L.) [13]. Additionally, *E. servus* populations are greater in cotton fields with adjacent woods containing many oak species (*Quercus* spp.) and black cherry [18].

This study also addressed the influence of adjacent habitats on stink bug density and quantified differences in density at various distances from the field edges in soybean and corn crops. We found that adjacent habitats, particularly wooded habitats, influenced the densities of *H. halys* and other stink bugs. In both corn and soybeans, fields adjacent to wooded habitats, across all study sites and distances from field edge, consistently harbored

Table 3. Statistical results of LMMs for analyzing the relationship between stink bug density and various soybean seed damage categories and yield.

Response variable	Data Transformation	Intercept	Intercept SE	Estimate	SE	DF	Wald t	Pval	psuedo r²
% normal seeds	None	75.8	8.04	−2.11	0.25	145	8.28	<0.001	0.30
% stink bug damaged seeds	Square Root	3.41	0.5	0.07	0.01	145	4.58	<0.001	0.12
% purple damaged seeds	log	1.39	0.14	0.09	0.01	145	9.99	<0.001	0.44
% moldy + shriveled + immature seeds	Square Root	2.59	0.47	0.09	0.02	148	5.87	<0.001	0.19
% all damaged seeds	Square Root	4.78	0.63	0.18	0.02	145	9.03	<0.001	0.35
Total Yield (grams/20 plants)	Square Root	17.1	1.04	−0.20	0.04	140	4.67	<0.001	0.13

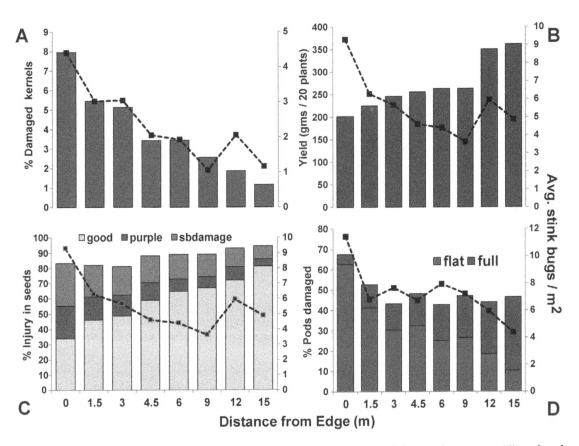

Figure 3. Patterns of kernel damage in field corn (A), soybean yield (B), soybean seed damage by category (C), and soybean pod development (D) in relation to mean stink bug density at different distance from field edge. The proportions of soybean seeds in each seed quality category (stink bug damaged, purple damaged, and normal seeds) and pod types (flat and full) are also provided. Mean stink bug abundance are denoted by the dashed lines which represent the second y axis.

significantly higher densities of stink bugs than in fields adjacent to open habitats. Also, stink bug density in fields adjacent to wooded habitats was consistently higher than in fields adjacent to buildings and corn habitats. These results suggest that wooded habitats play an important role in serving as sources of stink bug populations that colonize crops.

Patterns of infestation along field edges of corn and soybean differed among study sites, partly influenced by overall stink bug density and other adjacent habitats. Stink bug density in both row crops at Keedysville was consistently (3–5 times) greater than the mean density at the other two study sites, and this could be attributed to the higher overall densities of *H. halys* observed in the Western Maryland region over the past 5 years. In addition to wooded habitats, building and other crop habitats served as sources of colonizing adults in row crops. *H. halys* often utilize buildings as overwintering sites and thus these structures would likely influence stink bug populations earlier in the spring when post-diapause adults are moving to host plants, thereby supporting high local populations that eventually invade soybean later in the season. At Keedysville, stink bug density was higher in corn fields adjacent to alfalfa and in soybean fields adjacent to corn than in corn or soybean fields adjacent to building habitats. However, at Clarksville and Beltsville adjacent building habitats showed a higher density of stink bugs than crop habitats.

As with other studies, our results highlight the role of other adjacent cultivated crops in influencing stink bug populations. For example, adjacent fields of alfalfa, field corn, and other cultivated borders have been reported as a sources contributing to higher

densities of stink bugs in tomato, cotton, sorghum, and peanut fields [12,16,19,23]. However, differences in the relative influence of adjacent habitats in our study could be related to overall stink bug density at each of the study sites. For example, stink bug density did not differ significantly among adjacent habitats in field corn at Beltsville where *H. halys* densities were lowest. Moreover, the influence of the landscape heterogeneity on stink bug density could extend to larger spatial scales beyond habitats adjacent to a crop. Since insect population dynamics and spatial patterns are affected by regional landscape context and species traits such as dispersal ability [61], distribution, and density of *H. halys* may depend on habitat and other environmental characteristics at spatial scales greater than the local agricultural field [62,63]. Differences in landscape structure among our study sites may explain the higher density of stink bugs at Keedysville, and specifically the role of adjacent crop habitats as sources of stink bugs in field corn and soybean.

We related the various corn and soybean damage measurements to stink bug density within plots. As expected, stink bug damage to corn kernels increased with stink bug density. The percentage of damaged kernels reached levels up to 8% at the field edge to less than 3% between 9–15 m from field edge, and was positively correlated with the stink bug density within plots. The percentage of collapsed kernels was negligible and not significantly influenced by stink bug density. Based on findings by earlier studies [25,64], neither kernel damage, ear weight or grain weight are affected beyond tasseling stage (VT) from feeding damage by *E. servus* or *N. viridula*. Although *H. halys* density can be high along edges of corn fields, our result suggests that *H. halys* kernel quality loss is restricted to about 10 m from the field edge.

Halyomorpha halys populations in soybean had a significant impact on pod development, with the percentage of flat pods significantly increasing with increasing stink bug density. Concomitantly, the proportion of fully developed pods significantly decreased with increasing stink bug density. Changes in the development and maturation of soybean pods due to *H. halys* feeding have been recently documented [35], showing that most severe pod loss occurred at the R4 (full pod) growth stage. Observed effects on pod and seed development with higher stink bug density were similar to damage caused by other stink bug species [26–31]. Our finding that higher stink bug density is related to increased proportions of moldy and purple stained seeds suggest a potential role of *H. halys* in transmitting or facilitating various pathogens; however, this needs to be further investigated experimentally. Our study found a significant, yet weak negative association between soybean yield and stink bug density per plot. In contrast, recent field research using cages addressed the effects of *H. halys* feeding on soybean growth, and did not detect a significant relationship between H. *halys* stink bug densities and yield loss [35]. Soybean field studies to compare yields of insecticide-treated and untreated plots are needed to establish the relationship between soybean yield losses and stink bug density.

Knowledge of how adjacent habitats influence *H. halys* populations and the within-field distribution has several implications in stink bug management. First, our results indicate that scouting corn and soybean fields for decision-making is more efficient if efforts initially concentrated on field edges bordering wooded habitats where there is a greater likelihood of colonization and higher risk of infestation. Secondly, the infestation patterns of stink bug communities dominated by *H. halys* are primarily edge-centric, and population densities beyond 12 m are generally low ($<1/m^2$). Based on our results, if insecticides are to be applied, edge-only applications, particularly along wooded and building habitats, could reduce the cost of control while preventing damage caused by stink bugs in field corn and soybean. Preliminary studies show that treating just 12 m into soybean field prevented further invasion by *H. halys* and other stink bugs (Ames Herbert, personal communication). The edge-only treatment prevented reinvasion and also resulted in an 85–95% reduction in insecticide used compared with whole-field treatments [11]. Results presented here showing highest stink bug density and associated damage limited to the immediate field edge provide validity for the edge-only

treatment. While experimental research on effective insecticide application strategies are currently underway, based on our findings, we suggest that integrated pest management programs for the stink bug complex in field crops should include farmscape-level planning - crop location with regards to adjacent habitats, and targeted interventions in the form of edge-only treatments to prevent seed quality and yield losses.

Supporting Information

Figure S1   GLMM estimated mean stink bug densities in field corn (bold lines) and 95% CI (shaded region) among adjacent habitats and distance from the field edge.

Figure S2   Site and year wise raw stink bug averages in field corn among adjacent habitats and distance from the field edge.

Figure S3   GLMM estimated mean stink bug densities in soybean (bold lines) and 95% CI (shaded region) among adjacent habitats and distance from the field edge.

Figure S4   Site and year wise raw stink bug averages in soybean among adjacent habitats and distance from the field edge.

Acknowledgments

The farm managers and staff at each of the research facilities - Kevin Conover (Beltsville), Tim Ellis (Keedysville), David Justice and Mike Dwyer (Clarksville), and David Swain (USDA–BARC) - all extended excellent support and assistance for conducting the study. We thank Terry Patton for his help to identify sampling fields and for data collection. We thank the summer technicians, Jesse Ditillo, Jessie Saunders, Jake Bodart, Taylor Schulden, Kyle Runion, and Emily Zobel for data collection.

Author Contributions

Conceived and designed the experiments: PDV GPD WOL. Performed the experiments: PDV GPD PLC. Analyzed the data: PDV. Contributed reagents/materials/analysis tools: PDV. Contributed to the writing of the manuscript: PDV PLC GPD WOL.

References

1. Nestel D, Carvalho J, Nemny-Lavy E (2004) The spatial dimension in the ecology of insect pests and its relevance to pest management. In: Horowitz PDAR, Ishaaya PDI, editors. Insect pest management. Springer Berlin Heidelberg. 45–63.

2. Ekbom BS, Erwin ME, Robert Y (2000) Interchanges of insects between agricultural and surrounding landscapes. Kluwer Netherlands. 266 p.

3. Stinner RE, Barfield CS, Stimac JL, Dohse L (1983) Dispersal and movement of insect pests. Annu Rev Entomol 28: 319–335. doi:10.1146/annurev.en.28.010183.001535.

4. Kennedy GG, Margolies DC (1985) Mobile arthropod pests: management in diversified agroecosystems. Bull Entomol Soc Am 31: 21–35.

5. Kennedy GG, Storer NP (2000) Life systems of polyphagous arthropod pests in temporally unstable cropping systems. Annu Rev Entomol 45: 467–493. doi:10.1146/annurev.ento.45.1.467.

6. Panizzi AR (1997) Wild hosts of Pentatomids: ecological significance and role in their pest status on crops. Annu Rev Entomol 42: 99–122. doi:10.1146/annurev.ento.42.1.99.

7. Panizzi AR, Slansky F (1985) Review of phytophagous Pentatomids (Hemiptera: Pentatomidae) associated with soybean in the Americas. Fla Entomol 68: 184–214.

8. Nielsen AL, Hamilton GC, Shearer PW (2011) Seasonal phenology and monitoring of the non-native *Halyomorpha halys* (Hemiptera: Pentatomidae) in soybean. Environ Entomol 40: 231–238. doi:10.1603/EN10187.

9. Hooks CRR (2011) Stink bugs and their soybean obsession. Agron News 2: 1–4.

10. CABI (2014) *Halyomorpha halys* [original text by Leskey TC, Hamilton GC, Biddinger DC, Buffington ML, Dicekhoff C, et al.]. In: Invasive Species Compendium. Wallingford, UK: CAB International. www.cabi.org/isc. Available: http://www.cabi.org/isc/?compid=5&dsid=27377&loadmodule=datasheet&page=481&site=144. Accessed 19 March 2014.

11. Leskey TC, Hamilton GC, Nielsen AL, Polk DF, Rodriguez-Saona C, et al. (2012) Pest status of the brown marmorated stink bug, *Halyomorpha halys* in the USA. Outlooks Pest Manag 23: 218–226. doi:10.1564/23oct07.

12. Toscano NC, Stern VM (1976) Dispersal of *Euschistus conspersus* from alfalfa grown for seed to adjacent crops. J Econ Entomol 69: 96–98.

13. Jones WA, Sullivan MJ (1982) Role of host plants in population dynamics of stink bug pests of soybean in South Carolina. Environ Entomol 11: 867–875.

14. Outward R, Sorenson CE, Bradley JR (2008) Effects of vegetated field borders on arthropods in cotton fields in eastern North Carolina. J Insect Sci 8. Available: http://www.ncbi.nlm.nih.gov/pmc/articles/PMC3061576/.

15. Tillman PG, Northfield TD, Mizell RF, Riddle TC (2009) Spatiotemporal patterns and dispersal of stink bugs (Heteroptera: Pentatomidae) in peanut-

cotton farmscapes. Environ Entomol 38: 1038–1052. doi:10.1603/022.038.0411.

16. Toews MD, Shurley WD (2009) Crop juxtaposition affects cotton fiber quality in Georgia farmscapes. J Econ Entomol 102: 1515–1522. doi:10.1603/029.102.0416.

17. Pease CG, Zalom FG (2010) Influence of non-crop plants on stink bug (Hemiptera: Pentatomidae) and natural enemy abundance in tomatoes. J Appl Entomol 134: 626–636. doi:10.1111/j.1439-0418.2009.01452.x.

18. Reay-Jones FPF (2010) Spatial and temporal patterns of stink bugs (Hemiptera: Pentatomidae) in wheat. Environ Entomol 39: 944–955. doi:10.1603/EN09274.

19. Reeves RB, Greene JK, Reay-Jones FPF, Toews MD, Gerard PD (2010) Effects of adjacent habitat on populations of stink bugs (Heteroptera: Pentatomidae) in cotton as part of a variable agricultural landscape in South Carolina. Environ Entomol 39: 1420–1427. doi:10.1603/EN09194.

20. Reisig DD (2011) Insecticidal management and movement of the brown stink bug, Euschistus servus, in corn. J Insect Sci 11: Article 168. doi:10.1673/031.011.16801.

21. Herbert JJ, Toews MD (2011) Seasonal abundance and population structure of brown stink bug (Hemiptera: Pentatomidae) in farmscapes containing corn, cotton, peanut, and soybean. Ann Entomol Soc Am 104: 909–918. doi:10.1603/AN11060.

22. Olson DM, Ruberson JR, Zeilinger AR, Andow DA (2011) Colonization preference of Euschistus servus and Nezara viridula in transgenic cotton varieties, peanut, and soybean. Entomol Exp Appl 139: 161–169. doi:10.1111/j.1570-7458.2011.01116.x.

23. Tillman PG (2011) Influence of corn on stink bugs (Heteroptera: Pentatomidae) in subsequent crops. Environ Entomol 40: 1159–1176. doi:10.1603/EN10243.

24. Lee D-H, Short BD, Joseph SV, Bergh JC, Leskey TC (2013) Review of the biology, ecology, and management of Halyomorpha halys (Hemiptera: Pentatomidae) in China, Japan, and the Republic of Korea. Environ Entomol 42: 627–641. doi:10.1603/EN13006.

25. Ni X, Da K, Buntin GD, Cottrell TE, Tillman PG, et al. (2010) Impact of brown stink bug (Heteroptera: Pentatomidae) feeding on corn grain yield components and quality. J Econ Entomol 103: 2072–2079. doi:10.1603/EC09301.

26. Daugherty DM, Neustadt MH, Gehrke CW, Cavanah LE, Williams LF, et al. (1964) An evaluation of damage to soybeans by brown and green stink bugs. J Econ Entomol 57: 719–722.

27. Todd JW, Turnipseed SG (1974) Effects of southern green stink bug damage on yield and quality of soybeans. J Econ Entomol: 421–426.

28. McPherson RM, Newsom LD, Farthing BF (1979) Evaluation of four stink bug species from three Genera affecting soybean yield and quality in Louisiana. J Econ Entomol 72: 188–194.

29. Brier HB, Rogers DJ (1991) Susceptibility of soybeans to damage by Nezara viridula (L.) (Hemiptera: Pentatomidae) and Riptortus serripes (f.) (Hemiptera: Alydidae) during three sages of pod development. Aust J Entomol 30: 123–128. doi:10.1111/j.1440-6055.1991.tb00403.x.

30. McPherson RM, Douce GK, Hudson RD (1993) Annual variation in stink bug (Heteroptera: Pentatomidae) seasonal abundance and species composition in Georgia soybean and its impact on yield and quality. J Entomol Sci 28: 61–72.

31. Corrêa-Ferreira BS, De Azevedo J (2002) Soybean seed damage by different species of stink bugs. Agric For Entomol 4: 145–150. doi:10.1046/j.1461-9563.2002.00136.x.

32. Clarke RG, Wilde GE (1971) Association of the green stink bug and the yeast-spot disease organism of soybeans. III. Effect on soybean quality. J Econ Entomol: 222–223.

33. Mitchell PL (2004) Heteroptera as vectors of plant pathogens. Neotrop Entomol 33: 519–545. doi:10.1590/S1519-566X2004000500001.

34. Medrano EG, Esquivel J, Bell A, Greene J, Roberts P, et al. (2009) Potential for Nezara viridula (Hemiptera: Pentatomidae) to transmit bacterial and fungal pathogens into cotton bolls. Curr Microbiol 59: 405–412. doi:10.1007/s00284-009-9452-5.

35. Owens DR, Herbert Jr DA, Dively GP, Reisig DD, Kuhar TP (2013) Does feeding by Halyomorpha halys (Hemiptera: Pentatomidae) reduce soybean seed quality and yield? J Econ Entomol 106: 1317–1323. doi:10.1603/EC12488.

36. Hoebeke ER, Carter ME (2003) Halyomorpha halys (Stål) (Heteroptera: Pentatomidae): a polyphagous plant pest from Asia newly detected in North America. Proc Entomol Soc Wash 105: 225–237.

37. USDA National Agricultural Statistics Service (2014) Crop production. Available: http://www.usda.gov/nass/PUBS/TODAYRPT/crop0614.pdf. Accessed 14 June 2014.

38. Hanway JJ (1963) Growth stages of corn (Zea mays, L.). Agron J 55: 487. doi:10.2134/agronj1963.00021962005500050024x.

39. Fehr WR, Caviness CE, Burmood DT, Pennington JS (1971) Stage of development descriptions for soybeans, Glycine Max (L.) Merrill. Crop Sci 11: 929. doi:10.2135/cropsci1971.0011183X001100060051x.

40. Schumann FW, Todd JW (1982) Population dynamics of the southern green stink bug (Heteroptera: Pentatomidae) in relation to soybean phenology. J Econ Entomol 75: 748–753.

41. Walters HJ (1980) Soybean leaf blight caused by Cercospora kikuchii. Plant Dis 64: 961–962. doi:10.1094/PD-64-961.

42. Bolker BM, Brooks ME, Clark CJ, Geange SW, Poulsen JR, et al. (2009) Generalized linear mixed models: a practical guide for ecology and evolution. Trends Ecol Evol 24: 127–135. doi:10.1016/j.tree.2008.10.008.

43. Pinheiro JC, Bates DM (2000) Mixed effects models in S and S-Plus. New York: Springer. 560 p.

44. Zuur AF, Ieno EN, Walker N, Saveliev AA, Smith GM (2009) Mixed effects models and extensions in ecology with R. New York: Springer. 580 p.

45. Kéry M (2010) Introduction to WinBUGS for ecologists: bayesian approach to regression, ANOVA, mixed models and related analyses. Academic Press. 322 p.

46. Ver Hoef JM, Boveng PL (2007) Quasi-Poisson Vs. negative binomial regression: How should we model overdispersed count data? Ecology 88: 2766–2772. doi:10.1890/07-0043.1.

47. Nagelkerke NJD (1991) A note on a general definition of the coefficient of determination. Biometrika 78: 691–692. doi:10.1093/biomet/78.3.691.

48. R Development Core Team (2011) R: A language and environment for statistical computing. The R foundation for statistical computing, Vienna, Austria. Available: http://www.R-project.org/.

49. Bates D, Maechler M, Bolker B, Walker S (2013) lme4: Linear mixed-effects models using Eigen and S4. Available: http://cran.r-project.org/web/packages/lme4/index.html. Accessed 25 December 2013.

50. Pinheiro J, Bates DM, Debroy S, Sarkar D, R Core Team (2013) nlme: Linear and nonlinear mixed effects models R package version 3.1-117. Available: http://cran.r-project.org/web/packages/nlme/index.html. Accessed 25 August 2014.

51. Kuhn M, Weston S, Wing J, James JF, Thaler T (2013) contrast: A collection of contrast methods. Available: http://cran.r-project.org/web/packages/contrast/index.html. Accessed 25 December 2013.

52. Hothorn T, Bretz F, Westfall P (2008) Simultaneous inference in general parametric models. Biom J 50: 346–363.

53. Fox J, Weisberg S, Friendly M, Hong J, Andersen R, et al. (2013) effects: effect displays for linear, generalized linear, multinomial-logit, proportional-odds logit models and mixed-effects models. Available: http://cran.r-project.org/web/packages/effects/index.html. Accessed 25 December 2013.

54. Wickham H (2009) ggplot2: elegant graphics for data analysis. New York: Springer. Available: http://had.co.nz/ggplot2/book.

55. Harrell FE (2013) rms: Regression modeling strategies. Available: http://cran.r-project.org/web/packages/rms/index.html. Accessed 25 December 2013.

56. Bartoń K (2013) MuMIn: Multi-model inference. Available: http://cran.r-project.org/web/packages/MuMIn/index.html. Accessed 25 December 2013.

57. Todd JW (1989) Ecology and behavior of Nezara viridula. Annu Rev Entomol 34: 273–292.

58. Tillman PG (2010) Composition and abundance of stink bugs (Heteroptera: Pentatomidae) in corn. Environ Entomol 39: 1765–1774. doi:10.1603/EN09281.

59. BMSB IPM Working Group & Northeastern IPM Center (2013) Host plants of the brown marmorated stink bug in the U.S. Available: http://www.stopbmsb.org/where-is-bmsb/host-plants/. Accessed 2 January 2014.

60. Nielsen AL, Hamilton GC (2009) Life history of the invasive species Halyomorpha halys (Hemiptera: Pentatomidae) in northeastern United States. Ann Entomol Soc Am 102: 608–616. doi:10.1603/008.102.0405.

61. Tscharntke T, Brandl R (2004) Plant-insect interactions in fragmented landscapes. Annu Rev Entomol 49: 405–430. doi:10.1146/annurev.ento.49.061802.123339.

62. Thies C, Steffan-Dewenter I, Tscharntke T (2003) Effects of landscape context on herbivory and parasitism at different spatial scales. Oikos 101: 18–25. doi:10.1034/j.1600-0706.2003.12567.x.

63. Tscharntke T, Rand TA, Bianchi F (2005) The landscape context of trophic interactions: insect spillover across the crop-noncrop interface. Ann Zool Fenn 42: 421–432.

64. Negrón JF, Riley TJ (1987) Southern green stink bug, Nezara viridula (Heteroptera: Pentatomidae), feeding in corn. J Econ Entomol 80: 666–669.

15

Estimation of Agricultural Water Consumption from Meteorological and Yield Data: A Case Study of Hebei, North China

Zaijian Yuan[1,2], Yanjun Shen[2]*

1 School of Economics & Management, Hebei University of Science and Technology, Shijiazhuang, China, **2** Center for Agricultural Resources Research, Institute of Genetics and Developmental Biology, Chinese Academy of Sciences, Shijiazhuang, China

Abstract

Over-exploitation of groundwater resources for irrigated grain production in Hebei province threatens national grain food security. The objective of this study was to quantify agricultural water consumption (AWC) and irrigation water consumption in this region. A methodology to estimate AWC was developed based on Penman-Monteith method using meteorological station data (1984–2008) and existing actual *ET* (2002–2008) data which estimated from MODIS satellite data through a remote sensing *ET* model. The validation of the model using the experimental plots (50 m^2) data observed from the Luancheng Agro-ecosystem Experimental Station, Chinese Academy of Sciences, showed the average deviation of the model was −3.7% for non-rainfed plots. The total AWC and irrigation water (mainly groundwater) consumption for Hebei province from 1984–2008 were then estimated as 864 km^3 and 139 km^3, respectively. In addition, we found the AWC has significantly increased during the past 25 years except for a few counties located in mountainous regions. Estimations of net groundwater consumption for grain food production within the plain area of Hebei province in the past 25 years accounted for 113 km^3 which could cause average groundwater decrease of 7.4 m over the plain. The integration of meteorological and satellite data allows us to extend estimation of actual *ET* beyond the record available from satellite data, and the approach could be applicable in other regions globally where similar data are available.

Editor: Zhi Zhou, National University of Singapore, Singapore

Funding: The Natural Science Foundation of China (NSFC) and Chinese Academy of Sciences (CAS) supported this work through grants 40901130, 40871021, and KSCX2-EW-J-5. The funders had no role in study design, data collection and analysis, decision to publish, or preparation of the manuscript.

Competing Interests: The authors have declared that no competing interests exist.

* E-mail: yjshen@sjziam.ac.cn

Introduction

The most critical resource for agroecosystems in China is water [1]. The total annual water resources available in China are around 2,800 km^3. With a population of 1.3×10^9, the available water per capita is only 2,100 m^3/y. Thus, China is a nation with high water scarcity compared to a global average of 6,466 m^3/y [2]. Water resources in the northern parts of China account for less than 20% of the national total, whereas arable land accounts for 65% of the total [3], and the grain production in the North has exceeded to 50% since 2005. As 80% of China's food is produced on irrigated farmland, irrigation water plays an important role in feeding the large population [4,5]. The North China Plain (NCP) is one mostly important granary of China. It has 140,000 km^2 of arable land and produces about 20% of the nation's grain food.

The natural rainfall cannot meet crop water requirements in NCP, supplementary irrigation is therefore widely applied to increase yields and to secure the food supply for the nation [3]. However, excessive use of diverted river ows and groundwater has caused severe environmental problems. For example, since 1972 the lower reaches of the Yellow River has frequently dried up during the dry seasons for several years. During the droughts of 1997 it didn't reach the sea for even 228 days. However, it must be mentioned that since the beginning of the 2000s, after a river basin

management plan approach was adopted in Yellow River Basin, no drying up has occurred so far [2].

On the other hand, in most places of the NCP, such as Hebei province, groundwater is the primary source of water for irrigation. Grain production in Hebei province totaled 2.9×10^{10} kg in 2008, accounting for 5.5% of the country's total, while the production of wheat and corn shared for 10.9% and 8.7% of the national total, respectively. Due to continually over-pumping, groundwater resource has been greatly depleted and facing to great challenges in sustainability. The water table at the piedmont plain for example has declined rapidly from ~10 m below ground surface in the 1970s to ~30 m in 2001 [6], and to ~40 m in 2010 [7].

It is extremely important for a sustainable agricultural water management to explicitly estimate the groundwater consumption for agriculture in recent decades in NCP. FAO Penman-Monteith equation combined with crop coefficient was widely used for estimation crop water requirement over the world. For the NCP region, Liu et al. [8] calculated the crop water requirement for winter wheat and summer maize in North China in the past 50 years and found a widely decreasing trend of $-0.9 \sim -19.2$ mm per decade for wheat and $-8.3 \sim -24.3$ mm per decade for maize, respectively. Li et al. [9] successfully estimated the water consumption and crop water productivity of winter wheat in NCP using remote sensing for a growing season in 2003–2004. Their calculation suggested the average water consumption (i.e. ET) by winter wheat in 83 counties was 424 mm, which was 118 mm higher than the precipitation. Yang et al. [10] estimated that the crop water requirement for five major crops (wheat, maize, cotton, fruit trees, vegetables) in NCP using crop models DSSAT and COTTON2K, and found wheat accounted for over 40% of total irrigation water requirement in the plain, while maize and cotton together accounted for 24% of the total irrigation water requirement. They also estimated that the annual averaged irrigation requirement for grain crops was 6.16 km^3 during the period of 1986–2006. This estimation is of great importance to make regional water resources planning. Though the crop model with careful calibrations can provide relatively accurate estimation of crop water consumption, the difficulties in collecting huge amount of information on soil profiles and crops biology and phenology together with the complicated parameterization restrict the wide application of crop model to regional water resources management, especially for the regions with limited data. Moreover, even in some developed countries, actual evapotranspiration (ET) has been observed only in recent 1–2 decades, mostly at field scale. Simple methods to estimate agricultural water consumption (AWC) at larger spatial and temporal scales are urgently needed for water resources assessment and planning. In the present study, we attempt to propose a simple method to calculate long-term regional AWC by using limited meteorological and census data.

Therefore, the main objectives of the present study are to estimate 1) the AWC changes over past decades in Hebei province; and 2) irrigation water consumption for agriculture and related groundwater depletion. The results from this study will provide critical information for the future development of sustainable agricultural water resources management practices for local governments.

Materials and Methods

Hebei province (36°05′N-42°40′N, 113°27′E-119°50′E, Figure 1) is 190,000 km^2 in area with a population of 69 million (2009), and is divided into 11 prefectures (including 138 counties). The topography consists of mountains, hills, and plateaus in the north and west part, and a broad plain in the central and southeastern region. 34% of the provincial land area is cultivated with grain crops such as wheat, maize, rice, soybean, potato and millet, and among them the yield of winter wheat and summer maize account for 85% of the total grain yield (winter wheat is cultivated from early October to early June, summer maize grows from mid-June to late September). In plain area, most arable lands are irrigated except for the eastern part where the saline shallow groundwater restrains the irrigation but irrigation increased gradually in recent 3 decades due to technology evolution.

The study area is located in a temperate and continental monsoon climate zone with a mean annual precipitation of 500 mm (1984–2008), 70% of which occurs between June and September. Mean annual temperature is 10°C (1984–2008). Precipitation and temperature decrease from southeast to northwest.

Data

The meteorological data for 1984–2008, including daily average temperature, relative humidity, precipitation, sunshine duration, atmospheric pressure, vapor pressure, wind speed, were obtained from 55 national weather stations (Figure 1). The economic statistics data for each county from 1984 to 2008, including grain yield, sowing area and effective irrigation area, were obtained from Hebei economic statistical yearbooks. The meteorological data were used to calculate reference evapotranspiration, and economic data were employed to estimate the actual evapotranspiration.

An independent remote sensing ET dataset was employed to analyze the relationship between grain yield and ET and to calibrate a key parameter, i.e. K_f, of the model we proposed. The remote sensing ET data were produced based on moderate-resolution imaging spectroradiometer (MODIS) data by combining meteorological records and an scheme called ETWatch. There are 7 years (2002–2008) ET data available for Hebei province with a 1 km spatial resolution. Wu et al. [11] presented the details of the algorithm of ETWatch and its validation.

Validation data are collected from five years (2007–2011) field experiments on irrigation and water productivity at Luancheng Agro-ecosystem Experimental Station (35°53′N, 114°41′E), the Chinese Academy of Sciences, which is located at the piedmont, with an elevation of 50 m above sea level. The experiments have been conducted in 16 water balance plots with an area of 50 m^2 each. Irrigation was applied as five treatments to control the soil moisture at different levels (see Sun et al. [12] for details). The data of annual irrigation amount, annual total yield of the double crops wheat and maize, actual ET calculated from soil water balance for each treatment were collected as well as the daily meteorological data and groundwater depth monitoring data. The meteorological data was used to calculate the reference ET at this station, other data were employed to validate and evaluate the model's applicability.

Reference Evapotranspiration

Reference evapotranspiration was estimated through FAO56-PM model [13],

$$ET_0 = \frac{0.408\Delta(R_n - G) + \gamma \frac{900}{T+273} u_2 (e_s - e_a)}{\Delta + \gamma(1 + 0.34 u_2)} \qquad (1)$$

where ET_0 is reference evapotranspiration (mm d^{-1}) and annual ET_0 was accumulated from daily ET_0; R_n is the net radiation at the crop surface (MJ m^{-2}d^{-1}); G is the soil heat flux density (MJ m^{-2}d^{-1}); T is daily average temperature (°C); u_2 is the wind speed at 2 m height (m s^{-1}); e_s is the vapor pressure of the air at saturation (kPa); e_a is the actual vapour pressure (kPa); Δ is the slope of the vapor pressure curve (kPa °C^{-1}) and γ is the psychrometric constant (kPa °C^{-1}). A complete set of equations is proposed by Allen et al. [13] to compute the parameters of Eq. (1) according to the available weather data and the time step computation, which constitute the so-called FAO-PM method. G can be ignored for daily time step computation. Using Eq. (1), we firstly calculated ET_0 for the 55 weather stations based on conventional meteorological observation data, and then estimated ET_0 for 138 counties through Kriging interpolation.

Actual evapotranspiration (ET) of croplands. Actual evapotranspiration of croplands was calculated by using the following equation,

$$ET = K_c \times K_f \times ET_0 \qquad (2)$$

where ET is actual evapotranspiration (mm); K_c is crop coefficient; K_f is soil moisture correction coefficient. The crop coefficient is largely dependent on crop varieties and planting patterns such as sowing density, fertilizer management, etc. So it varies largely in space and time and difficult to be collected, especially for the past, because information on grain varieties and growing observation

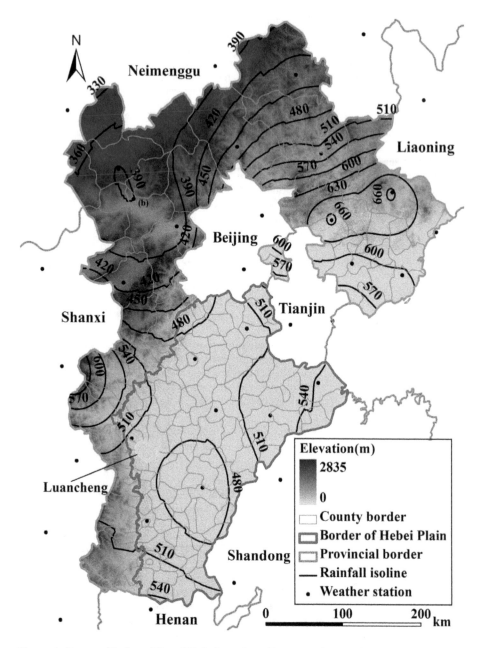

Figure 1. Geographical position of Hebei province. The contour lines and the points indicate average precipitation (1984–2008) and locations of weather stations used in this study, respectively.

data are not available. Alternatively, we assume that the temporal change of crop coefficient can be reflected by the grain yield coefficient (GY_c) without distinguishing crop species in this study.

The GY_c is based on our analysis of the relationship between grain yield (GY) and water consumption, i.e. ET, in 121 counties of the all 138 counties by using the statistical yield for 2002–2008 and independent source of remote sensing derived ET data (thereafter, ET_{rs}) for the same period. There are 17 counties, where the cultivated croplands mostly grow cotton and the grain croplands shares little to their total cultivated land, were removed from the correlation analysis of observed GY and ET_{rs} at county level. Figure 2 illustrated that the annual ET from the remote sensing ET products is significantly linearly correlated to the grain yield.

The grain yield coefficient GY_c is calculated as follows,

$$GY_{ci} = \frac{GY_i}{\overline{GY}} = \frac{GY_i}{\sum\limits_{i=1984}^{2008} GY_i/25} \tag{3}$$

where GY_{ci} is grain yield coefficient of year i, GY_i is grain yield of year i (t/ha), and $\overline{GY}$ is the mean yield from 1984 to 2008 (t/ha). Therefore, Eq. (2) can be modified as,

$$ET = GY_c \times K_f \times ET_0 \tag{4}$$

Then, the soil moisture correction coefficient K_f can be expressed

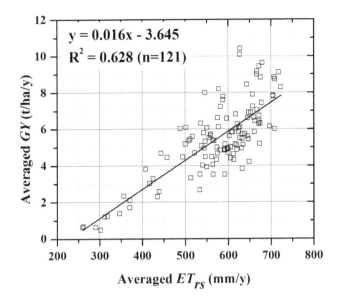

Figure 2. Relationship between remote sensing derived annual ET (ET_{rs}) and grain yield (GY) from 2002 to 2008 of 121 major grain production counties.

as below,

$$K_f = ET / (GY_c \times ET_0) \qquad (5)$$

Determination of K_f

The annual K_f of farmland was calculated using Eq. (5) on the basis of annual remote sensing ET_{rs} products (2002–2008) for each county, combined with annual ET_0 calculated for the same period. In this calculation, we used the areal weighted ET_{rs} from grain crop lands for each county since the grain yield coefficient GY_c only presents the water consumption and productivity from grain lands. Then we got the K_f parameter for all the counties during the period 2002–2008.

There are no long-term soil moisture data available for the region, but it should reflect the annual precipitation and irrigation. So, we analyzed the correlations of annual K_f with annual precipitation (P) and annual irrigation rate (Irr_{rate}) of the 121 counties during the 7 years (totally, 847 samples), and resulted in an empirical equations below,

$$K_f = 0.00033P + 0.2754\, Irr_{rate} + 0.0818$$
$$(R^2 = 0.54, n = 847, F = 489.4) \qquad (6)$$

where P is annual precipitation for each county (mm), Irr_{rate} is annual irrigation rate for each county which was defined as the effective irrigation area (EIA, km^2) of a county to its total cultivated area (A, km^2),

$$Irr_{rate} = EIA / A \qquad (7)$$

With assumption of the regression coefficients in Eq. (6) keep stationary during the study period, we can get K_f for each year in

each county by using Eq. (6) from the mean annual precipitation and irrigation rates from 1984 through 2008.

Agricultural Water Consumption and Irrigation Water Consumption

Agricultural water consumption (WC_{ag}, km^3) for each county was estimated as follows,

$$WC_{ag} = ET \times A / 1000000 \qquad (8)$$

According to the water balance equation, total ET for the study period also can be estimated by the following formula,

$$ET = P + Irr_n + (SM_0 - SM_1) - (R_o - R_i) \qquad (9)$$

where Irr_n is effective irrigation water (mm); SM_0 is initial and SM_1 is final soil moisture (mm), when water balance for a relatively long period were calculated, $SM_0 - SM_1 \approx 0$; R_o is outflow runoff (mm), R_i is inflow runoff (mm) of croplands. In counties on the plain R_i is basically equal to R_o, and in mountainous counties, we estimate the difference between annual R_o and R_i using the method proposed by Ji et al. [14].

$$R_o - R_i = 15.782\, e^{0.0035P} \qquad (10)$$

According to the above analysis, the annual agricultural irrigation water of a county on plain area can be estimated as follows,

$$Irr_n = ET - P \qquad (11)$$

while for the counties in mountainous region, it can be expressed as,

$$Irr_n = ET + 15.782\, e^{0.0035P} - P \qquad (12)$$

The total irrigation water consumption Irr_{nc} (km^3), or net groundwater mining for each county can be estimated as follows,

$$Irr_{nc} = Irr_n \times EIA / 1000000 \qquad (13)$$

So, the annual decline rate of the groundwater table affected by agricultural irrigation can be simply estimated through,

$$D_g = (Irr_{nc} \times 1000 / LA) / P_e \qquad (14)$$

where D_g is the decline depth of groundwater (m); LA is land area (km^2); P_e is effective porosity, which ranges from 10 to 30% in the piedmont area, 5 to 20% in the middle alluvial plain, and 5 to 7% in the coastal plain, respectively [15]. In this study, uniform effective porosity of 25% was used across the plain area of the province.

Validation and Sensitivity Analysis

The 5 years experimental data from Luancheng Agro-ecosystem Experimental Station as introduced earlier were employed to validate and evaluate the model's performance. We assumed the five different irrigation treatments for the five years as different irrigation rate. Firstly, according to the different irrigation levels, such as rainfed, fully irrigation, 80% irrigation, 75% irrigation and 70% irrigation, we set the irrigation rates (Irr_{rate}) of the 5 treatments as 0, 1.0, 0.8, 0.75 and 0.7, respectively. Then, the key parameters of GY_c and K_f were calculated according to Eq. (3) and (6). GY_c ranged from 0.44 to 1.30 and K_f from 0.20 to 0.55, respectively. Finally, we applied all the yield data, P, and ET_0 to the model and calculated the ET for different treatments in each year.

The comparison of calculated ET with field observed ET through soil water balance demonstrated a quite good consistency (Figure 3) except for 3 rainfed treatments in dry years. The relative bias for the 22 samples is only -3.7% and RMSE is 78.9 mm. But for the rainfed treatment in dry years, without supplementary irrigation the grain yield will be largely dependent the occurrence of rainfall on by both amount and timing, which induces uncertainty of the grain yield response to rainfall. In our studies, the main purpose is to give a good projection of the groundwater depletion because of irrigation pumping in past decades. We judge that the bias happened in rainfed cropland will have minor effects on this objective.

In order to evaluate the effectiveness of parameter K_f, we conducted a sensitivity analysis of the estimated ET to the key variables in Eq. 6, i.e. precipitation (P) and irrigation rate (Irr_{rate}). Figure 4 illustrated the responses of ET change to changes in P under different irrigation rates (Figure 4a) and to changes in irrigation rate under different annual precipitations (Figure 4b). $\Delta ET/\Delta P$ varies from 0.32–0.67 when irrigation rate varies from 100% to zero (Figure 4a); the dependence of ET on P decreases as irrigation rate increases. While, the dependence of ET on irrigation rate shows a smaller range, $\Delta ET/\Delta Irr_{rate}$ varies from 0.35–0.53 when annual precipitation decreases from 700 mm to 200 mm. So, the soil wetness parameter K_f is sensitive enough to annual P and irrigation rate, and the model can reflect good responses of estimated ET to precipitation and irrigation at an annual base.

Results

ET_0, ET and AWC

Mean annual ET_0 (1984–2008) ranged from 1,294 mm to 1,365 mm, decreasing gradually from southeast to northwest and showing a similar spatial pattern to air temperature (Figure 5a). And mean annual ET of croplands in each county ranged from 286 to 674 mm (Figure 5b). ET has significantly increased during the past 25 years except for a few counties located in mountainous regions. Increasing ET from croplands is attributed mainly to intensified agricultural activity, such as changes in sowing density, irrigation rate, etc., especially in the plain areas, and partly to increasing temperature (Figure 5c). Decreased ET was detected in some mountainous counties as was shown in Figure 5c, this phenomenon may reflect the effects of the state policy so-called 'Grain to Green', which was launched in the end of 1990s to prevent the land desertification and sand storm through returning cropland to forest or grassland. Mean annual agricultural water consumption (AWC) for the counties ranged from 50 million m^3 to 550 million m^3 in Hebei province during the period from 1984 to 2008. The total water consumption for agricultural grain production was estimated as much as 864 km^3 in Hebei province during the 25 years.

Net consumption of Irrigation Water

Mean annual net irrigation water (mainly groundwater, Irr_n) for each county ranged from 16 to 214 mm (Figure 6a) during the study period, in other words, the groundwater table changes would response to these numbers. The counties at the southeast part of the low plain region showed large increase in irrigation water consumption during the period of 1999–2008 compared with that in 1984–1993 (Figure 6b). That region used to be saline soil and shallow groundwater. The grain productivity has increase greatly during past 3 decades due to the efforts in drainage system construction and irrigation technology evolution. The total net groundwater consumption for irrigation (Irr_{nc}), calculated through Eq. (11 & 12) for the plain area during the study period was projected as 113 km^3 with the mean annual value for each county ranging from 2.7 million m^3 to 140 million m^3.

Discussion and Conclusions

Water shortage has become a major limiting factor for the sustainable development of agriculture in Hebei. The estimation of agricultural water and groundwater irrigation net consumption will provide scientific information for developing efficient irrigation practices to improve crop water productivity. During the study period from 1984 to 2008, the 138 counties in Hebei province produced a total of 6.1×10^8 Mg of grain, and consumed 864 km^3 of water (with an average of 34.6 km^3/y), including 139 km^3 of groundwater. The AWC estimating result was close to the fresh water footprint of agriculture, which was calculated by using of Gini coefficient and Theil index accounting for 33.4 km^3 in Hebei province in 2007 [16]. Figure 7 shows the variations of annual grain yield (GY), water from precipitation (Q_p) and groundwater irrigation net consumption (Irr_{nc}) in Hebei from 1984 to 2008. In terms of spatial distribution, the grain yield and AWC in the southeastern part of the province are significantly higher than those in the northwest.

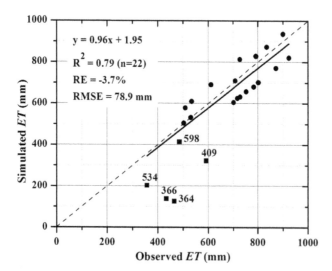

Figure 3. Validation of the model by using field experimental data from Luancheng Agro-ecosystem Experimental Station, Chinese Academy of Sciences. The numbers associated with the 5 points, representing rain-fed treatments, at the lower part refer to annual precipitation. The data for 3 filled blocks (409 mm, 366 mm, and 364 mm) are not included in the regression line because of the large deviations to the observed ET, indicating the model will underestimate actual ET for the non-irrigated lands when annual P less than around 400 mm.

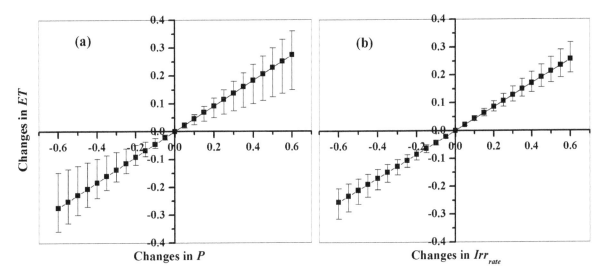

Figure 4. Sensitivities of estimated ET to the changes in annual precipitation (a), and to the changes in irrigation rates (b). The error bars indicates the range of different irrigation rates in (a), and range of different annual precipitation in (b), respectively.

Based on the linear correlation of ET and grain yield of each county (Figure 2), we estimated the grain yield without groundwater consumption, and subtracted this rain water fed yield from the actual statistical yield to obtain the grain yield gain (GYG) benefited from groundwater irrigation. It is found that the accumulated grain yield gain in the 25 years would be 1.9×10^8 Mg, which accounts for 31% of the province's total grain production during the same period.

In addition, we took Luancheng County (location shown in Figure 1) as a typical example to analyze the trade-off between groundwater consumption and grain yield gain. The irrigation rate of Luancheng County has reached to more than 90% since the beginning of 1980s. Although exploitation of groundwater ensured a stable increase in grain production, the groundwater table in

Luancheng fell 20.82 m from 1984 to 2008 due to continual over-pumping (Figure 8). The total groundwater consumption in Luancheng County estimated by the model accounted for 1.2 km^3 in the 25 years, which could cause the underground water table falling of 13.5 m in Luancheng area during the same period. Our estimation attributes the agricultural irrigation for grain production contributed 65% of the groundwater depletion in this county.

Large-scale mining of groundwater in Hebei Province began in the 1970s, the rapid socio-economic development consumed a large amount of groundwater in recent decades, the consumption of agricultural irrigation accounted for 77% of the total. Due to over-exploitation of groundwater, the underground water level was steadily declining. The total groundwater consumption in the

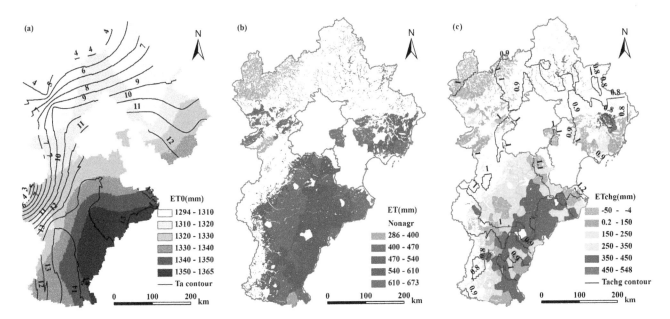

Figure 5. Distribution of annual mean ET_0 (a), actual ET (b), and changes in averaged ET for the period of 1984–1993 to 1999–2008 (c). The contour lines in (a) indicate the distribution of annual mean temperature (Ta); contour lines in (c) indicate the change of annual air temperature (Tachg) for the same time slices.

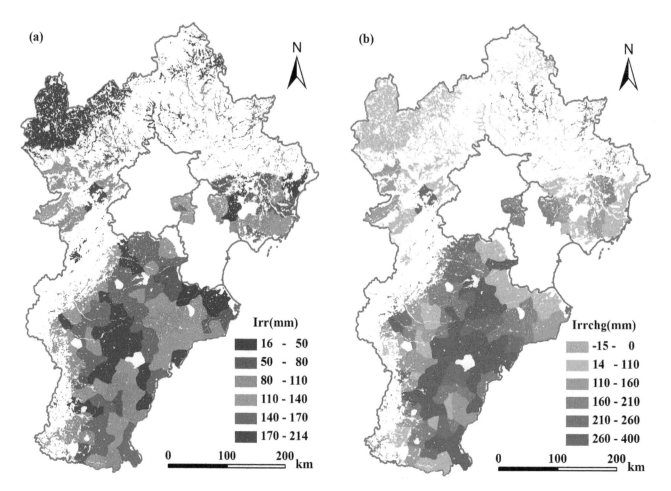

Figure 6. Annual mean net groundwater irrigation in 1984–2008 (a), and its change (b) from the periods of 1984–1993 to 1999–2008.

plain area during the study period was estimated as 113 km³, which could cause an average groundwater falling of 7.4 m over the plain. This estimation is greatly agree with the results reported by Cao [17], who used a numerical groundwater flow model to simulate the groundwater pumping and water table decline over the Hebei plain.

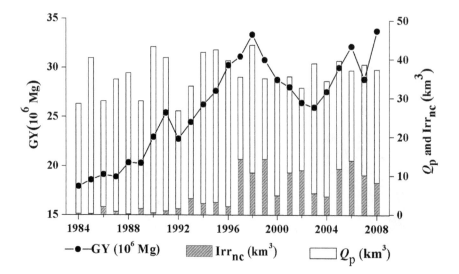

Figure 7. Inter-annual variations of precipitation (Q_p), net groundwater consumption (Irr_{nc}), and grain yield (GY) in Hebei Province (1984–2008).

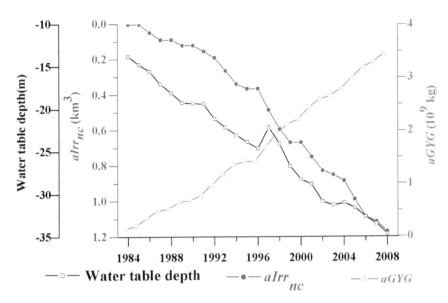

Figure 8. Accumulated net groundwater consumption (a*Irr*nc), grain yield gain (a*GYG*), and observed groundwater table depth change for Luancheng county (1984–2008).

In this study, we aimed to estimate the AWC and groundwater irrigation consumption of Hebei province in recent decades using a simple model. The model proposed in this study need only the basic meteorological data and annual grain yield data. Based on some important assumptions the model can give good estimates of agricultural water consumption and net groundwater consumption for grain food production, and meet the study objectives well. However, we would like to call the audience attention to the uncertainties included in this study. First of all, we used a grain yield coefficient to substitute the crop coefficient in calculation of actual *ET*. This assumption ignored the differences in crop varieties, planting and field managements, irrigation methods, etc. and might cause some deviations of the results. Second, soil moisture of each year is different, but for any region it remains basically unchanged over the long term. The *ET* calculated by Eq. (1) for each year therefore varies, but it is reasonable to use Eq. (1) to calculate the sum and mean annual *ET* over the 25 years period. The *ET* products derived from remote sensing data and economic statistics data also contain some uncertainties [18], these sources of uncertainty may affect the accuracy of this study.

However, through comparing our estimations with the observed *ET* and groundwater depth at Luancheng county and further with the independent simulation over Hebei Plain [17] we have great confidence to believe the method proposed in this study could be extrapolated and applied to other regions where limited data such as meteorological and yield census data are available. Also it may be used in those regions for assessing the water footprint or aiding better water management for sustainable development.

Acknowledgments

Hongwei Pei helped calculating water balance for validation using the experimental data from Luancheng Agro-ecosystem Experimental Station, Chinese Academy of Sciences. We are also grateful to the constructive comments from 3 anonymous reviewers and editors.

Author Contributions

Conceived and designed the experiments: YJS. Performed the experiments: ZJY. Analyzed the data: ZJY. Wrote the paper: ZJY YJS.

References

1. Heilig GK (1999) China food: Can China feed itself? IIASA, Laxenburg (CD-ROM Vers. 1.1).
2. FAO's Information System on Water and Agriculture (2011) Available: http://www.fao.org/nr/water/aquastat/countries_regions/CHN/index.stm.Accessed 2013 Feb 9.
3. Deng XP, Shan L, Zhang HP, Turner NC (2006) Improving agricultural water use efficiency in arid and semiarid areas of China. Agricultural Water Management 80: 23–40.
4. Zhang H, Wang X, You M, Liu C (1999) Water-yield relations and water use efficiency of winter wheat in the North China Plain. Irrigation Science 19: 37–45.
5. Yang H, Zhang XH, Zehnder JB (2003) Water scarcity, pricing mechanism and institutional reform in northern China irrigated agriculture. Agricultural Water Management 61: 143–161.
6. Shen YJ, Kondoh A, Tang C, Zhang Y, Chen J, et al. (2002) Measurement and analysis of evapotranspiration and surface conductance of a winter wheat canopy. Hydrological Processes 16: 2173–2187.
7. Zhang YC, Shen YJ, Sun HY, Gates J (2011) Evapotranspiration and its partitioning in an irrigated winter wheat field: A combined isotopic and micrometeorologic approach. Journal of Hydrology 408: 203–211.

8. Liu XY, Li YZ, Hao WP (2005) Trend and causes of water requirement of main crops in North China in recent 50 years. Transactions of the CSAE 21: 155–159 (in Chinese with English abstract).
9. Li HJ, Zheng L, Lei YP, Li CQ, Liu ZJ, et al. (2008) Estimation of water consumption and crop water productivity of winter wheat in North China Plain using remote sensing technology. Agricultural Water Management 95: 1271–1278.
10. Yang YM, Yang YH, Moiwo JP, Hu YK (2010) Estimation of irrigation requirement for sustainable water resources reallocation in North China. Agricultral Water Management 97: 1711–1721.
11. Wu BF, Yan NN, Xiong J, Bastiaanssen WGM, Zhu WW, et al. (2012) Validation of ETWatch using field measurements at diverse landscapes: A case study in Hai Basin of China. Journal of Hydrology 436–437: 67–80.
12. Sun HY, Shen YJ, Yu Q, Flerchinger GN, Zhang YQ, et al. (2010) Effect of precipitation change on water balance and WUE of the winter wheat-summer maize rotation in the North China Plain. Agricultural Water Management 97: 1139–1145.
13. Allen RG, Pereira LS, Raes D, Smith M (1998) Crop evapotranspiration-Guidelines for computing crop water requirements. FAO Irrigation and drainage paper 56. FAO, Rome, Italy.
14. Ji ZH, Yang CX, Qiao GJ (2010) Reason analysis and calculation of surface runoff rapid decrease in the northern branch of Daqing River. South-to-North

Water Transfers and Water Science & Technology 8: 73–75, 79 (in Chinese with English abstract).

15. Chen W (1999) Groundwater in Hebei Province. Beijing: Seismological Press (in Chinese).

16. Sun CZ, Liu YY, Chen LX, Zhang L (2010) The spatial-temporal disparities of water footprints intensity based on Gini coefficient and Theil index in China. Acta Ecologica Sinica 30: 1312–1321 (in Chinese with English abstract).

17. Cao GL (2011) Recharge estimation and sustainability assessment of groundwater resources in the North China Plain. Ph.D. Thesis, Tuscaloosa: the University of Alabama.

18. Long D, Sing VP (2011) How sensitivity is SEBAL to changes in input variables, domain size and satellite sensor? Journal of Geophysical Research- Atmospheres 116: D21107.

16

Effects of Local and Landscape Factors on Population Dynamics of a Cotton Pest

Yves Carrière[1]*, Peter B. Goodell[2], Christa Ellers-Kirk[1], Guillaume Larocque[3¤], Pierre Dutilleul[3], Steven E. Naranjo[4], Peter C. Ellsworth[1]

1 Department of Entomology, The University of Arizona, Tucson, Arizona, United States of America, 2 Kearney Agricultural Center, University of California Cooperative Extension, Parlier, California, United States of America, 3 Department of Plant Science, McGill University, Sainte-Anne-de-Bellevue, Quebec, Canada, 4 Arid-Land Agricultural Research Center, USDA-ARS, Maricopa, Arizona, United States of America

Abstract

Background: Many polyphagous pests sequentially use crops and uncultivated habitats in landscapes dominated by annual crops. As these habitats may contribute in increasing or decreasing pest density in fields of a specific crop, understanding the scale and temporal variability of source and sink effects is critical for managing landscapes to enhance pest control.

Methodology/Principal Findings: We evaluated how local and landscape characteristics affect population density of the western tarnished plant bug, *Lygus hesperus* (Knight), in cotton fields of the San Joaquin Valley in California. During two periods covering the main window of cotton vulnerability to *Lygus* attack over three years, we examined the associations between abundance of six common *Lygus* crops, uncultivated habitats and *Lygus* population density in these cotton fields. We also investigated impacts of insecticide applications in cotton fields and cotton flowering date. Consistent associations observed across periods and years involved abundances of cotton and uncultivated habitats that were negatively associated with *Lygus* density, and abundance of seed alfalfa and cotton flowering date that were positively associated with *Lygus* density. Safflower and forage alfalfa had variable effects, possibly reflecting among-year variation in crop management practices, and tomato, sugar beet and insecticide applications were rarely associated with *Lygus* density. Using data from the first two years, a multiple regression model including the four consistent factors successfully predicted *Lygus* density across cotton fields in the last year of the study.

Conclusions/Significance: Our results show that the approach developed here is appropriate to characterize and test the source and sink effects of various habitats on pest dynamics and improve the design of landscape-level pest management strategies.

Editor: Dorian Q. Fuller, University College London, United Kingdom

Funding: This work was funded by USDA-CSREES RAMP, Developing and Implementing Field and Landscape Level Reduced-Risk Management Strategies for Lygus in Western Cropping Systems (Project #0207436). The funders had no role in study design, data collection and analysis, decision to publish, or preparation of the manuscript.

Competing Interests: The authors have declared that no competing interests exist.

* E-mail: ycarrier@ag.arizona.edu

¤ Current address: Quebec Centre for Biodiversity Science, McGill University, Montreal, Quebec, Canada

Introduction

Landscape transformation resulting from increases in the extent and intensity of agricultural activities is often associated with greater pest pressure and use of environmentally-disruptive pesticides [1]. As increased demand for food will continue to favor agricultural intensification for decades, the vulnerability of intensively managed agro-ecosystems may increase in the future [2]. Accordingly, the need for sustainable pest management is increasing interest in manipulating agricultural landscapes to disrupt the capacity of pests to infest crops [3–5]. Much work has been done at the field scale to understand how spatial arrangements of vegetation affect pest movement and population dynamics [6,7]. However, much less information is available on effects of landscape heterogeneity on pest metapopulation dynamics [8,9].

The demographic impact on population density of particular habitats for other patches in the landscape has been characterized based on the local balance between birth and death rates and immigration and emigration [10,11]. Here, we define source habitats as areas that increase pest density in fields of a specific crop, while sink habitats are areas that reduce pest density in fields of that crop. Many significant polyphagous pests exploit a wide array of crops and uncultivated habitats that may act as sources or sinks for focal crops at some time during the growing season [12].

Lygus spp. (Hemiptera: Miridae) provide classical examples of source-sink dynamics resulting in crop damage through spatial subsidies [13,14]. After overwintering in uncultivated habitats with weedy host plants, adults colonize crops such as alfalfa (*Medicago sativa* L.) and safflower (*Carthamus tinctorius* L.), where large populations develop in the spring and early summer. Adults move to cotton when alfalfa is harvested or safflower matures and becomes less suitable [13,14]. Insecticides are typically applied in cotton following such dispersal because cotton is highly vulnerable to *Lygus* feeding during fruit formation. Landscape-based management to reduce *Lygus* populations in cotton has included lessening the source potential of certain crops and uncultivated habitats, or planting alternative hosts in cotton fields to divert *Lygus* feeding from cotton [15,16]. Another management practice could involve manipulating the spatial arrangements of source and sink habitats [16,17]. However, the source or sink potential of habitat patches can vary dramatically in space and time, and the consequences of this variation on the spatial structure of pest populations remain largely unknown [12,18].

The goals of this study were to characterize temporal variation in effects of local and landscape characteristics on the population density of *L. hesperus* in cotton, and to assess whether the spatial pattern of *L. hesperus* populations can be predicted despite this temporal variation in the San Joaquin Valley of California. The main period of cotton vulnerability to *L. hesperus* attack occurs between June and August. The source and sink characteristics of particular habitats may vary during this period, due to changes in suitability of host plants or harvest. Among-year variation in abundance of crops could also influence the source and sink characteristics of habitats if habitat choice of migrants is affected by the relative availability of habitats. We thus used geographic information system (GIS) technology combined with spatial statistics to evaluate: 1) within- and among-year variations in effects of cotton field characteristics and of certain crops and uncultivated habitats, 2) the spatial scale of the associations between abundance of the habitats and *L. hesperus* density in cotton, and 3) how the local and landscape factors combine to determine *L. hesperus* population density in cotton.

Methods

Ethics Statement

No specific permits were required for the described field studies.

Field Sites and GIS Mapping of Agricultural Fields

In 2007, 2008, and 2009, we sampled *L. hesperus* in cotton fields once a week between June and August (Table 1), which is the main period of cotton vulnerability to *L. hesperus* attack. Most fields were sampled in only one year ($n = 128$), although two fields sampled in 2007 were resampled in 2008 and two other fields sampled in 2007 were sampled again in 2009. These cotton fields were located in the Fresno and Kings Counties of the San Joaquin Valley (see Fig. S1). The study area was larger in 2007 than in subsequent years because cotton fields were more extensively distributed in 2007 than in 2008 and 2009. Location and shape of agricultural fields were determined with U.S. Department of Agriculture's Farm Service Agency common land unit maps [19] and validated from the ground with a Global Positioning System (GPS) at a resolution of 5 m. Over the three years, the average shortest distance between pairs of sampled cotton fields and area of sampled cotton fields varied between 2.5 and 3.9 km and 62.8 and 91.4 ha, respectively (Table 1). Pima cotton (*Gossypium barbadense* L.) was more frequently planted than Upland cotton (*G. hirsutum*) in the study area.

Winters in the San Joaquin Valley are moist and foggy but summers are hot and dry. Non-reproductive *L. hesperus* adults overwinter in uncultivated habitats and move into crops in late winter and spring when uncultivated vegetation starts to dry up [20]. Here we focused on the source and sink potential of uncultivated habitats and crops known to harbor significant *L. hesperus* populations in the study area [21]: cotton, forage alfalfa, safflower, seed alfalfa, sugar beet (*Beta vulgaris* L.), and tomato (*Solanum lycopersicum* L.) (see Fig. S2). Crops in fields <3 km from the edge of sampled cotton fields were identified by visual survey from the ground. Uncultivated habitats within this distance were identified from geographically referenced data (see below).

Within-fields Variables

Lygus spp. prefer to feed on developing cotton flower buds and young fruits [22]. We thus investigated the impact of cotton flowering date in addition to effects of landscape composition. Date of initiation of flowering was determined with planting dates obtained from producers and a model based on accumulation of degree-days [23]. We also evaluated effects of insecticide sprays applied during the sampling periods. Insecticide data were provided by cotton producers.

Sampling Method

Sampled cotton fields were divided in four quadrants with each quadrant sampled weekly. Samples were collected starting at least 25 m inside each quadrant and consisted of 100 sweeps. The upper part of plants was sampled because it is a preferred feeding and oviposition area for *L. hesperus* [24]. The number of adults from the 400 sweeps was recorded for each week and field.

Landscape Analysis

Fields were mapped using ArcGIS version 10.0 [25]. Roads and urban areas were overlaid on field maps. We drew twelve concentric rings around the edge of each sampled cotton field. The first ring had a distance from the field edge of 250 m and the distance of each subsequent ring increased by 250 m; the largest ring had a distance of 3000 m. The area of each crop type (m^2) between the edge of a sampled cotton field and a ring was calculated with ArcGIS. Uncultivated vegetation within rings was primarily found along irrigation canals and roads, in riparian areas, near urban developments, or in rangelands. Inspection from the ground and with high-resolution imagery in Google Earth [26] showed that such uncultivated habitats mainly comprised grass or weeds and shrubs, and thus plausibly contained *L. hesperus* hosts. Area of uncultivated habitats in each ring was calculated by subtracting the area occupied by agricultural fields, roads and urban development from the area of the ring.

Data Analysis

Source and sink effects. Here we use the slope of the statistical association between *L. hesperus* density in sampled cotton fields and the area of a habitat type surrounding the cotton fields to infer source or sink effects, whereby a significant negative association indicates a sink effect and a positive association a source effect. The source and sink potential of particular habitats could vary during the cotton vulnerability period (i.e., June to August), due to changes in host suitability or harvest. To evaluate potential variation in the associations between areas of habitat types and *L. hesperus* density during this period, the mean number of *L. hesperus* adults sampled per week in each field was averaged over two successive periods. Duration of the first and second periods varied between four and six weeks depending on year (Table 1). *L. hesperus* density did not differ significantly between Pima and Upland cotton in any period (2-sample *t*-tests, *P*-values >0.1). Furthermore, preliminary analyses conducted at all scales (method described below) indicated qualitatively similar effects of Pima and Upland cotton. The two species of cotton were thus considered as a single crop for analysis.

Table 1. Characteristics of cotton fields sampled for western tarnished plant bug, *Lygus hesperus*.

Year*	Field Area†	Closest distance†	First period‡	Second period‡	Flowering date¶	% Pima cotton	Insecticide sprays†	*Lygus* density†§
2007	67.8 (7.1)	3.9 (0.3)	10 Jun (6)	22 Jul (5)	30 Jun	85	1.2 (0.10)	4.3 (0.5)
2008	62.8 (4.5)	2.7 (0.2)	15 Jun (5)	20 Jul (4)	10 Jul	92	7.8 (0.4)	13.3 (1.4)
2009	91.4 (6.6)	2.5 (0.1)	9 Jun (5)	14 Jul (5)	23 Jun	89	3.9 (0.4)	5.3 (0.4)

Variables shown are average field area (ha), average closest distance between pairs of sampled cotton fields (km), date of initiation of first and second sampling periods, average date of initiation of flowering, percentage of sampled fields planted to Pima cotton, average number of insecticide sprays and average *Lygus* density (calculated per 100 sweeps to facilitate comparison with thresholds) for combined sampling periods in each year.
*Number of fields sampled: 41 in 2007; 39 in 2008; 56 in 2009.
†Standard error in parentheses.
‡Date is for onset of sampling period; number of weeks sampled per period is in parentheses.
¶Range associated with average flowering date was 22 Jun–7 Jul in 2007, 7 Jul–12 Jul in 2008, and 17 Jun–2 Jul in 2009.
§Suggested thresholds for *Lygus* spraying depend on cotton phenology [48]. Number of individuals per 100 sweeps that would trigger spraying is: >4–8 adults (early squaring); >14–20 individuals with at least two nymphs (bloom); and >20 individuals with nymphs present (boll filling).

The number of potential explanatory variables (i.e., nine: areas of six crops and uncultivated habitats, cotton flowering date, and number of insecticide applications) was relatively high, compared to the number of experimental units (Table 1, between 39 and 56 fields sampled per year). Therefore, we first used stepwise regression (with forward selection and backward elimination) to select a subset of relevant explanatory variables. Average *L. hesperus* density in a field was the response variable, and area of each crop, area of uncultivated habitats, cotton flowering date (number of days since January 1 of each year), and number of insecticide sprays in cotton were the candidate explanatory variables. For each period of the three years, we performed an analysis at each of the 12 spatial scales (ring distance from 250 to 3000 m). Variables with significant explanatory effect ($P<0.05$) at one or more scales were retained for subsequent analysis.

Multiple regression was then used to evaluate the association between *L. hesperus* density and the explanatory variables selected in the stepwise procedure. For a given period and year, the same multiple regression model was fit for all 12 scales. Partial F-tests were used to assess significance of explanatory variables included in the model. As in Carrière et al. [17,27], we used rank-based statistics in stepwise and multiple regression analyses because assumptions of normality and homogeneity of variance were not met by the raw data. Statistical analysis was adjusted for spatial autocorrelation when required (see below).

Scale of source and sink effects. A significant association between area of a habitat type and *L. hesperus* density is expected if the area in a ring comprises patches that affect *L. hesperus* density in sampled cotton fields, but statistical significance is expected to decline once the scale of analysis exceeds the distance at which patches affect *L. hesperus* density [28]. Therefore, we used the largest ring at which a significant effect was found for a habitat type to infer the scale of source or sink effects. Because larger rings included patches present in smaller rings, this procedure may overestimate the scale of source and sink effects through "carry-over effects". To assess this possibility, we performed additional multiple regression analyses with two adjacent rings of increasing width (from 250 m to 1500 m). Pairs of adjacent rings do not share patches because the larger rings do not include patches in the smaller rings (e.g., a 250 m-wide ring includes patches from edge of field up to a distance of 250 m, while a 500 m-wide ring includes patches at distance between 250 m and 500 m), so the maximum scale at which a significant association is observed in two-ring analyses is not affected by carry-over effects [17,27]. Nevertheless, two-ring analyses may have lower statistical power

than single-ring analyses because the number of explanatory variables required to investigate source and sink effects in the former is doubled (e.g., with six crops, twelve explanatory variables are used in multiple regression, instead of six). As expected, two-ring analyses detected a lower number of significant effects than single-ring analyses. However, across periods and years, the scale of effects detected in both single-ring and two-ring analyses did not differ significantly (paired *t*-test, $P = 0.33$), indicating that carry-over effects were not important. Here we only report results from single-ring analyses across the 12 scales. Among-habitat differences in average scale of source and sink effects (across periods and years) were assessed with one-way ANOVA [29].

Prediction of L. hesperus density. The goal of the predictive model was to show that factors with consistent effects were sufficient to predict spatial variation in *L. hesperus* density, even when other important factors were not considered in the predictive model (see **Discussion**). Analyses of data from the first two years revealed that the areas of cotton, uncultivated habitats and seed alfalfa, and date of flower initiation had consistent effects on *L. hesperus* density (see **Results**). Safflower also had consistent effects during the first two years, but its effects changed in the third year. Therefore, only the first four variables were included in a predictive model derived from data obtained in the first two years (see **Discussion**).

Before pooling data from the first two years to formulate the predictive model, one-way ANOVAs with year as the classification factor were performed to remove between-year variations in the response and explanatory variables. Standardized residuals from these ANOVAs (i.e., centered data divided by the standard deviation within each year) provided the response and explanatory variables in rank-based multiple regression analyses, which evaluated the association between *L. hesperus* density (calculated over the main period of cotton vulnerability between June and August) and areas of the cultivated and uncultivated habitats and flowering date at each of the 12 scales. Thus, data from 2007–2008 were used to analyze the association between among-site variations in *L. hesperus* density and among-site variations in areas of habitat types near each sampled field and date of cotton flower initiation.

The regression model with the highest coefficient of determination (R^2) was selected for prediction of *L. hesperus* density in 2009. Values of the explanatory variables at the corresponding scale for each sampled field in 2009 were substituted in the multiple regression model to calculate predicted values of ranks for *L. hesperus* density. A rank-based simple linear regression was then used to assess the association between predicted and observed values of *L. hesperus* density in 2009.

Spatial autocorrelation. In each stepwise and multiple regression analysis, and in the analysis of predicted versus observed *L. hesperus* density, semivariograms were computed to quantify and analyze spatial autocorrelation in *L. hesperus* density and other variables across fields [30]. By assessing spatial patterns in residuals (response variable) and partial residuals (explanatory variables), we evaluated spatial autocorrelation at all scales and corrected for potential non-independence of observations. Spatial autocorrelation was accounted for in tests of significance through the use of effective sample sizes, in modified *t*- and *F*-tests performed in simple linear correlation analysis, stepwise regression, and multiple regression [30–32]. Programs for these statistical analyses were written in Matlab [33].

Results

Composition of the landscape in rings surrounding the sampled cotton fields varied during the three years. The main changes involved a decrease in the area occupied by cotton and uncultivated habitats from 2007 to 2008 and 2009, and an increase in the area occupied by safflower and tomato in 2008 compared with 2007 and 2009 (Table S1). Other differences included greater *L. hesperus* population density and use of insecticides, and later cotton flowering dates in 2008 than in 2007 and 2009 (Table 1). Abundance of cotton, seed alfalfa, and uncultivated habitats were frequently and consistently associated with *L. hesperus* density in sampled cotton fields across sampling periods and years (Table 2). The significant negative associations for cotton and uncultivated habitats indicate that these habitats were sinks for *L. hesperus*. On the other hand, the significant positive associations for seed alfalfa indicate that this crop was a source of *L. hesperus* for cotton fields.

Abundance of safflower was frequently associated with *L. hesperus* density in cotton fields. However, the significant coefficients were positive in 2007 and 2008 and negative in 2009, indicating that this crop was a source of *L. hesperus* for cotton in the first two years but a sink in the last year. Forage alfalfa also had variable effects, as the significant coefficients were positive in 2007 and negative in 2008. Areas of sugar beet and tomato were associated with *L. hesperus* density only once, although sugar beet was only included in analyses in 2007 because it was rare in other years (Table 2).

The scale at which abundances of cultivated and uncultivated habitats were significantly associated with *L. hesperus* density differed among habitats ($F = 3.35$, *d.f.* $= 4$, 13, $P = 0.043$). Cotton, safflower and seed alfalfa affected *L. hesperus* density over larger spatial scales than forage alfalfa and uncultivated habitats (Fig. 1). The associations between date of flower initiation and *L. hesperus* density were positive and significant in 2007 and 2008 (Table 2). The number of insecticide sprays was negatively associated with *L. hesperus* density only once, in 2009.

The coefficient of determination of the multiple regression model including the abundances of cotton, seed alfalfa, uncultivated habitats and flowering date varied from 16.7 to 24.4% across the 12 scales. The R^2 value was highest at the 2750-m scale, which was thus the scale used to test the predictive model. For fields sampled in 2009, the association between predicted and observed ranks of *L. hesperus* density was positive and significant (Fig. 2, $R^2 = 33.2\%$, $F = 26.9$, *d.f.* $= 1$, 54, $P < 0.0001$, spatial autocorrelation was not significant in this analysis).

Discussion

Landscapes dominated by annual crops represent networks of ephemeral patches for multivoltine generalist pests that track the availability and suitability of resources during the growing season. Although control of generalist pests has typically hinged on within-field management, migration among patches often affects pest population dynamics locally and regionally [8,12,34]. Here we found that the abundance and distribution of cotton, uncultivated habitats and seed alfalfa surrounding the monitored cotton fields, and date of cotton flower initiation had consistent within- and among-year effects on density of *L. hesperus*. These factors were sufficient to predict *L. hesperus* density across cotton fields in the Fresno and Kings Counties in the San Joaquin Valley of California in 2009. Flowering date of cotton may have affected *L. hesperus* density because it synchronized dispersal of individuals with presence of the most suitable cotton phenological stages [12]. Thus, each factor included in the predictive model probably influenced among-patch migration, suggesting that a landscape-

Table 2. Average regression coefficient for the association between *Lygus* density in sampled cotton fields and abundance of crops and uncultivated habitats, estimated for two sampling periods in three years.

Habitat	First period			Second period		
	2007*	2008	2009	2007	2008	2009
Cotton	−0.39 (0.03, 11)	−0.27 (NA†, 1)	−0.38 (0.01, 9)	−0.40 (0.03, 11)	−0.41 (0, 2)	−0.46 (0.01, 12)
Forage alfalfa	0.31 (0.005, 5)	−0.4 (NA, 1)				
Uncultivated habitats	−0.21(0.02, 5)			−0.45 (0.05, 4)		−0.31 (0.06, 2)
Safflower	0.14 (NA, 1)	0.34 (0.02, 9)	−0.34 (0.006, 5)			−0.31 (0.005, 2)
Seed alfalfa	0.47 (0.06, 8)	NA†		0.74 (0.02, 12)	NA	0.59 (0.06, 7)
Sugar beet		NA	NA	0.12 (0.13, 4)	NA	NA
Tomato	−0.22 (0.03, 3)					
Flowering date		0.44 (0.01, 12)		0.25 (0, 3)		
Insecticide sprays						−0.34 (0.01, 8)

Effects of flowering date and insecticide sprays are also shown.
*After correcting for spatial autocorrelation, criterion for assessing significance of regression coefficients was P<0.1. Number reported is average of the significant regression coefficients in analyses performed at the 12 scales. Parentheses contain standard error followed by number of significant regression coefficients.
†NA: Standard error was not calculated because a single coefficient was significant, or a crop was not included in analyses because it was rare in rings.

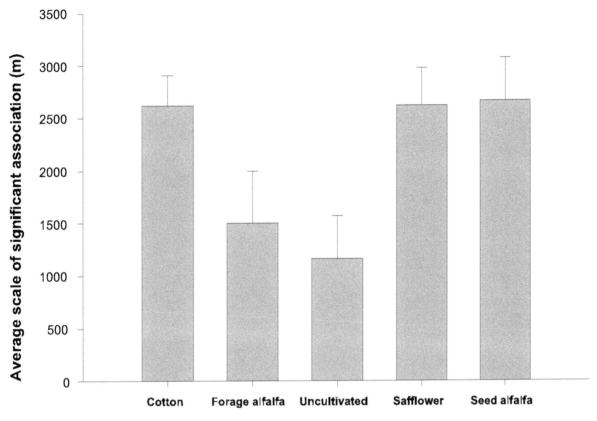

Figure. 1. Scale of association between *Lygus* density in cotton fields and abundance of surrounding habitats. Average scale (mean + SE) for habitats found to have significant effects in a least two of the six analyses are shown. Standard errors were derived from the ANOVA.

based approach will be useful to manage *L. hesperus* populations in cotton.

The consistent negative association between abundance of cotton and *L. hesperus* density in sampled cotton fields may be explained by the low density of *L. hesperus* in cotton compared to other habitats [17,21,35] and the low attractiveness of cotton compared to other hosts [14,35,36]. Control of *L. hesperus* populations with insecticides during the fruiting period (Table 1) could account at least in part for the low *L. hesperus* densities in cotton. Conversely, the consistent positive association between abundance of seed alfalfa and *L. hesperus* density in cotton likely occurred because seed alfalfa is an attractive and suitable *L. hesperus* host and many individuals disperse from this crop when irrigation is terminated before harvest. Similar sink and source effects of cotton and seed alfalfa were respectively found in a one-year study conducted in an arid agricultural landscape of Central Arizona [17].

A negative association between abundance of uncultivated habitats and *L. hesperus* density occurred in both parts of the cotton vulnerability period. The reasons for this pattern are not clear. Most uncultivated habitats harbor sparse *L. hesperus* populations in years with low rainfall, suggesting that uncultivated hosts should not be significant sources during these years [16]. However, large *L. hesperus* populations can develop in uncultivated habitats in years with high rainfall [37]. In such years, uncultivated habitats may attract and retain *L. hesperus* until July because high moisture availability postpones weed dry-up and *L. hesperus* prefers some weeds over cotton [35,37]. Yet, uncultivated habitats should become sources when weeds eventually dry up at the end of July. Accordingly, it seems that seasonal changes in suitability and

attractiveness of uncultivated hosts may not account for the negative associations of *L. hesperus* with uncultivated habitats found in the second sampling period.

Our finding that abundances of forage alfalfa and safflower were frequently associated with *L. hesperus* density in cotton indicates that a better understanding of landscape effects of management practices in these crops could greatly contribute in managing *L. hesperus* populations in cotton. Forage alfalfa and safflower have been managed to reduce *L. hesperus* movement to cotton since the mid 1960s in the San Joaquin Valley [13,21]. To reduce movement from forage alfalfa to cotton, strips of alfalfa are left at harvest to retain adults in the uncut portions of fields [38]. Insecticides can also be applied to safflower before harvest to limit adult emigration [39]. Although these practices do not significantly improve yield or quality of the treated crop, they increase insecticide use and complexity of crop management. Consequently, they are only profitable for producers that also grow cotton, or when cotton producers compensate alfalfa and safflower producers for extra costs and difficulties associated with *L. hesperus* management. On average, about 50% of producers manage alfalfa and safflower to reduce *L. hesperus* migration to other crops in the San Joaquin Valley, although use of these management practices varies across years and counties [16,40,41]. Accordingly, the inconsistent source and sink effects of forage alfalfa and safflower observed here may have been due to spatial and temporal variations in implementation of practices to reduce the source effect of these crops. Indeed, there is evidence that many safflower fields were treated with insecticides before harvest in 2009 but few in 2007 and 2008 [41]. This may explain why safflower was a source in 2007 and 2008 but a sink in 2009.

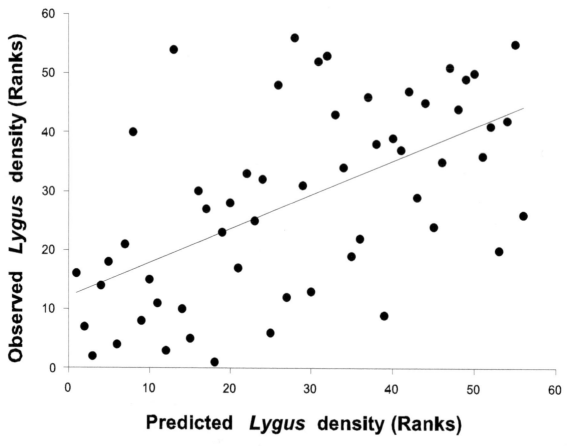

Figure. 2. Association between observed and predicted density of *L. hesperus* **in cotton fields.** Rank-based regression analysis was used to evaluate the association across 56 cotton fields sampled in 2009. The model used to calculate predicted values of ranks for *Lygus* density was: *Lygus* density = 44.3−0.41 (area of cotton) +0.096 (flowering date) +0.25 (area of seed alfalfa) −0.073 (area of uncultivated habitats).

The influence of other ecological factors such as natural enemy induced mortality on these observed landscape patterns is unclear. Although many species of parasitoids and generalist predators attack *Lygus* spp. in a variety of habitats in the U.S. [42,43], the impact of these natural enemies on pest dynamics is not well understood [42]. Recent work shows that abundance of *Geocoris* spp. in cotton is associated with reductions in immature stages of *L. hesperus* [44]. Species of parasitoids from Europe have become established in limited areas of central California [45], but their impacts appear restricted to strawberry production in coastal regions. Ongoing work is examining the influence of landscape factors on the dynamics of natural enemies. This work may allow us to better explain the spatial patterns observed here, including the sink effects of uncultivated habitats on *L. hesperus* populations in cotton.

Number of insecticide applications in cotton was rarely associated with *L. hesperus* density across cotton fields. Birth and immigration contribute to population growth in a patch while death and emigration reduce it [10,11]. Insecticides are generally applied when *L. hesperus* density exceeds a specific threshold (Table 1). If immigration rates varied among cotton fields and fields with high *L. hesperus* influx received more insecticides, immigration and mortality from insecticides may have often compensated each other. Thus, insecticides may contribute in reducing *L. hesperus* damage to cotton, especially by sedentary nymphs, but not in reducing populations of the mobile adult stage over time, as observed here in most sampling periods and years.

Independent sets of data were used to select factors included in the predictive model (i.e., data from 2007–2008) and evaluate accuracy of this model (i.e., data from 2009). However, safflower was excluded from the predictive model because analyses revealed that effects of this crop on *L. hesperus* density changed in 2009. The change in effects of safflower in 2009 was likely due to changes in safflower management in that year (see above). Thus, in the strictest sense, formulation and evaluation of the predictive model were not accomplished with independent data. Nevertheless, for the four factors included in the predictive model, an independent set of data was used to evaluate the quality of model predictions, in accord with recommended practices for development of predictive distribution models [46]. Importantly, excluding safflower from the predictive model here is appropriate because the goal of this model was to show that the four factors with consistent effects were sufficient to predict spatial variation in *L. hesperus* density, even when other important factors such as the abundance of safflower and forage alfalfa were not considered. The statistical approach used here was recently applied to predict spatial variation in the evolution of resistance to an insecticide in *Bemisia tabaci* [28]. Taken together, these studies indicate that such approach will be useful for the development of spatially-explicit integrated pest management.

The associations between abundances of particular habitats (e.g., forage alfalfa and safflower) and *L. hesperus* density in cotton varied among years. In the absence of knowledge on the cause of such temporal variation, manipulation of the spatial arrangement of these habitats is difficult. Thus, a fundamental question in the design and implementation of landscape-level pest management strategies is whether the modification of a limited number of factors with consistent effects will be sufficient to produce the desired outcome. A positive answer to this question is suggested when a statistical model including these factors provides accurate prediction of pest population dynamics [46,47]. Specifically, our demonstration that the spatial structure of *L. hesperus* populations in cotton was predicted with a model built on one local and three landscape factors with consistent effects across years increases the credibility of a landscape-based approach to manage this pest in the San Joaquin Valley.

The findings of this study indicate that patches of cotton, uncultivated habitats and seed alfalfa affects *L. hesperus* population density in cotton. Because increased abundance of cotton was associated with lower *L. hesperus* density in sampled fields, clumping cotton fields could contribute in reducing *L. hesperus* populations in cotton. The maximum scale of the significant negative associations between abundance of uncultivated habitats and *L. hesperus* density varied between 500 and 2000 m across periods and years. This indicates that groups of cotton fields at a distance <500 m from uncultivated habitats could harbor the lowest *L. hesperus* densities. Conversely, separating groups of cotton fields from seed alfalfa by more than 3 km (i.e., the maximum spatial scale of source effects of seed alfalfa observed here) could contribute in reducing *L. hesperus* populations in cotton. Cotton producers generally spread planting of fields over a few weeks. Because late flowering was associated with increased *L. hesperus* populations in cotton, fields located where *L. hesperus* immigration is expected to be high (e.g., near unmanaged safflower or seed alfalfa) could be planted earlier than fields in locations where *L. hesperus* dispersal is expected to be lower.

Pest infestations triggering applications of insecticides in specific crops often occur because polyphagous pests migrate from other source habitats [12]. Furthermore, the population dynamics of many polyphagous pests are likely affected by sink habitats, as these pests commonly prefer specific hosts or plants in particular phenological stages, and recurring application of insecticides in crops highly sensitive to damage and other management practices can drastically reduce their populations [12]. Accordingly, the metapopulation dynamics of many polyphagous pests depends on characteristics of the surrounding landscape and crop manage-ment practices applied to individual fields. Our results suggest that a systematic, spatially-explicit statistical approach taking into account the distribution of source and sink habitats and management practices in crop fields of interest can provide strong insights for designing landscape-level management strategies for such polyphagous pests.

Supporting Information

Fig. S1   Cotton fields sampled for *Lygus hesperus* in the Fresno and Kings Counties of the San Joaquin Valley in 2007, 2008 and 2009. The insert shows location of the San Joaquin Valley in California (bottom left, dark area) and location of the study area in the San Joaquin Valley.

Fig. S2   Location of sampled cotton fields, crops assessed for source and sink effects, unidentified crops, uncultivated habitats, and urban areas in 2008. Rings with a distance from the field edge of 3000 m are shown. Across the three years, the largest uncultivated areas surrounding sampled cotton fields were range-lands (shown here in top-left ring), periphery of an airport (five center-right rings), and riparian zones (four lower-right rings).

Table S1   Mean % area occupied by crops assessed for source and sink effects, uncultivated habitats, and urban development in 3000-m rings surrounding sampled cotton fields. Standard errors are in parentheses.

Acknowledgments

We would like to thank Doug Cary, Nathan Cannell, Idalia Orellana, and Ashley Pedro for their assistance in field sampling and sorting the data, and Frances Sivakoff and Jay Rosenheim for providing comments on an earlier version of the MS.

Author Contributions

Conceived and designed the experiments: YC PD PG PE SN. Performed the experiments: PG CK. Analyzed the data: YC CK PD GL. Contributed reagents/materials/analysis tools: YC PD GL. Wrote the paper: YC PD PG PE SN.

References

1. Meehan TD, Werling BP, Landis DA, Gratton C (2011) Agricultural landscape simplification and insecticide use in the Midwestern United States. Proc Natl Acad Sci USA 108: 11500–11505.

2. Godfray HCJ, Beddington JR, Crute IR, Haddad L, Lawrence D, et al. (2010) Food Security: the challenge of feeding nine billion people. Science 327: 812–818.

3. Gardiner MM, Landis DA, Gratton C, DiFonzo CD, O'Neal M, et al. (2009) Landscape diversity enhances biological control of an introduced crop pest in the north-central USA. Ecol Appl 19: 143–154.

4. Bianchi FJJA, Schellhorn NA, Buckley YM, Possingham HP (2010) Spatial variability in ecosystem services: simple rules for predator-mediated pest suppression. Ecol Appl 20: 2322–2333.

5. Jonsson M, Wratten SD, Landis DA, Tompkins J-ML, Cullen R (2010) Habitat manipulation to mitigate the impacts of invasive arthropod pests. Biol Inv 12: 2933–2945.

6. Bommarco R, Banks JE (2003) Scale as modifier in vegetation diversity experiments: effects on herbivores and predators. Oikos 102: 440–448.

7. Letourneau DK, Armbrecht I, Rivera BS, Lerma JM, Carmona EJ, et al. (2011) Does plant diversity benefits agroecosystems? A synthetic review. Ecol Appl 21: 9–21.

8. Tscharntke T, Brandl R (2004) Plant-insect interactions in fragmented landscapes. Ann Rev Entomol 49: 405–430.

9. Zaller JG, Moser D, Drapela T, Schmöger C, Frank T (2008) Insect pests in winter oilseed affected by field and landscape characteristics. Bas Appl Ecol 9: 682–690.

10. Pulliam HR (1988) Sources, sinks, and population regulation. Am Nat 132: 652–661.

11. Thomas CD, Kunin W (1999) The spatial structure of populations. J An Ecol 68: 647–657.

12. Kennedy GG, Storer NP (2000) Life system of polyphagous arthropod pests in temporally unstable cropping systems. Ann Rev Entomol 45: 467–493.

13. Mueller AJ, Stern VM (1974) Timing of pesticide treatments on safflower to prevent *Lygus* from dispersing to cotton. J Econ Entomol 67: 77–80.

14. Sevacherian V, Stern VM (1974) Movements of *Lygus* bugs between alfalfa and cotton. Environ Entomol 4: 163–165.

15. Snodgrass GL, Scott WP, Abel CA, Robbins JT, Gore J, et al. (2006) Suppression of tarnished plant bug (Heteroptera: Miridae) in cotton by control of early season wild host plants with herbicides. Environ Entomol 35: 1417–1422.

16. Goodell PB (2009) Fifty years of the integrated control concept: the role of landscape ecology in IPM in San Joaquin valley cotton. Pest Manag Sci 65: 1293–1297.

17. Carrière Y, Ellsworth P, Dutilleul P, Ellers-Kirk C, Barkley V, et al. (2006) A GIS-based approach for area-wide pest management: the scales of *Lygus hesperus* movements to cotton from alfalfa, weeds and cotton. Entomol Exp Appl 118: 203–210.

18. Hunter M (2002) Landscape structure, habitat fragmentation, and the ecology of insects. Agr Forest Entomol 4: 159–166.

19. USDA Farm Service Agency website. Aerial Photography Field Office, Imagery Products. Available: http://www.fsa.usda.gov/FSA/apfoapp?area=home&subject=prod&topic=clu. Accessed 2012 Jun 7.

20. Godfrey LD, Leigh TF (1994) Alfalfa harvest strategy effects on *Lygus* bug (Hemiptera:Miridae) and insect predator population density: implications for use as trap crop in cotton. Environ Entomol 23: 1106–1118.

21. Goodell PB, Lynn K, McFeeters SK (2002) Using GIS approaches to study western tarnished plant bug in the SJV of California. Proc Belt Cott Prod Res Conf. Atlanta GA.

22. Layton MB (2000) Biology and damage of the tarnished plant bug, *Lygus lineolaris*, in cotton. Southwest Entomol 23: 7–20.

23. University of CA Cotton Web Site. Cotton guidelines. Heat unit averages and time to mature bolls. Available: http://cottoninfo.ucdavis.edu/files/133240.pdf. Accessed 2012 Jun 7.

24. Wilson LT, Leigh TF, Gonzalez D, Forestiere C (1984) Distribution of *Lygus Hesperus* (Miridae: Hemiptera) on cotton. J Econ Entomol 77: 1313–1319.

25. ESRI (2011) ArcGIS Desktop: Release 10.0. Redlands, California.

26. Google Earth web site. Google Earth Version 6. Available: http://earth.google.com. Accessed 2012 Jun 7.

27. Carrière Y, Dutilleul P, Ellers-Kirk C, Pedersen B, Haller S, et al. (2004) Sources, sinks, and zone of influence of refuges for managing insect resistance to Bt crops. Ecol Appl 14: 1615–1623.

28. Carrière Y, Ellers-Kirk C, Harthfield K, Larocque G, Degain B, et al. (2012) Large-scale, spatially-explicit test of the refuge strategy for delaying insecticide resistance. Proc Natl Acad Sci USA. 109: 775–780.

29. JMP (2008) JMP 8.0. SAS Institute, Cary.

30. Dutilleul P (2011) Spatio-temporal heterogeneity: Concepts and analyses. Cambridge: Cambridge University Press. 416 p.

31. Alpargu G, Dutilleul P (2006) Stepwise regression in mixed quantitative linear models with autocorrelated errors. Com Stat Simul Comp 35: 79–104.

32. Dutilleul P, Pelletier B, Alpargu G (2008) Modified F tests for assessing the multiple correlation between one spatial process and several others. J Stat Plan Infer 138: 1402–1415.

33. The Mathworks Inc. (2008) MATLAB Version R2008a. Natick.

34. Hanski I (1999) Metapopulation ecology. New York: Oxford University Press. 313 p.

35. Barman AJ, Parajulee MN, Carroll SC (2010) Relative preference of *Lygus hesperus* (Hemiptera: Miridae) to selected host plants in the field. Ins Sci 17: 542–548.

36. Stern VM, Mueller AJ, Sevacherian V, Way M (1969) *Lygus* bug control through alfalfa interplanting. Calif Agric 23: 8–10.

37. Goodell PB, Ribeiro B (2006) Measuring localized movement of *Lygus hesperus* into San Joaquin valley cotton fields. pp 1375–1379 Proc Belt Cot Conf, San Antonio, Texas.

38. Summers CG, Goodell PB, Mueller SC (2004) *Lygus* bug management by alfalfa harvest manipulation. In: Pimentel D, editor. Encyclopedia of Pest Management Vol. 2. Boca Raton: CRC Press, Taylor & Francis Group. pp 322–325.

39. Sevacherian V, Stern VM, Mueller AJ (1977) Heat accumulation for timing *Lygus* control measures in a safflower-cotton complex. J Econ Entomol 70: 399–402.

40. Brodt SB, Goodell PB, Krebil-Prather RL, Vargas RN (2007) California cotton growers utilize integrated pest management. Cal Agric 16: 24–30.

41. Goodell PB (2010) Managing the ecosystem for IPM: effects of reduced irrigation allotments. pp 26–31 in Proc Amer Soc Agron, Tulare, California.

42. Ruberson JR, Williams LH (2000) Biological control of *Lygus* spp.: A component of areawide management. Southwest Entomol 23: 96–110.

43. Hagler JR (2006) Development of an immunological technique for identifying multiple predator-prey interactions in a complex arthropod assemblage. An Appl Biol 149: 153–165.

44. Zink AG, Rosenheim JA (2008) Stage-specific predation on *Lygus hesperus* affects its population stage structure. Entomol Exp Appl 126: 61–66.

45. Pickett CH, Swezey SL, Nieto DJ, Bryer JA, Erlandson M, et al. (2009) Colonization and establishment of *Peristenus relictus* (Hymenoptera: Braconidae) for control of *Lygus* spp. (Hemiptera: Miridae) in strawberries on the California Central Coast. Biol Contr 49: 27–37.

46. Guisan A, Zimmermann NE (2000) Predictive habitat distribution models in ecology. Ecol Model 135: 147–186.

47. Rykiel EJ Jr (1996) Testing ecological models: the meaning of validation. Ecol Model 90: 229–244.

48. UC IPM Online website. Statewide Integrated Pest Management Program. Available: http://www.ipm.ucdavis.edu/PMG/r114301611.html. Accessed 2012 Jun 7.

PERMISSIONS

LIST OF CONTRIBUTORS

Liang Wu, Xinping Chen, Zhenling Cui, Weifeng Zhang and Fusuo Zhang
Center for Resources, Environment and Food Security, China Agricultural University, Beijing, People's Republic of China

Caroline H. Orr and Stephen P. Cummings
Faculty of Health and Life Sciences, Northumbria University, Newcastle-Upon-Tyne, United Kingdom

Carlo Leifert and Julia M. Cooper
Nafferton Ecological Farming Group, Newcastle University, Nafferton Farm, Stocksfield, Northumberland, United Kingdom

Xiu-liang Jin
Institute of Crop Science, Chinese Academy of Agricultural Sciences/Key Laboratory of Crop Physiology and Production Ministry of Agriculture, Beijing, China
Beijing Research Center for Information Technology in Agriculture, Beijing, China

Wan-ying Diao and Chun-hua Xiao
Key Laboratory of Oasis Ecology Agriculture of Xinjiang Construction Crops, Shihezi, China

Fang-yong Wang and Bing Chen
Institute of Cotton, Xinjiang Academy of Agricultural Reclamation Sciences, Shihezi, China

Ke-ru Wang and Shaokun Li
Institute of Crop Science, Chinese Academy of Agricultural Sciences/Key Laboratory of Crop Physiology and Production Ministry of Agriculture, Beijing, China
Key Laboratory of Oasis Ecology Agriculture of Xinjiang Construction Crops, Shihezi, China

Eva L. M. Figuerola, Leandro D. Guerrero, Silvina M. Rosa and Leandro Simonetti
Instituto de Investigaciones en Ingeniería Genética y Biología Molecular (INGEBI-CONICET) Vuelta de Obligado 2490, Buenos Aires, Argentina

Matías E. Duval and Juan A. Galantini
CERZOS-CONICET Departamento de Agronomía, Universidad Nacional del Sur, Bahía Blanca, Argentina

José C. Bedano
Departamento de Geología, Universidad Nacional de Río Cuarto, Río Cuarto, Córdoba, Argentina

Luis G. Wall
Departamento de Ciencia y Tecnología, Universidad Nacional de Quilmes, Roque Sáenz Peña 352, Bernal, Argentina

Leonardo Erijman
Instituto de Investigaciones en Ingeniería Genética y Biología Molecular (INGEBI-CONICET) Vuelta de Obligado 2490, Buenos Aires, Argentina
Facultad de Ciencias Exactas y Naturales, Universidad de Buenos Aires, Ciudad Universitaria, Pabellón 2, Buenos Aires, Argentina

Rajan Ghimire, Jay B. Norton and Peter D. Stahl
Department of Ecosystem Science and Management, University of Wyoming, Laramie, Wyoming, United States of America

Urszula Norton
Department of Plant Sciences, University of Wyoming, Laramie, Wyoming, United States of America

Richard V. Scholtz III and Allen R. Overman
Agricultural & Biological Engineering Department, University of Florida, Gainesville, Florida, United States of America

Sarina Macfadyen
CSIRO Ecosystem Sciences and Sustainable Agriculture Flagship, Canberra, Australia

Darryl C. Hardie
Department of Agriculture and Food, Irrigated Agriculture and Diversification, Perth, Australia
The University of Western Australia, School of Animal Biology, Perth, Australia

Laura Fagan and Helen Spafford
The University of Western Australia, School of Animal Biology, Perth, Australia

Katia Stefanova
The University of Western Australia, The UWA Institute of Agriculture, Perth, Australia

Kym D. Perry and Helen E. DeGraaf
South Australian Research and Development Institute, Entomology Unit, Urrbrae, Australia

Joanne Holloway
New South Wales Department of Primary Industries, Wagga Wagga Agricultural Institute, Wagga Wagga, Australia

Paul A. Umina
The University of Melbourne, Department of Zoology, Melbourne, Australia
Cesar, Melbourne, VIC, Australia

Jean-Noël Aubertot
Institut National de la Recherche Agronomique, Unité Mixte de Recherche 1248 Agrosystémes et Agricultures, Gestion des Ressources, Innovations et Ruralitès, Castanet-Tolosan, France
Université Toulouse, Institut National Polytechnique de Toulouse, Unité Mixte de Recherche 1248 Agrosystémes et Agricultures, Gestion des Ressources, Innovations et Ruralités, Castanet-Tolosan, France

Marie-Héléne Robin
Institut National de la Recherche Agronomique, Unité Mixte de Recherche 1248 Agrosystémes et Agricultures, Gestion des Ressources, Innovations et Ruralite´ s, Castanet-Tolosan, France
Université de Toulouse, Institut National Polytechnique de Toulouse, Ecole d'Inge´nieurs de Purpan, Toulouse, France

Xuelin Zhang, Qun Wang, Yilun Wang and Chaohai Li
The Incubation Base of the National Key Laboratory for Physiological Ecology and Genetic Improvement of Food Crops in Henan Province, Zhengzhou, China; Agronomy College of Henan Agricultural University, Zhengzhou, China

Feina Cha
Meteorological Bureau of Zhengzhou, Zhengzhou, China

Frank S. Gilliam
Department of Biological Sciences, Marshall University, Huntington, West Virginia, United States of America

Anders S. Huseth
Department of Entomology, Cornell University, New York State Agricultural Experiment Station, Geneva, New York, United States of America

Russell L. Groves
Department of Entomology, University of Wisconsin-Madison, Madison, Wisconsin, United States of America

Marijn van der Velde, Linda See, Juraj Balkovič, Steffen Fritz, Nikolay Khabarov and Michael Obersteiner
International Institute for Applied Systems Analysis (IIASA), Ecosystem Services and Management Program, Laxenburg, Austria

Liangzhi You
International Food Policy Research Institute (IFPRI), Washington D.C., United States of America

College of Economics and Management, Huazhong Agricultural University, Wuhan, China

Stanley Wood
Global Development Program, Bill & Melinda Gates Foundation, Seattle, Washington, United States of America

Syed Tahir Ata-Ul-Karim, Xia Yao, Xiaojun Liu, Weixing Cao and Yan Zhu
National Engineering and Technology Center for Information Agriculture, Jiangsu Key Laboratory for Information Agriculture, Nanjing Agricultural University, Nanjing, Jiangsu, P. R. China

Jun Wang
College of Urban and Environmental Sciences, Northwest University, Xian, Shaanxi Province, China

Upendra M. Sainju
U.S. Department of Agriculture, Agricultural Research Service, Northern Plains Agricultural Research Laboratory, Sidney, Montana, United States of America

P. Dilip Venugopal, Peter L. Coffey, Galen P. Dively and William O. Lamp
Department of Entomology, University of Maryland, College Park, Maryland, United States of America

Zaijian Yuan
School of Economics & Management, Hebei University of Science and Technology, Shijiazhuang, China
Center for Agricultural Resources Research, Institute of Genetics and Developmental Biology, Chinese Academy of Sciences, Shijiazhuang, China

Yanjun Shen
Center for Agricultural Resources Research, Institute of Genetics and Developmental Biology, Chinese Academy of Sciences, Shijiazhuang, China

Yves Carrière, Peter C. Ellsworth and Christa Ellers-Kirk
Department of Entomology, The University of Arizona, Tucson, Arizona, United States of America

Peter B. Goodell
Kearney Agricultural Center, University of California Cooperative Extension, Parlier, California, United States of America

Guillaume Larocque and Pierre Dutilleul
Department of Plant Science, McGill University, Sainte-Anne-de-Bellevue, Quebec, Canada

Steven E. Naranjo
Arid-Land Agricultural Research Center, USDA-ARS, Maricopa, Arizona, United States of America

Index

Printed in the USA
CPSIA information can be obtained
at www.ICGtesting.com
JSHW051446221024
72173JS00006B/1594

9 781682 863794